Sustaining Biodiversity in Today's World

Sustaining Biodiversity in Today's World

Edited by **Neil Griffin**

New York

Published by Callisto Reference,
106 Park Avenue, Suite 200,
New York, NY 10016, USA
www.callistoreference.com

Sustaining Biodiversity in Today's World
Edited by Neil Griffin

International Standard Book Number: 978-1-63239-588-7 (Hardback)

Printed in the United States of America.

Contents

Preface

Every book is initially just a concept; it takes months of research and hard work to give it the final shape in which the readers receive it. In its early stages, this book also went through rigorous reviewing. The notable contributions made by experts from across the globe were first molded into patterned chapters and then arranged in a sensibly sequential manner to bring out the best results.

This book presents a comprehensive study regarding biodiversity, its conservation and utilization. It discusses the regulation and consumption of resources to meet human demands in different spheres varying from agriculture to wildlife and exhaustible to inexhaustible resources to fulfill human necessities and the range of life is assessed according to the variety of the species that exist. Diversity is an essential feature of environment. But a few factors have led environmentalists into a dilemma where they are bound to allow a variety of species to grow and flourish together even at the cost of vulnerability of a few other species to preserve overall biodiversity. The economic benefits of biodiversity are well discussed in the book at local, national and international levels.

It has been my immense pleasure to be a part of this project and to contribute my years of learning in such a meaningful form. I would like to take this opportunity to thank all the people who have been associated with the completion of this book at any step.

Editor

Agricultural Science

Agricultural Systems and the Conservation of Biodiversity and Ecosystems in the Tropics

F.F. Goulart, T.K.B. Jacobson, B.Q.C. Zimbres,
R.B. Machado, L.M.S. Aguiar and G.W. Fernandes

Additional information is available at the end of the chapter

1. Introduction

One quarter of the terrestrial surface is composed of cultural systems, while in the tropics, 70% of the land has already been converted into pastures, agriculture, or a mixture of managed landscapes [1,2]. Agricultural expansion is recognized as the most significant human alteration of the global environment, with the addition of fertilizers in the agricultural sector accounting for high input of nitrogen and phosphorus in terrestrial ecosystems. The conversion of natural ecosystems in agricultural areas has increased fire frequency, and caused profound rupture in nutrient cycles. Furthermore, agricultural expansion has modified landscapes, making them more vulnerable to invasion by exotic species.

In spite of these facts, there is enough evidence that anthropogenic systems managed using agroecological principles can support high levels of biodiversity [3,4], contribute to the maintenance of a healthy environment and its services, as well as depend less on costly external inputs of pollutant pesticides and fertilizers [5]. Therefore, there is a wide range of agricultural management strategies, and they differ greatly on their effect on biodiversity.

Today, agroforestry systems cover more than 16 million hectares, and they involve 1.2 billion people worldwide [6]. Traditional shade-cocoa [7], shade-coffee [3], and agroforestry home-gardens [8] are examples of agricultural systems that retain part of the natural habitat structure and ecosystems properties, providing habitat for rich and diverse fauna and flora including threatened and endemic species. On the other hand, intensive agricultural systems, such as pastures and extensive mono specific plantations, harbour low levels of biodiversity, hamper biological flux, and lead to soil leaching, and nutrient import/export. Intensive agriculture is one of the major drivers of change in some biogeochemical cycles such as nitrogen and phosphorus [9]. This "out of farm" nutrient input changes the coexistence and competition patterns between autotrophic organisms, changing the structure of natural ecosystems. The more intensive the agricultural systems, the less they are capable of harbouring biodiversity, maintaining landscape connectivity, and conserving ecosystems properties and services. Agricultural intensification is a process in which low-input agriculture (such as traditional mixed farming) becomes intensified in terms of input/output level, which in turn impacts negatively the associated biodiversity, and the natural ecological services.

2. Hunger and conservation in the tropics

Two of the most important issues in the political and scientific agendas are biodiversity conservation, and hunger. Of the world's 2 million formally described taxa, between 12% and 52% are threatened with extinction according to IUCN Red List of Endangered Species [10]. For example, 119 out of 273 species of turtles in the world are threatened, and 1,063 out of 4,735 mammal species are threatened [10]. At the same time, solving the hunger problem seems to be another great challenge for humanity. Global food production has increased 168% over the past 42 years. However, there is great inequality in food distribution. Only between 2000 and 2002, 852 million people suffered from malnutrition (96% in developing countries). Although biodiversity loss and hunger are global problems, nowhere are these problems more acute than in the tropical region. There is a clear pattern in the distribution of biodiversity and latitude, i.e. the closer it gets to the Equator, the larger the number of species. The Afrotropical and the Neotropical region account for 49% of the bird, 63% of the amphibian, and 45% of the mammal species of the world [10]. Only in the Neotropical region, more than 10,000 vertebrate species are found [10]. Hence, most world priority sites for biodiversity conservation are concentrated in the tropical region [11]. On the other hand, hunger problem are also much more intense in the tropics, where most underdeveloped countries are situated. South Asia alone accounts for 60% of the undernourished people in the world, and Subsaarian Africa also shows high starvation levels. In the Congo Democratic Republic, more than 60% of the population is unable to acquire sufficient calories to meet their daily caloric requirements. In India, one of the most populous countries, this value lies between 30 and 40% [6].

Because of the high level of biological diversity, and the great famine incidence, it seems clear that trade-off or win-win relationship between biodiversity and agriculture must be exacerbated. Therefore, global scientific and political concerns that address the issues of hunger and biodiversity conservation in the tropical region are necessary.

3. Myths and facts about conservation and agricultural production

Ecologists and biologists have focused their work primarily on 'pristine' or 'untouched' habitats [12,13,14]. This is done under the assumption that human modified ecosystems have virtually null or diminished importance for conservation, and that conservation efforts should go in the direction of establishing human-free reserves as large as possible to avoid species loss. What some ecologists and conservation biologists ignore is the fact that: even supposedly untouched places have actually moderate degree of human intervention [15]. Human-modified ecosystems vary greatly in their quality for biodiversity and maintenance of ecosystems properties. In the 'unaltered' habitat, biodiversity is often restricted to patches embedded in an anthropogenic matrix, which can serve as a conduit or barrier to species movement. Because connectivity is necessary for the long term maintenance of species in patchy landscapes [16], matrix management has deep effects on biodiversity, and functioning of the complex habitat mosaic [17,18]. Finally, human activities can reach far beyond anthropogenic environments, causing changes in several regional processes, hence affecting ecosystems and biodiversity at larger scales.

On the other hand, agricultural sciences are rarely aware of the effect of management on ecological patterns taking place in the agrienvironments and landscapes. The inverse is also true: they are unaware of how ecological patterns taking place in the agrienvironment and of the landscape affect agricultural production. Agribusiness and agricultural scientists generally aim at reaching the highest agricultural yields. There is an implicit assumption that the loss of ecosystems services will be overcome by biotechnological advances. It is common the thinking that if the weather is drier because of climatic alterations, resistant crop will be developed; that if soil is leached, higher fertilizer quantities can be applied; and so on.

A common argument, in which yield-maximization is based, is the poverty and hunger alleviation issue. The mostly accepted ideas are that agricultural managements should increase production at any cost, based on the hunger alleviation argument, and that conservation efforts, although being relevant to society, should never prevent food production from increasing. Facing these persuasive arguments, biological conservation is regarded as low priority, compared to productivist sectors in the stakeholder's agenda.

Even some conservation biologists accept such assumptions, so that they propose that agricultural areas should reach maximum yields in order to reduce the need to convert more natural areas into agricultural systems, but still maintain the production target [19,20]. This theory is called Land Sparing, and predicts that agricultural intensification would reduce deforestation by increasing productivity. This view has been criticized on the theoretical ground [13,14,21], as well as with empirical data from studies at both regional [22], and local scales [23]. For example, the agricultural product demand (mainly meat and soybean) has increased Amazonian ecosystems' conversion rates. Direct forest conversion into agricultural lands in 2003 accounted for 23% of forest and savannah deforestation in the Mato Grosso state of Brazil. While grazing areas remain the main deforestation cause in the Amazon, land conversion for the production of soybean crops for exportation is also leading to high deforestation rates [24].

Figure 1A. shows that, at global levels, food production has been steeply increasing since the sixties [25]. Food production per capita has also increased, although at lower rates, and food prices have been declining with some oscillation. Finally, the number of undernourished people has declined up to the mid 1990s, when it started increasing suddenly. Therefore, at global levels (in which agribusiness operates), the increase in food production per capita *per se* does not guarantee hunger alleviation. Hence even disregarding conservation issues, the argument that food production should increase to nurse the hunger problem, even if conservation policies are underprivileged, is a fallacy and lacks scientific base.

Using basic ecological principles, such as trophic webs, it is possible to maximize food by simply moving down to lower levels of the trophic pyramid. By changing our food habits so that we eat more vegetables and less meat, the quantity of food *per capita* will increase. Therefore, it is more reasonable to use actual food production in a more rational manner, rather than clearing forest for agricultural expansion, or increasing productivity at the cost of biodiversity loss. Another important issue that emerges from the hunger problem is that of food distribution. From 1980 to 2000, the number of obese adults have doubled in the United States [28], and tripled in the United Kingdom [29]. Obesity is growing around the world and affects mostly high income countries, but it is also epidemic of many low income countries. For instance, in some cities of China, 20% of the population is overweight [30]. In some countries in Africa, Latin America, Asia, and the Pacific, there is a double burden of diet-related diseases caused by obesity and undernourishment. Figure 1.B shows the proportion of the population which is overweight in the last years in some countries.

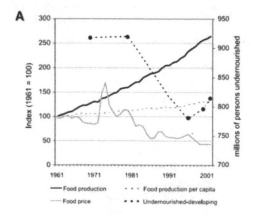

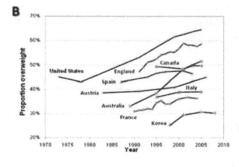

Figure 1. A) Trends in key indicators of world's food production 1961-2002 [26]. (B) Proportion of the population which is overweighed in the last years in some countries [27].

Another wrong aspect of the Land Sparing Theory is that it does not account for some important political and social aspects [8]. For instance, the rural poor comprise 80% of those hungry worldwide [6], and any management that addresses the hunger issue must therefore focus on poor farmers. Studies in Brazil [31], Central America [32], and India [33] showed that agricultural intensification leads to social disasters. In Brazil, agricultural intensification has lead to an increase or to the maintenance of rural poverty levels, and to a drastic increase of poverty in the cities. It has reduced prematurely the labour demand, inflated land prices, expelling small landholders from their lands [31]. In Andra Pradesh, in India, 16,000 farmers committed suicide between 1995 and 1997, mainly because of farm failure. Most failure was caused by conversion of traditional mixed farming systems into monocultures of a high yield variety of cotton, which was highly dependent on external inputs [33]. Hence the assumption that agricultural intensification can solve the problem of hunger and poverty is naïve. Furthermore, the idea that large scale agriculture produces more than at smaller scales is erroneous for most countries [34], as is the idea that organic systems are generally less productive than conventional ones [35,36].

4. Agriculture and biodiversity conservation

Great part of the world's terrestrial surface is in the agrienvironment, so that most of the world's terrestrial biodiversity inhabits the agrienvironment (known as farm biodiversity), or inhabits patches of natural habitat embedded in an agricultural matrix. In the last 40 years, most of these agricultural landscapes have gone through deep changes in farming practices, which have negatively affected farm biodiversity [37,38,39], as well as the metapopulation dynamics of the species inhabiting patches embedded in an agriculture matrix [13,40,21]. "Agricultural intensification" is the general term given to changes in farming practices that have begun after the Green Revolution. Intensification includes pesticides, irrigation systems, machinery, an increase in farm size, and a decrease of spatial and temporal heterogeneity [37].

Concerning agriculture and biodiversity, it is important to distinguish "planned biodiversity" or "agrobiodiversity", which are the species intentionally introduced in the agricultural systems for the proposes of production, from "associated biodiversity" defined as the biological components that exist in an agricultural system by chance, without being actively introduced [41]. From an ecological point of view, it is also useful to distinguish the associated biodiversity that inhabits the agrienvironment (that feeds, reproduces and roosts in it) from species using the agricultural matrix simply for dispersion.

Agricultural intensification leads to declines at the species level, through conversion of mixed crop systems into monocultures, but also at the genetic level though the replacement of highly diverse traditional cultivated varieties for single high-yield varieties [42,43]. This has caused the extinction of many traditional varieties worldwide, leading to homogenization of cultivated species at the genetic level. Because traditional varieties have gone through centuries or millenniums of adaptation and selection, many traditional varieties are much more adaptable and less demanding in terms of external inputs than modern ones.

Moreover, agriculture intensification leads to the widespread use of pesticides, which cause the loss of biodiversity through direct poisoning. The famous Rachel Carson's Silent Spring [44] is a keystone in the environmental cause, and describes the effect of pesticides on birds. Pesticides increase the risk of egg breakage by reducing egg shell thickness [44, 46]. They also cause changes in brain activity of the birds [47]. Consequently, pesticides may not only cause population decline of birds inhabiting agrienvironments, but it may also negatively affect species inhabiting adjoining habitat forest areas [48]. Pesticides exposure associated to trematode infection can cause morphological deformities in amphibians [49]. Another vertebrate species negatively affected by pesticides are human beings. The WHO [51] estimates that 220,000 people die annually because of unintended pesticide poisoning. The majority of cases occur in low-income countries, where knowledge of health risks and safe use of pesticides is limited [6].

Another effect of agricultural intensification on farm biodiversity is the loss of spatial and temporal heterogeneity which seems to be the major cause of farm associated biodiversity decline worldwide [37]. Studies conducted in 'natural landscapes' suggest that spatial heterogeneity increases the number of possible niches in a given habitat [51], as well as the possibility of co-existence among species that share the same niche [52]. Heterogeneity also increases the chance of co-existence of predator-prey dynamics [53]. Bird nest predation rates in homogeneous environments are higher than in heterogeneous ones [55,37]. Also, heterogeneity can affect perceived risks for bird species, so that even if there is no real predation pressure (for instance, because top predators abundance has decreased due of habitat alteration), birds will not establish in homogeneous habitats, where perceived risk is high [55,56]. Therefore agriculture intensification, by affecting predation pressure via habitat homogenization deeply reduces bird diversity associated with agricultural habitats [37,55].

For animals that disperse through the agricultural matrix, heterogeneity can increase matrix permeability to species flux, reducing (re)colonization in patches in agricultural landscapes [37]. Regarding forest birds, individuals face great actual and perceived risk when dispersing in open habitats [57], which suggest that intensification of the agricultural matrix may reduce avian dispersion rates. Hence, agriculture intensification leads to a loss of temporal and spatial heterogeneity among farm plots and regions via homogenization of farming practices [37]. Heterogeneity is a key concept in agroecology, and can go beyond the biological sphere, reaching important aspects of social spheres, such as cultural (diversity of agricultural practices and knowledge in the community), gender (woman participation in farming activities), and individual (individual empowerment in rural communities) levels. Hence, heterogeneity must be a flagship in the management of agricultural landscapes [58].

Following, we exemplify how agricultural practices affect biodiversity in the tropical region, by highlighting the effects of agricultural intensification on biodiversity and production of shade-cocoa (Fig 2.A), shade-coffee (Fig 2.B) and home gardens (Fig 2.C).

Figure 2. Highly diverse agroforestry systems in the Atlantic Forest Hotspot of Brazil. Figure 2.a is a shade-cocoa (*cabruca*) in the south of the Northeast region in the state of Bahia; Fig 2.b is a traditional rustic coffee agroforest in the southeast region in the state of Minas Gerais; and Fig 2.c is a home-garden in the southeast in the state of São Paulo. Photos by F.F. Goulart.

4.1. Shade-cocoa plantations

Shade-cocoa systems are the largest agroforestry system in the world, accounting for 7 million hectares worldwide [6]. Cocoa is planted under a shade canopy in many tropical countries, such as Indonesia [59], Costa Rica [61], Mexico [62], Camerron [63], and Brazil. The cabruca is the local name of traditional rustic shade-cocoa plantations in the state of Bahia, in the northeast Brazil. The system involves growing cocoa under the canopy of native Atlantic Forest. In the 1960s, the Brazilian military government implemented a policy of promoting the intensification of the cabruca to increase production. The program involved the reduction of shade in a way to reduce the incidence of the witches broom fungi (*Moniliophthora perniciosa*), a cocoa pest responsible for the collapse of the productivity in the region. Additionally, the use of fertilizers and pesticides was suggested due to an increase in insect pests in the low-shade management strategy. The government conceded credit loans to farmers who removed trees from the system. This involved the use of arboricides in order to facilitate the 'deforestation' of the high-shade cabruca. Fortunately, many farmers did not adopt the program, while many that obtain the loan did not remove the trees. The reason for this is that fertilizer and insecticide expenditures outweighed the gains of the low-shade management. Also farmers considered that tree removal would increase risk under the condition of price uncertainty.

Today, the rustic cabruca is what was left of the Atlantic Forest in the region, and it harbours high levels of forest biodiversity. High richness of forest ants [64], bats, birds [4], frogs, lizards, ferns [7], and trees have been reported in the cabruca [65]. Furthermore, many threatened species, such as the golden lion tamarin (*Leontopithecus rosalia*), one of the rarest monkeys of the world [66], the golden headed lion tamarin (*Leontopithecus chrysomelas*), the yellow breasted porcupine (*Chaetomys subspinosus*) [67], the white-necked hawk (*Leucopternis lacernulatus*), and the pink-legged graveteiro (*Acrobatornis fonsecai*) [4] live in the cabrucas. This *Acrobatornis* is a mono-specific genus, first described in the cabruca [68], and has never been reported outside of this environment [69]. Because all of these organisms are forest dwelling species, it seems plausible that, if all cabruca were converted into low-shade system, as proposed by the government, these species would be locally extinct. In the case of *Acrobatornis* it is would be globaly extinct, as it is restricted to high shade systems. Hence, if farmers were risk takers, this species would have disappeared without science ever knowing about it, as it was described 30 years after the implementation of the intensification policy. We consider the *cabruca* one of the most biological important agroecosystem in the world.

Faria and coworkers [4,7] found that the richness of species in the rustic agroforestry is even higher than the found in the forest. Despite this great importance of shade plantations for biodiversity conservation, most studies indicate that many forest dwelling species are absent or found at a much lower abundance in the agroforest, compared to the primary forest, suggesting that shade plantations cannot substitute the forest in its ecological function. In a study in Costa Rica, two species of sloths (*Bradypus variegatus* and *Choloepus hoffmannis*) were radio-tracked to understand the use of the agrienvironment by the individuals. The results indicated that the shade-cocoa, the riparian forest, and live fences provided habitat and increased connectivity for these species [70].

Concerning the relationship between biodiversity and cocoa production, a recent study conducted in Sulawesi concluded that, for most taxa (Fig. 3), including endemic species (Fig. 4), there is no correlation between both variables [71]. This suggests that the relationship between conservation importance and yield is not a trade-off, but can be win-win, as high production can be coped with biological conservation. Additionally, authors found no relationship between forest distance and biodiversity, so that that species richness is related to management structure rather than landscape patterns. Therefore Landsparing Theory, which assumes a trade-off between conservation and production [20], cannot be applied to these cocoa systems.

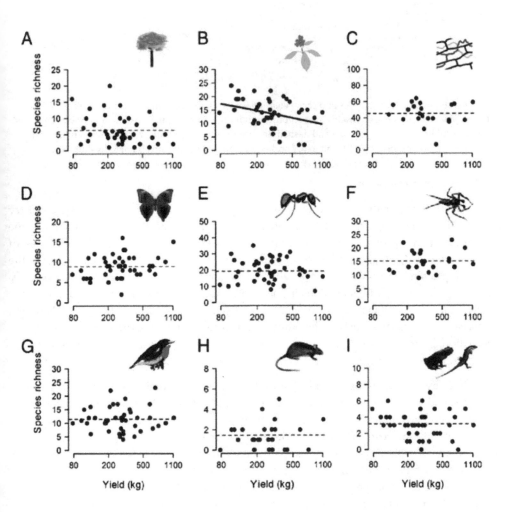

Figure 3. Associated biodiversity in smallholder cacao agroforestry in relation to cacao yeild in Sulawesi, Indonesia, for (A) trees, (B) herbs (C) endophytic fungi, (D) butterflies, (E) ants, (F) spiders, (G) birds, (H) rats, and (I) amphibians and reptiles. Broken lines are intercept-only linear models. Source: [71]

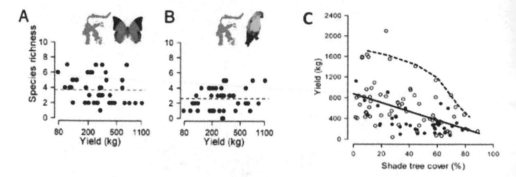

Figure 4. Endemic species richness and cacao yield relationship for (A) butterflies and (B) birds in Sulawesi, and the influence on shade cover on productivity (C). In all figures dashed lines are linear adjust, and in C solid line is the linear adjust and dashed line are maximum son-linear adjust. Broken lines are intercept-only linear models (A, B), and in (C) broken line are the maximum simulated values. Source: [71]

4.2. Shade-coffee systems

Just as in the case of the shade-cocoa, coffee is planted in high-shade rustic agroforestry systems in Mexico [3], Jamaica [72], Guatemala [73] and Brazil [74] among others. In the Mexican systems (probably the most studied agroforest in the world), high richness of species is found [4,75]. The conversion of these systems into sun monocultures, as part of the worldwide intensification policy, has called the attention of ornithologists because many migrant birds used shade plantations in the winter. Consequently, the substitution of high-shaded for sun systems leads to a decline in migrant bird populations. Therefore, conservationist agencies, such as the Smithsonian Bird Migratory Center, the Conservation International, and the Nature Conservancy launched a campaign to conserve this agrienvironment by certifying coffee farms as "bird friendly", or "biodiversity friendly".

High diversity of trees, birds, frogs [75], ants, butterflies [41], orchids [76], bats, dung-beetles [77], bees, and wasps [78] are found in shade-coffee plantations in comparison with sun coffee systems. Birds are the most well known taxa in coffee plantations, as in other agricultural systems [79]. By the year 2006, forty studies had been published in well known international scientific journals on birds in shade-coffee plantations [80]. This review suggests that more than 50 North-American migrant species use shade-coffee plantations. Concerning endangered birds, eight species that use shade-coffee are considered to be threatened at some degree [80]. Some other aspects of bird ecology are worth noting. For example, migrants to show high winter site fidelity (individuals return to the same area in consecutive years) in shade plantations, which suggests that these areas are highly suitable for wintering [81]. Many threatened taxa of mammals are found in shaded-coffee systems in Mexico, such as the tamandua anteater (*Tamandua mexicanus*), the river otter (*Lutra longicaudis*), the mexican porcupine (*Shiggurus mexicanus*), and the margay (*Leopardus wiedii*) [82].

Agriculture intensification acts by reducing both planned and associated biodiversity, and coffee systems are not an exception. For instance, in the southeast of Brazil, traditional coffee farmers cultivated a traditional varieties of coffee (*Moca, Carolina, Cravinho* among others), which is shade tolerant, and has a long productive life cycle (according to farmers, the *Moca* plants can produce for more than a hundred years). In the 1970s, agricultural intensification policies introduced a sun-variety of coffee (locally called *Catuaí*). This variety starts producing at earlier life stages than the traditional varieties, but also stops producing earlier and it is much more dependent on fertilizer and pesticides [74]. The substitution of traditional varieties for modern ones leads to a widespread failure of coffee production in the region, because farmers could not keep up with the high costs of necessary inputs. The result was the conversion of coffee plantations into pastures. Many traditional varieties, as the *Carolina*, are at risk of extinction due to the substitution for the modern varieties. We once heard a statement from a local farmer, saying in a mix of disappointment and anger: "Agronomists have ruined coffee production in the region". Although we believe that the problem lies beyond simple government agriculture agencies' technicians, this is a good example of the misleading efforts of the agriculture intensification programs.

Regarding the pattern of conservation-production of coffee systems, studies conducted in Central America, points out to a trade off. This occurs because the reduction in tree density may, depending on the shade values, decrease biodiversity and increase yields. Species richness and productivity negative relationship can be concave or convex. Convex pattern, means that significant amount biodiversity is present at intermediate levels of biodiversity, while in the concave, trade-off is steep and relatively low levels of biodiversity in systems with intermediate productivity. In the Mexican systems, butterflies show a convex, while ants show a concave pattern (Fig 5).

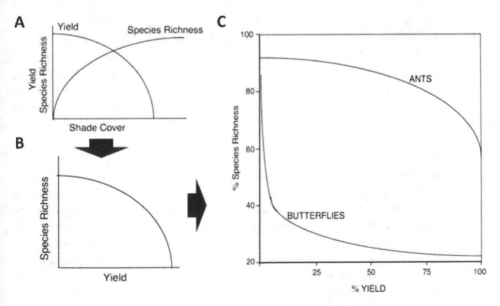

Figure 5. Construction of yield set from the functions of cover (A and B) and the yield-richness relationship for butterflies and ants in coffee systems of Mexico (C). Source: [41]

4.3. Home-gardens and other agricultural systems

Home-gardens are the oldest agroecosystems in the world [84], and may be more than 10,000 years old [85]. These systems are generally characterized by a high heterogeneity within plots and regions. Because home-gardens are a complex mosaic of orchards, live-fences, mixed farming practices (including animal raising), and they vary greatly on their temporal and spatial structure. Another aspect of home-gardens is that they individually occupy a small area, especially in the tropics, so that the management grain (minimal area in which a certain farming practice takes place) is small, and human density is high.

Home-gardens are a pool for agrobiodiversity, work as refuges, preventing the genetic erosion of cultivars, and are considered living gene banks [86]. For example, in the home-gardens of Nepal, twenty crop species have been lost in the last 10 to 15 years, and 11 species are threatened (their use has declined significantly over the last years) [87]. In an arid region of the northeast of Brazil, more than 50 woody plants were reported in 31 home gardens, including many native species [88]. In the tropical lowland forest of Indonesia, in only six traditional home-gardens (locally called *tembwang*), 144 of the tree species were found [89]. In Bangladesh, in 402 home-gardens, Kabir & Webb [90] recorded 419 tree species. Half of these species were native, and six are in the IUCN Red List.

The biodiversity associated with home-gardens is largely unknown. Mardsen and coleagues [91] have conducted bird diversity censuses in home-gardens of New Guinea, finding high richness of species in those systems. In spite of this, many forest species were found in lower numbers or were absent in the gardens. In the Brazilian Atlantic forest hotspot, two works [92, 93] were carried out in the Pontal of Paranapamena agroforestry home-gardens, comparing bird assemblages in pasture, forest, and gardens. Both concluded that agroforestry systems are very important for bird conservation in the region. However, one of them noted a great influence of the distance between gardens and the nearest forest on bird richness and abundance (Fig 6A). Two species of Psittacidae of conservation concern (threatened or near-threatened respectively), *Ara chloroptera* and *Primolius maracana*, were found in home-gardens [92]. Additionally, another study concerning the feeding ecology of frugivorous birds, including *Ara chloroptera*, reported that the feeding activity, and the diversity of food items consumed by this species were greater in home-gardens compared to forest. In spite of it, abundance was higher in the forest compared to gardens. Feeding bouts per abundance (FBPF), which describes the relative amount of feeding activities irrespectively of the frequency of habitat use, was greater in home gardens than in forest (Fig 6B). The study suggested that home-gardens have richer and more abundant food resources, but because they are more intensely disturbed, and perceived predator risk pressure is greater, birds spend less time in this habitat compared to forest [8] (Fig 6B). Figure 7 shows two species feeding in these home-gardens.

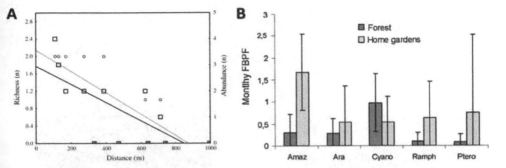

Figure 6. Variation in richness (circles and grey line) and number of individuals (squares and black line) as a function of distance to the closest large forest patch (A) [93], and Feeding bouts per frequency of *Amazona aestiva* (Amaz), *Ara chloroptera* (Ara), *Cyanocorax chrysops* (Cyano), *Ramphasto toco* (Ramph) and *Pterglossus castanotis* (Ptero) in forests and home gardens (B) in the Pontal of Paranapanema, Brazil [8]

Many other tropical agroforestry systems are known to harbour high biodiversity, such as the shade yerba mate [94], and the rubber jungle [95]. Live fences, isolated trees on pastures, and wind breaks also serve as habitat to many species, as well as increase landscape connectivity for many other species [96]. For example, a study conducted in Vera Cruz, Mexico, showed that pasture with isolated trees hosted 35 different orchid species [97]. Another study in the same region found that these systems harboured 58 species of vascular epiphytic and hemiepiphytic forest species [98]. Guevara and Laborde [99] recorded 73 species of bird species visiting four individuals of fig trees (*Ficcus* sp) in pastures of Veracruz. In Australia, isolated paddock trees in New South Wales are used by 31 bird species [99]. In Brazil, *Ara ararauna* has been seen foraging in pastures with high abundance of palms (*Syagrus romazoffiana*). This bird is considered threatened, and its population is declining in some states of Brazil [101]. Other studies show that fig trees (*Ficcus sp*) in pastures near forests are visited by bats [102], and primates [103].

5. Agriculture, biogeochemical cycles, and ecosystem services

Biogeochemical cycles represent the movement of chemical elements within and between several biotic and abiotic entities. These elements can be extracted from mineral sources, the atmosphere, or be recycled through conversion of organic to ionic form, returning to the atmosphere or soil. This cycle is performed by a wide variety of organisms, from a large number of nutrient compartments. The relative abundance of these compartments is specific for each ecosystem type [104]. Any biogeochemistry imbalance between compartments results in diversity loss through bottom-up effects, in which changes in nutrient levels trigger an imbalance in the whole ecosystem's trophic web. Ecosystem fertility is defined as the potential of soil, sediment or aquatic systems to provide nutrients in enough quantity, form or proportion to support optimal plant growth. Soil nutrient flows can be represented by the release of organic matter from microbial communities. However, the chemical balance and the maintenance of ecological processes (mainly carbon, nitrogen, phosphorus, and sulphur cycles) are strongly affected by agricultural and industrial activities.

Figure 7. Blue-fronted Amazon *(Amazona aestiva)* feeding on a mango tree *(Mangifera indica,* Fig. 3.A) and *Melia azedarach* (Fig 3.C). Chestnut-eared Acari *(Pteroglossus castanotis)* feeding on *Cecropia pachystachya* (Fig 3.b) and *Inga vera* (Fig 3.D). Photos by F.F.Goulart

Living organisms usually contain a relatively constant proportion of chemical elements, especially carbon, nitrogen, and phosphorus. In natural ecosystems, the regulation of biogeochemical cycles work at different space and time scales, allowing the adjustment of the nutrient flows from microbial activity to plant demand, reducing nutrient losses within ecosystems. The synchrony between nutrient release, plant use demand, and microorganisms is determined by complex chemical, biological, and physical interactions. These nutrient maintenance processes are rarely achieved in agroecosystems, in which nutrients are lost to the atmosphere and to aquatic ecosystems [105]. In spite of this, agricultural practices vary widely on their efficiency of conserving/losing nutrients. Agricultural intensification, by creating open systems, in which nutrients are lost, are highly dependent on constant inputs (mainly fertilization) in order to sustain production.

5.1. Carbon, nitrogen and phosphorous cycles

Ecosystem nutrient input to produce goods and services to humanity have amplified N and P global cycles by 100% and 400% after the industrial revolution, respectively [106]. Agriculture is responsible for approximately 15% of anthropogenic CO_2 emissions, 58% methane (CH_4) emissions, and 47% of N_2O emissions [107]. The global N cycle was changed by human activities, so that more N is fixed annually by human activities than by all natural means combined [108]. Furthermore, high N concentration in the biosphere interacts with the carbon and sulfur amplified cycles, affecting the global climate [109].

Ecosystem nitrogen increase has been recognized as an important cause for changes in plant species' composition, and for biodiversity loss in a wide range of ecosystems in the globe. According to Bobbink et al [110], an increase in N availability influences species composition and diversity due to changes in competitive interactions among plants, either through the direct effects of nitrogenous gases and aerosols toxicity, or by ammonium nitrate toxicity, which is the predominant N soil form. Increase in soil acidity, cation leaching, and Al concentration promotes ecosystem stress and susceptibility to disturbance, with direct effects on species diversity. Furthermore, changes in competitive interactions due to changes on N amount may be influenced by other edaphic conditions, such as P limitation.

The phosphorus cycle is also greatly affected by agriculture. Globally, 17 tetragrams of phosphorous are applied in the soil every year as fertilizers, and this element is the main driver of water eutrophication worldwide [111]. Menge and Field [111] have argued that an increase in N atmospheric deposition, coupled with increased CO_2 atmospheric concentration stimulates net primary productivity, increasing P demand or limitation. N limitation reduction can cause an increase in P limitation in many ecosystems where N is limiting. The increase in P limitation can modify plant communities by increasing organic P demand, increasing phosphatase enzyme levels in plants and microorganisms. This represents a phosphorus stress that induces limitation changes (from N to P), favouring species that may use P in its organic form [112].

However, effects due to changes in a particular nutrient concentration may be manifested not only in quantitative changes, but also in qualitative changes in the nature of nutrient limitation. According to Elser et al [114], and Davidson and Howarth [115], limitation by N and P are equivalent in terrestrial ecosystems, and the supply and demand of these nutrients are significantly correlated. Thus, the addition of a specific nutrient causes a modest limitation by another, which reduces both nutrients' limitation, causing synergistic positive effects in net primary production in several terrestrial ecosystems. The authors suggest that there is a stoichiometric relationship between N and P supply for autotrophic primary production due to the balance between cellular demand for protein synthesis, or by ATP and nucleotide synthesis. Jacobson et al [116] observed that simultaneous N and P addition affected density, dominance, richness, and diversity patterns in a central Brazilian Cerrado area, with increased rates of plant decomposition. Increased N levels resulted in a greater N loss, and a combined increase (N plus P) resulted in litter N immobilization.

Soil NO emissions were also higher when only N levels were increased, indicating that when increased P availability is not proportional to N increased availability, N losses are intensified. Nutrient cycling in the Brazilian Cerrado is very conservative [117], and increasing human disturbance may cause changes in chemical composition of the organisms' tissues, and also change nutrient cycling in an ecosystem adapted to low nutrient availability. Nutrient dynamic changes may lead to an environmental improvement for some species, increasing their competitiveness in relation to others, which may cause changes in species composition in response to long-term fertilization [118]. A large effect on the diversity of the soil microorganism should also be expected, principally on species rich ecosystems such as in some the Cerrado ecosystems [e.g., 119].

Despite these facts, agroforestry systems are sinks for many green house gases, such as carbon dioxide and atmospheric nitrogen. By using N-fixing leguminous species, agroforestry systems absorb nitrogen from the atmosphere, conserving this nutrient in the soil [119]. A study in USA concluded that agroforestry systems accumulated 530 kg of nitrogen per hectare. Agroforests are also sinks of carbon because of the wood density on these systems. On a global scale, agroforestry systems could potentially be established on 585 to 1275 million hectares of suitable land and these systems could store from 12 to 228 tons of CO_2 per hectare under current climate and soil conditions [120].

5.2. Water

Agroecosystems represent the planet's largest fresh water consumers, with 250 million hectares of irrigated agroecosystems accounting for 69% of the water withdrawing, and 84% of consumptive uses [6]. Forty percent of the world's food production derives from irrigated systems [2]. Water requirements from agriculture are high. For example, it takes 500 litters, 900 litters, and 2000 litters of transpired water to produce 1 kilogram of potatoes, wheat, maize, and rice, respectively [122]. Besides consuming large quantities of fresh-water, agricultural intensification can deeply affect the quality of water resources. Eutrophication of water bodies is mostly related to fertilizer use by intensive agricultural practices [123].

Agricultural development has historically been the main driver of inland water quality loss worldwide. It has been estimated that by 1985 56-65% of the suitable inland water had been used by intensive agriculture in Europe and North America, 27% in Asia, 6% in South America [124]. The nitrate concentration in the biggest rivers in the US increased from three to ten times since the beginning of the century. High quantities of this nutrient is responsible for the baby blue disease (Methemoglobinemia) [125].

Additionally, large quantities of money are spent on irrigation projects. In the 1970s, US investments had reached their peak of 1 billion dollars per year. In Brazil, there is great controversy involved in the project by the present government to transpose the São Francisco River. The ongoing project consists in deviating part of the river course into 600 kilometres of channels, and is costing 3.7 billion dollars to the Brazilian society. The main objective of the transposition is to increase water supply for shrimp farms, and agroindustry in the northeast region, so that only 4% of the water would be directed to the population. The ecological impacts of such enterprise will be enormous. The Rio São Francisco harbours 137 species, many of which are endemic or threatened with extinction [126]. The withdrawing of the river water will lower water levels, possibly causing profound impacts on fish assemblages.

One way of reducing the need for intensive irrigation systems is conserving the water in the agricultural systems, which is basically related to the presence of organic matter. Agricultural management that increases organic matter in the soil, such as no-till and specially agroforestry systems, increase water conservation in the soil. Many agroforestry management use trees (e.g. banana trees) that enhance soil moisture.

5.3. Ecosystems services

Several groups at the base of the food chain, from microorganisms to soil micro- and meso-fauna, present ecological roles that affect nutrient fixation, cycling, and mobilization in the soil. Arthropods are by far the most studied group regarding their ecological roles in agroecosystems. They have been shown to have crucial roles in pollination [127], and predation of or competition with pest species [128]. De Marco and Coelho [129] noted that the raw production of coffee systems near forest fragments increased 14%, due to an increase in pollination activity in the agroecosystem. Figure 4 shows the diversity of flower visiting insects in coffee plants in home-gardens of the Pontal of the Paranapanema, southwest of Brazil. Ecosystem services provided by arthropods are also negatively affected by the use of biocides [130]. The loss of base species, as well as the structural simplification inside farms, leads in turn to a decrease in richness and abundance of several vertebrate groups [131]. These processes that follow agricultural intensification cause agroecosystems to lose basic regulation processes, including soil fertility and pest control. The latter is the most well-known consequence of biodiversity reduction [131], with a great number of studies demonstrating the effects of native arthropod predation on pest species abundance and richness [e.g. 61,72,132,133].Vertebrate species, such as bats and birds, also present important ecological roles related to pollination, pest predation, and seed dispersal [61,72]. All these groups can be affected by agricultural intensification, which in turn could affect production in a feed-back fashion.

Figure 8. High diversity of flower visiting insects, such as bees (*Xylocopa sp*, (fig 4.A), *Epicharis flava* (fig 4.B), *Exomalopsis fulvofasciata* (fig 4.C), moths (Fig 4.D, 4.F), and butterflies (Fig 4.G, 4.E, 4.H) in coffee plants of home-gardens of Pontal do Paranapanema, southwest of Brazil (Photos: F.F.Goulart.)

It is possible measure the effect of the functional groups, such as predators and pollinator by assessing production or variables that are correlated to it (such as biomas, insect damage, herbivore abundance, etc..) with predator presence or abundance. With this respect, experiments that exclude pollination or predation action have been useful to estimate the value of these services.

A table showing selected studies on the influence of functional groups regarding predation of potential agronomic pests and pollination of cocoa and coffee systems are shown in the appendix A.

The mechanisms that lead to biodiversity simplification through agricultural intensification vary. Soil microorganisms and micro-fauna have been shown to decrease in richness and abundance with intensive soil tillage [135,136] due to the closing of soil cracks and pores, with the consequent drying of the soil surface [137].

The result of these losses is the dependence on external inputs, which increases the costs involved in food, fiber and fuel production, as well as a decrease in soil and water quality, quality of the food produced and of rural life [134]. As an example, Boyles et al. [138] have estimated that agricultural losses caused by the decline of bat populations in North America are worth more than 3.7 billion dollars/year. For the entire biosphere, estimates of the value of natural ecosystem services vary between 16 and 54 trillion dollars/year, with an average of 33 trillion dollars/year [127]. However, as Martis [139] has pointed out, although it is possible to evaluate objectively the ecological value of certain species within a complex net of relationships and connections, we currently do not appreciate the importance of certain species, until after their disappearance.

The ecological roles of a wide variety of taxa and functional groups are object of considerable research, but we still do not fully understand the roles and the links between them [140]. Anyway, since those ecological processes on which productive systems depend on are largely biological, feed-back loops caused by the modification of natural systems are bound to occur, affecting agricultural yield and stability. Therefore, there is a growing body of knowledge that demonstrates the beneficial economic effects of managing biodiversity so as to maintain ecosystems services functional [134, 141]. For instance, no-till practices generate more biologically complex soils, so that they could potentially enhance these groups' diversity [141]. Also, tropical crop productivity can be enhanced by the promotion of a more heterogeneous environment inside farms and at the landscape level, so that the occurrence of native pollinator species, pest species' natural predators [62], and competitors is promoted. Generally, diversified cropping systems harbour more arthropod populations, because these species respond to: 1) habitat heterogeneity; 2) higher predation, which facilitates species coexistence though density control mechanisms; 3) and higher stability and resource-partitioning, since production stability and predictability promotes temporal and spatial partitioning of the environment, permitting coexistence of species [131]. In a landscape perspective, management options which promotes biodiversity conservation by enhancing native or planned heterogeneity inside farms or in their surroundings, have been suggested [131,143].

Structurally complex landscapes, which involve mosaic formations and corridors, facilitate native species' dispersal between farms and natural strips of vegetation [144]. There is a strong relationship between pollination activity and distance from farms to the closer forests areas (Fig. 9). Therefore, forests are sources of pollinators, and agricultural landscapes that preserve landscape heterogeneity and the forest protection can increase agricultural yield.

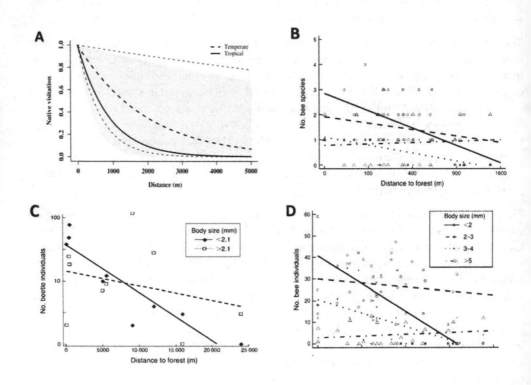

Figure 9. Decay curves for native visitation rates in cultivated systems of tropical and temperate region (A), so that solid line and shading are for tropical region, while dashed lines and lighter dashed lines concerns temperade region. (A). Source: Ricketts et., 2008. Distance from the nearest forest and the abundance of beetles (C), abundance of bees (B) and bees richness (D) of different body sizes in coffee farms of Indonesia. Source: [145]

6. Prospects for the future of farming and biodiversity

Farming practices are one of the greatest emitters of green house gases, such as carbon dioxide, methane, and atmospheric nitrogen, contributing greatly to climate change. Therefore agriculture is in part responsible for future climate changes. Climate change is affecting and will affect even more people's lives, biodiversity, and ecosystems. Climate change is already one of the major drivers of biodiversity loss worldwide [10]. Climate relates with many population processes, such as disease dynamics. In the highlands of Costa Rica, the outbreak of the disease caused by the fungus *Batrachochytrium dendrobatidis*, associated with global warming caused the extinction of many frog species, such as of the golden toad (*Bufo periglenes*) [146].

Predictions of the impact of future climate change on species, ecosystems and agriculture are alarming, if not catastrophic. Climate change is expected to be the major drivers of ecological shifts in a near future [147]. Simulations of International Panel of Climate Change (IPCC) suggest that the mean earth temperature will rise up to 5.8 C°, and the weather will become significantly drier [148].

This will involve several changes, and one of the most notable will be the shift in the range of species distributions. Using bioclimatic models, scientists have predicted the future distribution of several species. A review of 2,954 mammals, birds, and amphibians in the Western Hemisphere suggested that 10% of the extant fauna will be lost due to climatic change. Greater changes are predicted for the Tundra, Central America, and the Andes species, which will undergo over 90% of species turnover, assuming no dispersal [149]. For instance, one of the world's most species rich ecosystems, the high altitude rupestrian fields of Brazil, might have only 15-20% of species left by the year 2080 under the best IPCC scenario [150]. Because such models do not consider process such as biotic interactions the real scenario can be worse. In a study conducted at a smaller scale, Marini and colleagues [151] modelled the present and the future distribution of *Amazona pretreii*, a threatened parrot of the Atlantic Forest of Brazil. They conclude that the year-round distribution of this species will decrease 47% until 2060. A similar analysis was conducted with 120 bat species on the Brazilian Cerrado (woodland savanna) [152]. The study was aimed to evaluate possible responses by this group for climate change, considering the IPCC scenario for 2050. The results indicated that bat species would find, in average, similar climate conditions in the future 480 km away from current regions. For the majority of the 120 species modeled, suitable regions will be located to the South (80% of the species) and to the West (56% of the species). For two bat species there will be no suitable conditions on the Brazilian Cerrado on the future, and they will be locally extinct. For 96 species the models indicate a significant contraction on their distribution (41% of actual distribution in average) only due to climate change (not accounting habitat lost by deforestation). The region were the distribution shift is expected by bat species in the future is already extremely fragmented. According to the Ministry of Environment [153], the states of Parana and São Paulo, the region where the models indicated that most of bat species would find better climate conditions in the future, are two of the worst region in terms of natural vegetation coverage. The Cerrado areas were reduced to less than 10% of its original area. Due to its flight ability, perhaps bat species can cross large deforested area easily, but we can not say the same about the species that they depend on, such as plant species, insects, and small vertebrates. Species will have to disperse large distance to reach areas in which climate is the same as today. This suggests that future landscape connectivity will play a major role in the effectiveness of the species in reaching new areas. In this context, agroforestry systems will have a key importance in the maintenance of matrix permeability [154,93].

Climate change will affect not only species and natural ecosystems but also the agriculture. The Brazilian Center for Agriculture Research - EMBRAPA has already projected momentous changes for the regions where cultures such as soybean, sugarcane, cotton, coffee, cassava and corn are currently implemented. According to EMBRAPA and UNICAMP [155] all cultures cited above, except sugarcane and cassava, will have their areas decreasing due to climate change and global warming. Projections based on IPCC A2 scenario for 2050 verify that appropriate areas for soybean plantation will be mainly on the center and Southeastern regions of Brazil. The Southeastern region of Brazil is quite the projected area that will be sufficient to, for instance, birds and bat species of the Brazilian Cerrado. Thus, spatial competition for food production and species protection is a serious issue nowadays and for the future.

Regarding human aspects, the exponential growth of human population, and the increased *per capita* consumption reflected in the development of a highly expansive and intensive agriculture. It has been estimated that the human population will increase by 50% until 2050, with a higher expected proportion of individual meat consumption in the daily diet (feeding at higher levels of the trophic pyramid) [26]. Sustaining food production in the same magnitude of human growth is a challenge for all areas of human knowledge.

Already 1.2 billion people live in areas in which water is physically scarce, and this number should double by 2030 [6]. Projections of the proportion of total global food supply obtained from rain fed areas (non-irrigated) should decline from 65% currently to 48% in 2030 [156]. The total irrigated area is expected to grow from 254 million ha in 1995 to between to 280 and 350 million ha in 2025. Fertilizer use is expected to increase 188 million tons by 2030 [157], and the world's meat consumption is expected to grow by 70% in the 2000-2030 period, and 120% in the 2000-2050 period [6]. Concerning food production, future predictions are also alarming. Global cereal production is predicted to decline by more than 5%, but this value may reach more than 10%. The risk of hunger may rise up to almost 60% in the developing world [158]. In some countries, such as India, production of crops may decrease by 70% [159]. When food security, availability, stability, utilization, and access are considered, between 5 to 170 million additional people will be at risk of hunger by 2080 [160]. However, childhood malnutrition is projected to decline from 149 million children in 2000 to 130 million children by 2025, and 99 million children by 2050 [6]. In the Amazon, soybean yields will suffer a reduction of 44% by 2050 [161]. It is estimated that the average rate of atmospheric N deposition in 34 world biodiversity hotspots by 2050 will be twice the rate in all terrestrial ecosystems during the mid-nineties [162]. This will greatly affect plant assemblages through altered competition patterns.

According to the International Assessment of Agricultural Knowledge, Science and Technology for Development [6], an international report involving 800 well known scientists, the best way to nurse today's and future hunger problem is by fostering small scale agriculture based on agroecological principles. The problem of climate change is due to high quantities of green house gases that affect the earth surface temperature, and agroforestry systems are sinks of some of these gases, such as nitrogen and carbon [119]. Therefore, a possible alleviation for climate change is to 're-green' [163] the planet, using agroforestry systems over large areas. Agroforestry systems thus represent a keynote in this re-greening strategy. Because it enables the association between agriculture production and biodiversity and ecosystem conservation.

7. Conclusion

Overall, this chapter presents an overview on the agricultural systems and the effect of different types of management on biodiversity and ecosystems. We analyse data for shade-cocoa, shade-coffee and agroforestry home gardens in Brazilian atlantic forest region. In most situations, win-win relationship between conservation and production is possible, as farms with intermediate levels of yield are associated with high biodiversity. Also, the idea that there is a need to intensify agricultural systems to increase food production to feed the hungry does not apply to many tropical agriculture landscapes. Instead, changing food habits and promoting a more even food distribution using small scale eco-agriculture will guarantee a more resilient, social, and biodiversity friendly practices.

The future of farming and biodiversity depends on the type of agricultural management that will be applied in landscapes. If the agricultural intensification continues to expand, it is very likely that yields will increase, but with high variance and low resilience to environmental uncertainties, which are predicted to increase due to climatic changes and loss of ecosystem services. On the other hand, stakeholders may opt for more biodiversity friendly agricultural practices, which sometimes (but not always) are less productive than intensive systems, but have more productive stability and are more resilient. Additionally, these non-intensive systems can mitigate climate change by being sinks of green house gases.

Food security in the tropics depends on the recognition of the importance of the poor small holder agricultural systems, because they are the majority of the hungry people in the world. The *'business as usual'* strategy should increase economic inequality, increasing poverty and starvation, as well as causing deep ecological impacts. On the other hand, small-holder mixed-farming systems increase food security during times of ecological and economical instability. As we see it, heterogeneous agroforestry is the best option for biological conservation and social justice. It is expected that in the near future, millions of hectares of land will be occupied by agroforestry systems.

Appendix

Ecosystem service (taxa)	Plant species	Proportional yield loss or indirect effect caused by the reduction or absence of the functional group	Forest distance and functional group richness/abundance	Main findings	Country	Exclusion experiment	Ref.
Pollination (bees)	Coffee	0,15 (sites near forests)	Highly correlated	Pollination accounted for US$ 1860.55 ha per year.	Brazil	Yes	129
Pollination (bees)	Coffee	0,05 to 0,56 (mean = 0,17)	Not correlated	Appis melifera (honeybee) accounted for high proportion of the visits (95% of the visits)	Panama	Yes	164
Pollination (bees)	Coffee	0,27	Highly correlated.	Solitary bees generally show low abundance, but high pollination effectiveness. Diversity, rather than abundance explained variation in fruit set. Rare solitary bees are more important than abundant social bees.	Indonesia	Yes*	165, 166
Pollination (bees)	Coffee	0,8	Not addressed	Appis melifera showed high visitation rate.	Ecuador	No	169
Pollination (bees and ants)	Coffee	0,1 (low shade) to 0,41 (high shade)	Not addressed	Flying pollinators alone did not affect fruit set or fruit weight. In spite of it, the exclusion of ants and flying pollinators decrease fruit weight in shade plantations.	Mexico	Yes	168
Pollination (midges)	Cocoa	0,77	Not correlated	Ceratopogonidae midges have high effectiveness.	Ghana	No	169
Predation (birds)	Coffee	0,01 to 0,14	Landscape heterogeneity had significant effect	The migrant Dendroica caerulescens (Black throated Blue Warbler) showed high predation effectiveness.	Jamaica	Yes	72
Predation (birds)		Indirect effect: reduction in 64% to 80% of large arthropods caused by birds	Not addressed		Guatemala	Yes	62
Predation (birds, lizards, arthropods predator and parasitoids)	Coffee	Indirect effect: reduction in the abundance of the pests (Leucoptera coffella and Petrusa epilepis)	Not addressed	Birds and lizards had additive effects. Both groups fed on arthropods and parasitoids (intra-guild predation).	Porto Rico	Yes	170
Predation (birds)	Coffee	Indirect effect: Lepidoptera pest removal was smaller in the controls	Not addressed	The bird (Basileuterus rufifrons) showed high predation effectiveness.	Mexico	Yes	171
Predation (ants)	Coffee	Indirect effect: higher predation rates of coffee berry border	Not addressed	Higher removal rates were found in shade systems in the wet season.	Colombia	No	172
Predation (birds)	Cocoa	Indirect effect: Birds reduce leaf damage from 9.7% to 7.6%	Not addressed	Dendroica pensylvanica showed high predation effectiveness.	Panama	Yes	173
Predation (bats)	Coffee	Indirect effect: Bats and birds reduce arthropod abundance	Not addressed	Arthropods in the control were 46% higher than in the treatment in which both bats and birds were excluded.	Mexico	Yes	174
Predation Basileuterus rufifrons	Coffee	Indirect effect: birds can reduce arthropod abundance in 58%	Not addressed	When migrants are present, birds forage more frequently in the understory, where there is lower competition with migrants.	Mexico	Yes	175
Predation: (Azteca spp. ants)	Coffee	Indirect effect: Pieris rapae (a coffee pest) suffered higher removal	Not addressed	Contradictory results show that the ants can have potential as pests through their positive effect on scale, but also as biological control agents	Mexico	No	176
Predation: (Wasmannia spp. ants)	Cocoa	no significant effect:	Not addressed	This species have low effectiveness of predating on potential pests.	Brazil	No	177
Parasitoid	Cocoa	Indirect effects: parasitoids species correlated tree species	Not addressed	Higher diversity of shade trees maintains high parasitoid levels.	Brazil	No	178

Appendix A: Selected studies concerning ecosystem services in coffee and cocoa plantations in the tropics highlighting the main findings.

Author details

F.F. Goulart[*], B.Q.C. Zimbres and R.B. Machado
Laboratório de Planejamento para a Conservação da Biodiversidade,
Departamento de Zoologia, Universidade de Brasília, Campus Darcy Ribeiro, Brasília, DF, Brazil

T.K.B. Jacobson
Campus Planaltina, Universidade de Brasília/
Centro de Estudos UnB Cerrado, Universidade de Brasília, Brasília, DF, Brazil

L.M.S. Aguiar
Laboratório de Biologia e Conservação de Mamíferos,
Departamento de Zoologia, Universidade de Brasília, Campus Darcy Ribeiro, Brasília, DF, Brazil

G.W. Fernandes
Laboratório de Ecologia Evolutiva e Biodiversidade/DBG,
ICB/Universidade Federal de Minas Gerais, Belo Horizonte, MG, Brazil

Acknowledgement

We thank A. Azevedo (Biotrópicos- Instituto de Pesquisa em Vida Silvestre) for identifying the bee species, P. Saboya for the critiques on the model, and the farmers of Pontal do Paranapanema and Serro region. We would like to thank F. Takahashi for the critiques on the final manuscript. The Coordination for the Improvement of Higher Education Personel (CAPES) for a PHD fellowship for F.F.G.

We would like to thank the RedeCerrado and Graduate Program in Ecology of the University of Brasília." The name R.B. Machado appears twice. Where it says R.B. Machado refered to the Laboratório de Biologia e Conservação de Mamíferos, it should be L.M.S. Aguiar. Change the adress of T.K.B. Jacobson for: Universidade de Brasília, Campus UnB Planaltina, Área Universitária n.1, Planaltina-DF. Universidade de Brasília, Centro de Estudos UnB Cerrado da Chapada dos Veadeiros

8. References

[1] McNeely JA, Scherr SJ (2003) Ecoagriculture: strategies to feed the world and save wild biodiversity. Washington: Island Press. pp. 352

[2] Cassman K, Wood S (2005) Cultivated Systems. In: Milenium Ecosystem Assessment Report (available at http://www.maweb.org/en/index.aspx). Washington: Island Press. pp. 793

[3] Perfecto I, Rice RA, Greenberg R, Van der Voort ME (1996) Shade-coffee: a disappearing refuge for biodiversity. Biosci. 46: 598–608.

---

[*] Corresponding Author

[4] Faria D, Laps RR, Baumgarten J, Cetra M (2006) Bat and bird assemblages from forests and shade cacao plantations in two contrasting landscapes in the Atlantic Forest of southern Bahia, Brazil. Biodiv. cons. 15: 587–612.

[5] Altieri MA (1995) Agroecology: the Science of Sustainable Agriculture. Boulder: Westview Press. 433 p

[6] McIntyre BD, Herren HR, Wakhungu J, Watson RT (2009) Agriculture at a Crossroads. International assessment of agricultural knowledge, science and technology for development (IAASTD): global report. Synthesis Report. Washington: Island Press, 590

[7] Faria D, Paciencia ML, Dixo M, Laps RR, Baumgarten J (2007) Ferns, frogs, lizards, birds and bats in forest fragments and shade cacao plantations in two contrasting landscapes in the Atlantic forest, Brazil. Biodiv. and cons. 16: 2335–2357.

[8] Goulart FF, Vandermeer J, Perfecto I, da Matta-Machado RP (2011) Frugivory by five bird species in agroforest home-gardens of Pontal do Paranapanema, Brazil. Agrof. syst. 1–8. DOI10.1007/s10457-011-9398-z

[9] Matson PA, Parton WJ, Power A, Swift M (1997) Agricultural intensification and ecosystem properties. Sci. 277: 504.

[10] Mace G, Masundire H, Baillie J, (coord. authors) (2005) Biodiversity. In: Millenium Ecosystems Assessment, p. 77 – 122. Washington: Island Press

[11] Myers N, Mittermeier RA, Mittermeier CG, da Fonseca GA, Kent J (2000) Biodiversity hotspots for conservation priorities. Nat. 403: 853–858.

[12] Pimentel D, Stachow U, Takacs DA, Brubaker HW, Dumas AR, Meaney JJ, Onsi DE, Corzilius, DB (1992) Conserving biological diversity in agricultural/forestry systems. Biosci. 42: 354–362.

[13] Perfecto I, Vandermeer J (2008) Biodiversity conservation in tropical agroecosystems. Ann. n.y. acad. sci. 1134: 173–200.

[14] Goulart F, Vandermeer J, Perfecto I, Matta-Machado R (2009a) Análise agroecológica de dois paradigmas modernos. Revista Brasileira de Agroecologia 4: 76-85.

[15] Noble IR, Dirzo R (1997) Forests as human-dominated ecosystems. Sci. 277: 522.

[16] Taylor PD, Fahrig L, Henein K, Merriam G (1993) Connectivity is a vital element of landscape structure. Oikos 68: 571–573.

[17] Antongiovanni M, Metzger JP (2005). Influence of matrix habitats on the occurrence of insectivorous bird species in Amazonian forest fragments. Biol. cons. 122: 441–451.

[18] Vandermeer J, Carvajal R (2001) Metapopulation dynamics and the quality of the matrix. Am. Nat. 158: 211.

[19] Balmford A, Green R, Scharlemann JPW (2005) Sparing land for nature: exploring the potential impact of changes in agricultural yield on the area needed for crop production. Glob. chang. biol. 11: 1594–1605.

[20] Green RE, Cornell SJ, Scharlemann JP, Balmford A (2005) Farming and the fate of wild nature. Sci. 307: 550.

[21] Perfecto I, Vandermeer J (2010) The agroecological matrix as alternative to the land-sparing/agriculture intensification model. Proc. n. acad. sci. 107: 5786.

[22] Angelsen A, Kaimowitz D (2001) Agricultural technologies and tropical deforestation. CABI Publishing. pp. 411

[23] Makowski D, Dore T, Gasquez J, Munier-Jolain N (2007) Modelling land use strategies to optimise crop production and protection of ecologically important weed species. Weed res. 47: 202–211.

[24] Davidson EA, Howarth RW (2007) Nutrients in synergy. Nat. 449: 1000-1001.

[25] Wood S, Ehui S (2005) Food

[26] Levy M, Babu S, Hamilton K (2005) Ecosystem conditions and human well-being. In: Kakri AH, Watson R, editors. Millenium Ecossystem Assesment. Washington DC: Island Press. pp.794

[27] OECD, 2012. Available at : http://www.oecd.or (Organization of Economic Co-operation and Development).

[28] Bessesen DH (2008) Update on obesity. J. Clinic. Endocr. Metab. 93: 20-27.

[29] Kopelman PG, Caterson ID, Dietz WH (2005) Clinical obesity in adults and children. Wiley Online Library.

[30] WHO (2005a) Global database on Body Mass Index. Available at http://www.who.int/ncd _surveillance/infobase/web/InfoBaseCommon/.

[31] Guanziroli CE (2001) Agricultura familiar e reforma agrária no século XXI. Rio de Janeiro: Garamond. pp. 291

[32] Flynn DF, Gogol-Prokurat M, Nogeire T, Molinari N, Richers BT, Lin BB, Simpson N, Mayfield MM, DeClerck F (2009) Loss of functional diversity under land use intensification across multiple taxa. Ecol. Lett. 12: 22–33.

[33] Shiva V, Jafri AH, Shiva V, Bedi G (2002) Seeds of suicide: the ecological and human costs of globalization of agriculture. New Delhi: Sage Publications India Pvt Ltd. pp. 151

[34] Rosset M, Rosset PM, Write O (1999) The multiple functions and benefits of small farm agriculture. Policy Brief No 4, Washington DC: Institute for Food and Development Policy. pp. 22.

[35] Badgley C, Perfecto I (2007) Can organic agriculture feed the world? Renewable agric. food syst. 22: 80–85.

[36] Stanhill G (1990) The comparative productivity of organic agriculture. Agric. ecosys. envir. 30: 1–26

[37] Benton TG, Vickery JA, Wilson JD (2003) Farmland biodiversity: is habitat heterogeneity the key? Trends ecol. evol. 18: 182–188.

[38] Chamberlain DE, Fuller RJ, Bunce RGH, Duckworth JC, Shrubb M (2000) Changes in the abundance of farmland birds in relation to the timing of agricultural intensification in England and Wales. J. appl. ecol. 37: 771–788.

[39] Krebs JR, Wilson JD, Bradbury RB, Siriwardena GM (1999) The second silent spring? Nat. 611–612.

[40] Donald PF, Evans AD (2006) Habitat connectivity and matrix restoration: the wider implications of agrienvironment schemes. J. appl. ecol. 43: 209–218.

[41] Perfecto I, Vandermeer J, Mas A, Pinto LS (2005) Biodiversity, yield, and shade-coffee certification. Ecol. econom 54: 435–446.

[42] Altieri MA, Merrick L (1987) In situ conservation of crop genetic resources through maintenance of traditional farming systems. Econom. bot. 41: 86–96.

[43] Wood, D, Lenne J M (1997) The conservation of agrobiodiversity on-farm: questioning the emerging paradigm. Biodivers. cons. 6: 109–129.

[44] Carson R (1964) Silent spring. Boston: Mariner Books. pp. 368

[45] Ratcliffe DA (1970) Changes attributable to pesticides in egg breakage frequency and eggshell thickness in some British birds. J. Appl. ecol. 7: 67–115.

[46] Mellink E, Riojas-López ME, Luévano-Esparza J (2009) Organchlorine content and shell thickness in brown booby (Sula leucogaster) eggs in the Gulf of California and the southern Pacific coast of Mexico. Envir. poll. 157: 2184–2188.

[47] Busby DG, Pearce PA, Garrity NR, Reynolds LM (1983) Effect on an Organophosphorus Insecticide on Brain Cholinesterase Activity in White-Throated Sparrows Exposed to Aerial Forest Spraying. J. Appl. Ecol. 20: 255–263.

[48] Cooper RJ, Dodge KM, Martinat PJ, Donahoe SB, Whitmore RC (1990) Effect of diflubenzuron application on eastern deciduous forest birds. J. wildl. manag. 54: 486–493.

[49] Kiesecker JM (2002) Synergism between trematode infection and pesticide exposure: A link to amphibian limb deformities in nature? Proc. n. acad. sci. 99: 9900.

[50] WHO (2005b) Modern food biotechnology, human health and development: An evidence-based study.

[51] MacArthur RH, MacArthur JW (1961) On bird species diversity. Ecol. 42: 594–598.

[52] Begon M, Harper JL, Townsend CR (1996) Ecology: individuals, populations, and communities. Malden: Wiley-Blackwell. 1092 p.

[53] Paine RT (1969) A note on trophic complexity and community stability. The American Naturalist 103: 91–93.

[54] Bowman GB, Harris LD (1980) Effect of spatial heterogeneity on ground-nest depredation. J. Wildl. Manag. 44: 806–813.

[55] Whittingham MJ, Evans KL (2004) The effects of habitat structure on predation risk of birds in agricultural landscapes. Ibis 146: 210–220.

[56] Lima SL, Valone TJ (1991) Predators and avian community organization: an experiment in a semi-desert grassland. Oecologia 86: 105–112

[57] Desrochers A, Hannon SJ (1997) Gap crossing decisions by forest songbirds during the post-fledging period. Cons. biol. 11: 1204–1210.

[58] Marcelo C. Silva, pers. comm.

[59] Clough Y, Dwi Putra D, Pitopang R, Tscharntke T (2009) Local and landscape factors determine functional bird diversity in Indonesian cacao agroforestry. Biol. cons. 142: 1032–1041.

[60] Reitsma R, Parrish JD, McLarney W (2001) The role of cacao plantations in maintaining forest avian diversity in southeastern Costa Rica. Agrof. syst. 53: 185–193.

[61] Greenberg R, Bichier P, Angón AC (2000a) The conservation value for birds of cacao plantations with diverse planted shade in Tabasco, Mexico. Anim. cons. 3: 105–112

[62] Greenberg R, Bichier P, Angon AC, MacVean C, Perez R, Cano E (2000b) The Impact of Avian Insectivory on Arthropods and Leaf Damage in Some Guatemalan Coffee Plantations. Ecol. 81: 1750-1755.

[63] Sonwa DJ, Nkongmeneck BA, Weise SF, Tchatat M, Adesina AA, Janssens MJ (2007) Diversity of plants in cocoa agroforests in the humid forest zone of Southern Cameroon. Biodiversity and Conservation 16: 2385–2400.

[64] Delabie JH, Jahyny B, do Nascimento IC, Mariano CS, Lacau S, Campiolo S, Philpott SM, Leponce M (2007) Contribution of cocoa plantations to the conservation of native ants (Insecta: Hymenoptera: Formicidae) with a special emphasis on the Atlantic Forest fauna of southern Bahia, Brazil. Biodiv. cons. 16: 2359–2384.

[65] Sambuichi RH (2002) Fitossociologia e diversidade de espécies arbóreas em cabruca (mata atlântica raleada sobre plantação de cacau) na Região Sul da Bahia, Brasil. Acta Bot. Brasilica 16: 89–101.

[66] Mittermeier RA (1988) Primate diversity and the tropical forest. In: Wilson EO, Peter FM, editors. Biodiversity. Washington: National Academy Press. pp. 145–154.

[67] Oliver WLR, Santos IB (1991) Threatened endemic mammals of the Atlantic forest region of South-east Brazil. Jersey Wildl. Preserv. Trust, special Scientific Report (4): pp. 125.

[68] Pacheco LF, Whitney BM (1996) A new genus and species of furnariid (Aves: Furnariidae) from the cocoa-growing region of southeastern Bahia, Brazil. Wilson Bull. 108: 397–433.

[69] Ricardo R. Laps, pers comm.

[70] Vaughan C, Ramírez O, Herrera G, Guries R (2007) Spatial ecology and conservation of two sloth species in a cacao landscape in Limón, Costa Rica. Biodiv. cons. 16: 2293–2310

[71] Clough Y, Barkmann J, Juhrbandt J, Kessler M, Wanger TC, Anshary A, Buchori D, et al. (2011). Combining high biodiversity with high yields in tropical agroforests. Proc. nat. acad. sci. 108(20), 8311.

[72] Kellermann JL, Johnson MD, Stercho AMY, Hackett SC (2008) Ecological and economic services provided by birds on Jamaican Blue Mountain coffee farms. Cons. biol. 22: 1177–1185.

[73] Calvo L, Blake J (1998) Bird diversity and abundance on two different shade-coffee plantations in Guatemala. Bird cons. internat. 8: 297–308.

[74] Goulart FF, Monte AZL, Checoli CH, Saito CH, (2009b) Etnoecologia associada aos cafezais agroflorestais tradicionais da região do Serro. Annals of the 5th Congresso Nacional de Sistemas Agroflorestais, Luziânia, Brazil.

[75] Moguel P, Toledo VM (1999) Biodiversity conservation in traditional coffee systems of Mexico. Cons. biol. 13: 11–21.

[76] Solis-Montero L, Flores-Palacios A, Cruz-Angón A (2005) Shade-Coffee Plantations as Refuges for Tropical Wild Orchids in Central Veracruz, Mexico. Cons. biol. 19: 908–916]

[77] Pineda E, Moreno C, Escobar F, Halffter G (2005) Frog, bat, and dung beetle diversity in the cloud forest and coffee agroecosystems of Veracruz, Mexico. Cons. biol. 19: 400–410.

[78] Klein AM, Steffan-Dewenter I, Buchori D, Tscharntke T (2002) Effects of Land-Use Intensity in Tropical Agroforestry Systems on Coffee Flower-Visiting and Trap-Nesting Bees and Wasps. Cons. biol. 16: 1003–1014.

[79] Ormerod SJ, Watkinson AR (2000) Editors' introduction: birds and agriculture. J. Appl. ecol. 37: 699–705.

[80] Komar O (2006) Priority Contribution. Ecology and conservation of birds in coffee plantations: a critical review. Bird Conservation International 16: 1–23.

[81] Wunderle Jr JM, Latta SC (1996) Avian abundance in sun and shade-coffee plantations and remnant pine forest in the Cordillera Central, Dominican Republic. Ornit. neotrop. 7: 19–34.

[82] Somarriba E, Harvey CA, Samper M, Anthony F, González J, Staver C, Rice RA (2004): Biodiversity conservation in neotropical coffee (*Coffea arabica*) plantations. In: Schrotth G, Fonseca G, Harvey C, Gascon C,Vasconcelos H, Izac A. Agroforestry and biodiversity conservation in tropical landscapes. pp.198–226.

[83] Wunderle Jr JM, Latta SC (1996) Avian abundance in sun and shade-coffee plantations and remnant pine forest in the Cordillera Central, Dominican Republic. Ornit. neotrop. 7: 19–34.

[84] Kumar BM, Nair PKR (2004) The enigma of tropical home-gardens. Agrof. syst. 61: 135–152.

[85] Soemarwoto O, Conway GR (1992) The javanese home-garden. J. farm. syst. res. ext. 2: 95–118.

[86] Galluzzi G, Eyzaguirre P, Negri V (2010) Home-gardens: neglected hotspots of agro-biodiversity and cultural diversity. Biodiv. cons. 19(13): 3635–3654

[87] Sunwar S, Thornström CG, Subedi A, Bystrom M (2006) Home-gardens in western Nepal: opportunities and challenges for on-farm management of agrobiodiversity. Biodiv. conserv. 15: 4211–4238.

[88] Albuquerque UP, Andrade LHC, Caballero J (2005) Structure and floristics of home-gardens in Northeastern Brazil. J. arid envir. 62: 491–506.

[89] Marjokorpi A, Ruokolainen K (2003) The role of traditional forest gardens in the conservation of tree species in West Kalimantan, Indonesia. Biodiv. cons. 12: 799–822.

[90] Kabir ME, Webb EL (2008) Can home-gardens conserve biodiversity in Bangladesh? Biotropica 40: 95–103.

[91] Marsden SJ, Symes CT, Mack AL (2006) The response of a New Guinean avifauna to conversion of forest to small-scale agriculture. Ibis 148: 629–640.

[92] Goulart FF (2007) Aves em quintais agrflorestais do Pontal do Paranapanema: epistemologia, frugivoria e estrutura de comunidades. Dissertation. Universidade Federal de Minas Gerais.

[93] Uezu A, Beyer DD, Metzger JP (2008) Can agroforest woodlots work as stepping stones for birds in the Atlantic forest region? Biodiv. conserv. 17: 1907–1922.

[94] Cockle KL, Leonard ML, Bodrati AA (2005) Presence and abundance of birds in an Atlantic forest reserve and adjacent plantation of shade-grown yerba mate, in Paraguay. Biodiv. cons. 14: 3265–3288.

[95] Thiollay JM (1995) The role of traditional agroforests in the conservation of rain forest bird diversity in Sumatra. Conserv. biol. 9: 335–353.

[96] Harvey CA, Tucker NI, Estrada A (2004) Live fences, isolated trees, and windbreaks: tools for conserving biodiversity in fragmented tropical landscapes. Agroforestry and biodiversity conservation in tropical landscapes 261–289.

[97] Williams-Linera G, Sosa V, Platas T (1995) The fate of epiphytic orchids after fragmentation of a Mexican cloud forest. Selbyana 16: 36-40.

[98] Hietz-Seifert U, Hietz P, Guevara S (1996) Epiphyte vegetation and diversity on remnant trees after forest clearance in southern Veracruz, Mexico. Biol. cons. 75: 103–111.

[99] Fischer J, Lindenmayer DB (2002) The conservation value of paddock trees for birds in a variegated landscape in southern New South Wales. 1. Species composition and site occupancy patterns. Biodiv. cons. 11: 807–832.

[100] Guevara S, Laborde J (1993) Monitoring seed dispersal at isolated standing trees in tropical pastures: consequences for local species availability. Plant ecol. 107: 319–338.

[101] Rodrigues M, Goulart FF (2005) Aves regionais: de Burton aos dias de hoje. In: Goulart EM, editor. Navegando o Rio das Velhas das Minas aos Gerais. Belo Horizonte: Editora Guaycui. pp. 589-603

[102] Galindo-González J, Guevara S, Sosa VJ (2000) Bat-and Bird-Generated Seed Rains at Isolated Trees in Pastures in a Tropical Rainforest. Cons. biol. 14: 1693–1703.

[103] Slocum MG, Horvitz CC (2000) Seed arrival under different genera of trees in a neotropical pasture. Plant ecol.149: 51–62.

[104] Lavelle P, Berhe AA (2005) Nutrient Cycling. Ecosystems and human well-being: current state and trends: findings of the Condition and Trends Working Group of the Millennium Ecosystem Assessment 1, 331.

[105] Cadisch G, Giller KE (1997). Driven by Nature. CAB International, Wallingford, U.K.

[106] Falkowski PG, Scholes RJ, Boyle E, Canadell J, Canfield D, Elser J (2000) The global carbon cycle: a test of our knowledge of Earth as a system. Sci 290: 291-296.

[107] Smith P, Bertaglia M (2007) Greenhouse gas mitigation in agriculture. In: Cleveland CJ, editor. Encyclopedia of Earth. Washington: Available at http://www.eoearth.org/article/Greenhouse_gas_mitigation_in_agriculture.

[108] Vitousek PM (1994) Beyond global warming: Ecology and global change. Ecol. 75: 1861-1876.

[109] Gruber N, Galloway JN (2008) An Earth-system perspective of the global nitrogen cycle. Nat. 451: 293- 296.

[110] Bobbink R, Hicks K, Galloway J, Spranger T, Alkemade R, Ashmore M, Bustamante M, Cinderby S, Davidson E, Dentener F, Emmett B, Erisman JW, Fenn M, Gilliam F, Nordin A, Pardo L, de Vries W (2010) Global assessment of nitrogen deposition effects on terrestrial plant diversity: a synthesis. Ecol. appl. 20: 30-59.

[111] Lavelle P, Dugdale R, Scholes R (lead authors) (2009). Nutrient cycling. In: Millenium Ecosystems AssessmentIn: Kakri AH, Watson R, editors. Millenium Ecossystem Assesment. Washington DC: Island Press. , pp. 331 – 353.

[112] Menge, D. N. L.; Field, C. B. 2007. Simulated global changes alter phosphorus demand in annual grassland. Glob. chang. biol. 13: 1-10.

[113] Turner BL (2008) Resource partitioning for soil phosphorus: a hypothesis. J. ecol. 96: 698-702.

[114] Elser JJ, Bracken MES, Clelan, EE, Gruner DS, Harpole WS, Hillebrand H, Ngai JT, Seabloom EW, Shurin JB, Smith JE (2007) Global analysis of nitrogen and phosphorus limitation of primary producers in freshwater, marine and terrestrial ecosystems. Ecol. lett. 10: 1-8

[115] Davidson EA, Araújo AC, Artaxo P, Balch JK, Brown IF, Bustamante MMC, Coe MT, DeFries RS, Keller M, Longo M, Munger JW, Schroeder W, Soares-Filho BS, Souza Jr CM, Wofsy SC (2012) The Amazon basin in transition. Nat. 481: 321-328.

[116] Jacobson TKB, Bustamante MMC, Kozovits AR (2011) Diversity of shrub tree layer, leaf litter decomposition and N release in a Brazilian Cerrado under N, P and N plus P additions. Environ. poll. 159: 2236-2242

[117] Bustamante MMC, Medina E, Asner GP, Nardoto GB, Garcia-Montiel DC (2006) Nitrogen cycling in tropical and temperate savannas. Biogeochem. 79: 209-237.

[118] Fynn RWS, Morris CD, Kirkman KP (2005) Plant strategies and trait trade-offs influence trends in competitive ability along gradients of soil fertility and disturbamce. J. Ecol. 93: 384-394.

[119] Carvalho F, de Souza FA, Carrenho R, Souza FMS, Fernandes GW, Jesus EC (2012) The mosaic of habitats in the high-altitude Brazilian rupestrian fields is a hotspot for arbuscular mycorrhizal fungi. Appl. soil ecol. 52: 9-19

[120] Sharrow S, Ismail S (2004) Carbon and nitrogen storage in agroforests, tree plantations, and pastures in western Oregon, USA. Agrof. syst. 60: 123–130.

[121] Dixon RK, Schroeder PE, Winjum JK (1991) Assessment of promising forest-management practices and technologies for enhancing the conservation and sequestration of atmospheric carbon and their costs at the site level. Environmental Research Lab, Environmental Protection Agency, Corvallis.

[122] Klohn WE, Appelgren BG (1998) Challenges in the field of water resource management in agriculture. Sustainable Management of Water in Agriculture: Issues and Policies. Paris: OECD. pp. 33

[123] Carpenter SR, Caraco NF, Correll DL, Howarth RW, Shawley AN, Smith VH (1998) Nonpoint pollution of surface waters with phosphorus and nitrogen. Ecol. appl. 8: 559–56.

[124] Finlayson CM, D'Cruz R (cord authors) (2005) Inland Water Systems. In: Millenium Ecossystem Assesment. pp. 583. Washington: Island Press.

[125] Matson, P. A.; Parton, W. J.; Power, A. G.; Swift, M. J. 1997. Agricultural intensification and ecosystem properties. Sci. 277: 504-509.

[126] Petrere Jr M (1989) River fisheries in Brazil: A review. Regul. riv. res. manag. 4: 1–16.

[127] Costanza R, D'Arge R, De Groot R, Farber S, Grasso Ma, Hannon B, Limburg K, Naeem S, O'Neill RV, Paruelo J, Raskin RG, Sutton P, Van Den Belt M (1997) The value of the world's ecosystem services and natural capital. Nat. 387: 253-260.

[128] Starý P, Pike KS (1999) Uses of beneficial insect diversity in agroecosystem management. In: Collins W, Qualset CO, editors. Biodiversity in agroecosystems. Boca Raton: CRC Press. pp. 49-68.

[129] De Marco P, Coelho FM (2004) Services performed by the ecosystem: forest remnants influence agricultural cultures' pollination and production. Biodiv. cons. 13: 1245–1255.

[130] Yeates GW, Bamforth SS, Ross DJ, Tate KR, Sparling GP (1991) Recolonization of methyl bromide sterilized soils under four different field conditions. Biol. fert. soils 11: 181–189.

[131] Altieri MA (1994) Biodiversity and Pest Management in Agroecosystems. New York: Haworth Press. pp. 185

[132] Altieri MA, Letourneau DK (1982). Vegetation management and biological control in agroecosystems. Crop Protection 1: 405-430.

[133] Philpott SM, Greenberg R, Bichier P, Perfecto I (2004) Impacts of major predators on tropical agroforest arthropods: comparisons within and across taxa. Oecologia 140: 140- 149.

[134] Altieri MA (1999) The ecological role of biodiversity in agroecosystems. Agric. Ecosyst. envir. 74: 19-31.Oakland: Rowman & Littlefield Publishers Inc. pp. 175-182.

[135] Bamforth SS (1999) Soil microfauna: Diversity and applications of protozoans in soil. In: Collins W, Qualset CO, editors. Biodiversity in agroecosystems. Boca Raton: CRC Press. pp. 19-26.

[136] [136] Neher DA, Barbercheck ME (1998) Diversity and function of soil mesofauna. In: Collins W, Qualset CO, editors. Biodiversity in agroecosystems. Boca Raton: CRC Press. pp. 27–47.

[137] Klute A (1982) Tillage effects on the hydraulic properties of soil: a review. In: Unger PW, van Doren DC, editors . Predicting tillage effects on soil physical properties and processes. Madison: American Society of Agronomy. pp. 29-43.

[138] Boyles JG, Cryan PM, McCracken GF, Kunz TH (2011) Economic importance of bats in agriculture. Sci. 332: 41.

[139] Martis M (1988) Man vs. landscape. Prague: Horizont.

[140] Kremen C (2005) Managing ecosystem services: what do we need to know about their ecology? Ecology Letters 8, 468-479.

[141] Nelson E, Mendoza G, Regetz J, Polasky S, Tallis H, Cameron DR, Chan KM, Daily GC, Goldstein J, Kareiva PM, Lonsdorf E, Naidoo R, Ricketts TH, Shaw MR (2009) Modeling multiple ecosystem services, biodiversity conservation, commodity production, and tradeoffs at landscape scales. Frontiers in Ecol. Environ. 7: 4-11.

[142] Perdue JC, Crossley Jr. DA (1989) Seasonal Abundance of Soil Mites (Acari) in Experimental Agroecosystems: Effects of Drought in No-Tillage and Conventional Tillage. Soil till. res. 15: 117-124.

[143] Tscharntke T, Klein AM, Kruess A, Steffan-Dewenter I, Thies C (2005) Landscape perspectives on agricultural intensification and biodiversity - ecosystem service management. Ecol. Lett. 8: 857-874.

[144] Klein, A. M., Cunningham, S. A., Bos, M., & Steffan-Dewenter, I. (2008). Advances in pollination ecology from tropical plantation crops. Ecology, *89*(4), 935–943.

[145] Lacher TE, Slack RD, Coburn LM, Goldstein MI (1999) The role of agroecosystems in wildlife biodiversity. In: Collins W, Qualset CO, editors. Biodiversity in agroecosystems. Boca Raton: CRC Press. pp. 147-165

[146] Pounds JA, Bustamante MR, Coloma LA, Consuegra JA, Fogden MP, Foster PN, La Marca E, Masters KL, Merino-Viteri A, Puschendorf R, Ron SR, Sánchez-Azofeifa GA, Still CJ, Young BE (2006) Widespread amphibian extinctions from epidemic disease driven by global warming. Nat. 439: 161–167.

[147] Hannah L, Midgley G, Lovejoy T, Bond W, Bush M, Lovett J, Scott D, Woodward, FI (2002) Conservation of biodiversity in a changing climate. Cons. biol. 16: 264–268

[148] IPCC (2001). Report of the International Panel for Climate Change. Cambridge: Cambrige University Press.

[149] Lawler JJ, Shafer SL, White D, Kareiva P, Maurer EP, Blaustein AR, Bartlein PJ (2009) Projected climate-induced faunal change in the Western Hemisphere. Ecol. 90: 588–597

[150] NPU Barbosa & GW Fernandes, in prep

[151] Marini MÂ, Barbet-Mansin M, Martinez J, Prestes NP, Jiguer F (2010) Applying niche modelling to plan conservation actions for the Red-spectacled Amazon (*Amazona pretrei*). Biol. cons. 143: 102-112

[152] LMS Aguiar & RB Machado, in prep.

[153] MMA. 2009. Relatório técnico de monitoramento do desmatamento no bioma Cerrado, 2002 a 2008: dados revisados. Brasília-DF: Centro de Sensoriamento Remoto - Instituto Brasileiro do Meio Ambiente e dos Recursos Naturais Renováveis. Available at < http://siscom.ibama.gov.br/monitorabiomas/>.

[154] Perfecto I, Vandermeer J (2002) Quality of agroecological matrix in a tropical montane landscape: ants in coffee plantations in southern Mexico. Cons. biol. 16: 174–182.

[155] EMBRAPA and UNICAMP (2008). Aquecimento global e a nova geografia da produção agrícola no Brasil. Empresa Brasileira de Pesquisa Agropecuária – EMBRAPA. Brasília-DF. Technical report available at www.agritempo.gov.br/climateagricultura

[156] Bruinsma J (2003) World agriculture: Towards 2015/2030 — An FAO Perspective. London: Earthscan. pp. 432

[157] FAO (2004) The state of food insecurity in the World. Monitoring progress towards the World Food Summit and Millennium Development Goals. Rome: FAO. pp. 44

[158] Parry M, Rosenzweig C, Livermore M (2005) Climate change, global food supply and risk of hunger. Phil. trans. roy. soc. B: biol. sci. 360: 2125–2138.

[159] Rosenzweig C, Parry ML (1994) Potential impact of climate change on world food supply. Nat. 367: 133–138.

[160] Schmidhuber J, Tubiello FN (2007) Global food security under climate change. Proc. n. acad. sci. 104: 19703.

[161] Lapola DM, Schaldach R, Alcamo J, Bondeau A, Msangi S, Priess JA, Silvestrini R, Soares-Filho BS (2011) Impacts of climate change and the end of deforestation on land use in the Brazilian legal Amazon. Earth interact. 15: 1-30

[162] Phoenix GK, Hicks WK, Cinderby S, Kuylenstierna JCI, Stock WD, Dentener FJ, Giller KE, Austin AT, Lefroy RDB, Gimeno BS, Ashmore MR, Ineson P (2006) Atmospheric nitrogen deposition in world biodiversity hotspots: the need for a greater global perspective in assessing N deposition impacts. Glob. chang. biol. 12: 470-476.

[163] Thomas Lovejoy, pers. comm.

[164] Roubik, D. W. (2002). Feral African bees augment neotropical coffee yield. *Pollinating bees: the conservation link between agriculture and nature, Ministry of Environment, Brazilia, Brazil*, 255–266.

[165] Klein, A. M., Steffan-Dewenter, I., & Tscharntke, T. (2003). Pollination of Coffea canephora in relation to local and regional agroforestry management. J Appl. Ecol. 40(5), 837–845.

[166] Klein, A. M., Steffan–Dewenter, I., & Tscharntke, T. (2003). Fruit set of highland coffee increases .. h the diversity of pollinating bees. Proc. r. soc. lond. s B: biol. sci. 270(1518), 955–961.

[167] Veddeler, D., Olschewski, R., Tscharntke, T., & Klein, A. M. (2008). The contribution of non-managed social bees to coffee production: new economic insights based on farm-scale yield data. Agrof syst, 73(2), 109–114.

[168] Philpott, S. M., Uno, S., & Maldonado, J. (2006). The importance of ants and high-shade management to coffee pollination and fruit weight in Chiapas, Mexico. Arthrop. Divers. Cons. 473–487

[169] Frimpong, E. A., Gordon, I., Kwapong, P. K., Gemmill-Herren, B., & others. (2010). Dynamics of cocoa pollination: tools and applications for surveying and monitoring cocoa pollinators. Int. j. insect. sci. 29(2), 62.

[170] Borkhataria, R. R., Collazo, J. A., & Groom, M. J. (2006). Additive effects of vertebrate predators on insects in a Puerto Rican coffee plantation. *Ecol. Appl. 16*(2), 696–703.

[171] Perfecto, I., Vandermeer, J. H., Bautista, G. L., Nunez, G. I., Greenberg, R., Bichier, P., & Langridge, S. (2004). Greater predation in shaded coffee farms: the role of resident neotropical birds. Ecol. *85*(10), 2677–2681.

[172] Armbrecht, I., & Gallego, M. C. (2007). Testing ant predation on the coffee berry borer in shaded and sun coffee plantations in Colombia. Entomol. exp. appl. *124*(3), 261–267.

[173] Van Bael, S. A., Bichier, P., & Greenberg, R. (2007). Bird predation on insects reduces damage to the foliage of cocoa trees (*Theobroma cacao*) in western Panama. J. trop. ecol. *23, 715*–719.

[174] Williams-Guillén, K., Perfecto, I., & Vandermeer, J. (2008). Bats limit insects in a neotropical agroforestry system. Sci. 320(5872), 70–70.

[175] Jedlicka, J. A., Greenberg, R., Perfecto, I., Philpott, S. M., & Dietsch, T. V. (2006). Seasonal shift in the foraging niche of a tropical avian resident: resource competition at work? J. trop. ecol. 22(4), 385–395.

[176] Vandermeer, J., Perfecto, I., Ibarra Nuñez, G., Phillpott, S., & Garcia Ballinas, A. (2002). Ants (Azteca sp.) as potential biological control agents in shade coffee production in Chiapas, Mexico. Agrof. syst. 56(3), 271–27.

[177] Sperber, C. F., Nakayama, K., Valverde, M. J., & Neves, F. S. (2004). Tree species richness and density affect parasitoid diversity in cacao agroforestry. Bas. appl. ecol. 5(3), 241–251.

Chapter 2

Plant Diversity in Agroecosystems and Agricultural Landscapes

Dariusz Jaskulski and Iwona Jaskulska

Additional information is available at the end of the chapter

1. Introduction

Agricultural landscapes represent a cultural landscape group. Their origin, structure and ecological relations differ from natural landscapes considerably. By (Kizos and Koulouri 2005) they are defined as the visual result of land uses. They are nature systems developed with a great participation of the man, used by the man and maintained in the state of internal equilibrium. At present the role of rural areas does not mean only foodstuffs production. The sustainable rural areas development should involve maintaining the equilibrium between the productive, economic and social function of agricultural landscape and its ecological function, including maintaining the biodiversity. Those are the areas of numerous plant and animal organisms not connected directly with agricultural production, however, playing important environmental functions. The human activity performed in them should thus also consider the need of environmental protection [Millennium Ekosystem Assessment 2005, Fisher and Lindenmayer 2007].

The basic elements of the rural landscapes are the agroecosystems. Those are mainly grasslands and cultivated fields. Very important is their proportion in the agricultural landscape. The correct structure allows the agricultural production and maintain environmental values [Kovalev et al. 2004]. Biodiversity of agricultural fields is very small. Altieri [1999] citing Fowler and Mooney indicates that more than one billion hectares in the world are cultivated only about 70 species of plants. Therefore it is very important is the presence in the area of islands, corridors and other environmental elements.

1.1. The structure of agricultural landscapes and the biodiversity of plants

The biodiversity in agricultural landscape depends on its structure, including the share of natural components, land use structure and the intensity of farming. To evaluate the biodiversity in agricultural landscapes, there are applied various habitat and agricultural production parameters. For example Billeter et al. [2008] give:

- Land-use intensity parameters:
  - number of crops cultivated on a farm,
  - nitrogen input,
  - share of intensively fertilized arable area,
  - amount of livestock units per farm,
  - number of pesticide applications per field
- Landscape parameters:
  - area of semi-natural habitats,
  - number of semi-natural habitat types,
  - number of patches of woody and herbaceous semi-natural habitats,
  - average size of a semi-natural patch,
  - number of patches of woody and herbaceous semi-natural habitats per 100 ha,
  - semi-natural habitats edge density,
  - average Euclidean-nearest-neighbour distance between semi-natural landscape elements,
  - contagion index of woody and herbaceous semi-natural landscape elements,
  - proximity of woody and herbaceous semi-natural elements within a 5000 m radius.

In Poland [Jakubowski 2007] an attempt has been made to evaluate the biodiversity of agricultural landscape based on:

- the share of the landscape type with a varied little-mosaic use,
- the occurrence of protected habitats,
- the occurrence of rare field and meadow plant species,
- land relief enhancing the diversity of habitats,
- the occurrence of nature refuges connected with field or meadow habitats or species,
- the occurrence of large areas under extensive meadow or fen use,
- the occurrence of agrocenoses with numerous midfield woodlots and thickets, especially forming ecological corridors.

Agricultural landscapes are a significant component of the surface of the countries or regions. It can be shown that the diagram (Fig. 1). One of the conditions for its high biodiversity is multi-element structure and heterogeneity; the areas with low natural qualities, mostly due to strong anthropogenic impact on the environment and limited biodiversity: agricultural land. It also includes the areas of a high biodiversity, in general, however, small in size: forest islands and non-point woodlots, xerothermic grasses, fallow land and water ponds. Besides there is a network of ecological corridors, including: field boundaries, field margins, hedgerows, linear midfield woodlots, roads and shoulders. Those elements are supplemented with a settlement and transport network.

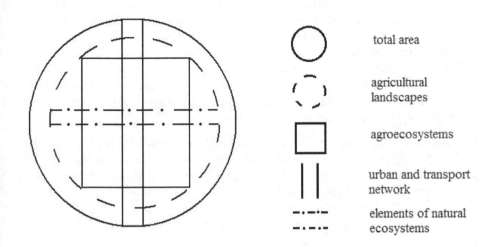

Figure 1. Diagram of the location and structure of agricultural landscape

Over the last decades the rural areas have undergone habitat homogenization and fragmentation [Jongman 2002]. In agricultural landscapes the structural diversity and heterogeneity are getting smaller and smaller. The development of agricultural production results in, one the one hand, a reduction in the number and diversity of natural elements of their structure and, on the other hand, the increase in the concentration and intensification of field crops, at the same time limiting the use of meadows and pastureland. In agricultural landscape there increases the number of large monoculture fields. Plant cultivation involves the application of technologies with high inputs of mineral fertilizers and plant protection agents. The plantation mechanization and large-size machines result in an elimination of midfield woodlots, water ponds and hollows. Striving for the consolidation of land and crops leads to the liquidation of wetlands, fallow land, and increasing the farm acreage is connected with giving up the field boundaries.

The key role in maintaining the biodiversity and biological equilibrium in agricultural landscapes is played by their elements with no direct effect on agricultural production. However, their indirect relationship through the impact on the biotope and the biocenosis of adjacent agricultural ecosystems is unquestionable and seen e.g., in the effect on the microclimate, soil properties and ecological relationships between organisms. Midfield and mid-meadow woodlots affect the biological and microclimatic conditions in the neighbouring arable fields. They limit the effects of water and wind erosion. Those areas act as a buffer, reduce non-point pollutions and the discharge of biogenes from the fields. They plan a crucial hydrological role. They create refuges for many species of fauna and flora non-specific for the neighbouring agricultural land. The organisms, by increasing the biodiversity in agricultural landscape, help maintaining its biological equilibrium. To maintain the richness of plant and animal species in the agricultural landscape, the following are of similar importance: the elements of natural landscapes, semi-natural land under use and fallow land, including: water ponds, swampy areas, wetland, peat bogs, dry turfs, field boundaries, slope, embankments, and others. Those, together with agricultural land, combined in the landscape join ecological corridors, thanks to which, numerous organisms can migrate between various ecosystems, which enhances the stability of their presence in the landscape. A high biodiversity occurs especially on the border of ecosystems. It is a result of varied habitat conditions in the zone of ecotone and the mutual penetration of organisms between the neighbouring habitats.

The development of agriculture with its economic and social function and, at the same time, the activity for the protection of the environment and landscape, are a springboard for the strategy of sustainable development of rural areas in many countries. The prevention of agricultural landscape degradation requires e.g. maintaining its multi-element, biologically-varied spatial structure, especially maintaining and revitalizing the landscape elements with a high plant biodiversity since the flora variation facilitates the development of zoocenosis.

Midfield woodlots and other woodland system elements in agricultural landscape support the production and ecological functions of agroecosystems [Benton et al. 2003]. Those are the elements which are non-homogenous in terms of origin, form, structure and nomenclature. In literature one can find various names: woodlots, shelterbelts, hedgerows, and also midfield clumps, water-edge hedgerows and avenues. Midfield woodlots occur as patches and linear forms. Woodlots, especially the linear ones, are also considered corridors found in the matrix of agricultural landscapes. They are mostly made up by woody vegetation with a share of herbaceous vegetation, and the total biodiversity is enhanced by abundant fauna. Linear woodlots, including hedgerows are a key ecological element in the countries of Western Europe; e.g. France, England [Baudry et al. 2000]. They are also present in Central and Eastern Europe [Ryszkowski et al. 2003, Lazarev 2006] as well as in North America [Brandle et al. 2004] and on other continents [Onyewotu et al. 2004, Tsitsilas et al. 2006].

An indirect effect of woodlots on the plant biodiversity in agricultural landscape involves the development of abiotic habitat conditions, which is seen e.g. from braking the wind speed, restricting the wind and water soil erosion, limiting water evaporation from soil, increasing air humidity, slowing-down the snow melting rate, decreasing daily and annual air temperature amplitudes, limiting the occurrence of ground frosts, restricting the mobility of harmful agrochemical compounds, which creates conditions favourable to the vegetation of many plant species, including crops which occur in agroecosystems.

Biotic elements of midfield woodlots, on the other hand, remain in a close ecologic relationship with agrophytocenosis. On that ecological island there are found, permanently or seasonally, pests and pathogens of crops as well as weeds which can migrate to arable fields. However, the species richness of those places is mostly made up by organisms favourable to crops; entomopathogenic fungi, predator beetles and flies, ladybirds feeding on aphids. The insects representing the family *Apidae* are of special importance since they pollinate many plants, including crops. Most herbaceous plants which occur in woodlots, are not, however, expansive weeds posing a threat to agroecosystems in nature. A complex character of the structure of midfield woodlots and their functions are seen from the environmental and agroecological research results reported by many authors from various research centres in the world and presented as a review by Mize et al. [2008]. Woodlot lanes are most frequently established with the use of 2-5 species of woody plants. Their biodiversity and effect on the landscape change with growth. At the initial stage a high share is accounted for by weeds, mono- and dicotyledonous plants. Their seeds are found in the soil seed bank and transferred with the wind and by animals. Later the trees and thicket vegetation start to dominate; their competition for light and water increases. Light-loving species give up. The vegetation of woodlot patches is also exposed to a strong human pressure resulting from the agrotechnical practises for crops, e.g. tillage, mineral fertilization and pesticide application and so it is, in general, less stable than in woodlands.

In Canada [Boutin et al. 2003] point to the diversity of the vegetation in hedgerows depending on their origin: natural woody, planted woody and herbaceous. Hedgerows made up of natural and planted woody plants demonstrated a greater diversity and richness of plant species. In natural hedgerows there were identified 31 woody species in the layer of trees > 5 m, 63 species of those plants in the layer shrubs < 5 m as well as 94 species of herbaceous plants. Planted hedgerows were mostly composed of ecotone vegetation, typical for the edges of arable fields.

Walker et al. [2006] point to a high biodiversity of plants and a complex nature of green lanes, composed of the external part with woody species and herbaceous plants, the inside verge and the central track. It was the inside verge which was richest in plant species. The area was most covered with *Urtica dioica, Rubus fruticosus, Arrhenatherum elatius*, while the central track - mostly with *Agrostis stolonifera, Ranunculus repens, Dactylis glomerata, Trifolium repens, Lolium perenne, Holcus lanatus, Plantago major*. Plant communities in respective parts of green lanes were developed due to habitat conditions, including light, moisture, reaction, nitrogen content, as well as the elements of agrotechnical practises in the adjacent arable fields.

In Poland, in Lower Silesia (south-western Poland), the biodiversity of plants of midfield woodlots depended on their type: midfield clumps, water-edge hedgerows and avenues. In total in 183 woodlots there were found 77 woody plant species; most occurred in midfield clumps, and least – in avenues. The greater the area of woody species in midfield woodlots or the greater their length, the greater their abundance [Orłowski and Nowak 2005].

Nevertheless, precious environmental islands to maintain the biodiversity in the landscape and agroecosystems include field boundaries, combining physical and functionally-different ecosystems of agricultural landscape. The smaller the arable fields and farms and the more extensive the farming, the greater the number of field boundaries. Le Coeur et al. [2002] quoting the results reported by many authors [Helenius, Hooper, McAdam et al., Pointereau and Bazile] demonstrates that along with the agricultural production intensification in the second half of the 20th century, those semi-natural landscape elements disappear. The scale of field boundaries loss is high; e.g. in the UK 5000 km annually, in Northern Ireland - 14%

of the field boundaries network between 1976 and 1982, in Finland 500 000 km, 740 000 km in France. The authors show, at the same time, a strong relationship between the diversity and the structure of vegetation which occurs in field boundaries and the effect of the interaction of many habitat and economic factors, e.g. the landscape structure, field management method, farm type as well as the nature of the field boundary itself.

Field boundaries, despite their small size, show a great richness of its organisms; mostly herbaceous plants, and sometimes also trees and shrubs. The flora is accompanied by abundant fauna. The species richness of field boundaries depends e.g. on their age and width. Czarnecka [2011], investigating along 4 field boundaries of a total length of 1000 m, identified 67 plant species. Symonides [2010] citing studies by many authors indicate that in Poland in the field boundaries and in the immediate vicinity may occur up to several hundred species of plants. Sometimes there is an expansion of those plants (weeds) into arable fields. The agroecological importance of field boundaries, however, mostly comes from the occurrence of pollinating insects and organisms entomophagous towards crop pests.

Field boundaries are often a part of field margins. Those are linear elements of agricultural landscape showing a complex structure and high biodiversity; e.g. aqueous, ruderal, woody vegetation. Depending on the margin structure and on the distance from the arable field, crops, herbaceous plants, shrubs, trees, and aqueous plants dominate. The flora of the area adjacent to fields is developed by agricultural activities; e.g. fertilization, herbicides application. The vegetation of field margins also affects agricultural vegetation, both directly and indirectly [Marshall and Moone 2002].

Other agricultural landscape elements showing high ecological qualities are midfield ponds, combining the biotopes of greater, open surface waters. They play a retention function and affect water relations in agroecosystems, which is crucial for the development of crops and other companion crops, especially when exposed to seasonal precipitation deficits. Midfield ponds are an essential component of biodiversity, including flora diversity in agricultural landscapes and agroecosystems. They serve as a habitat for many plant species representing various plant communities. The richness and the frequency of occurrence of the phytocenoses within water ponds, with an example of Wełtyń Plain (in Poland), are presented in the Table 1. by Gamrat [2009].

The diversity of plant species which occur in those habitats depends on their form of water ponds, changes which occur there; devastation, overgrowing, shallowing. The richness of plant species, their structure and biodiversity are much affected by agricultural and non-agricultural human activity, being an important cause of the eutrophization of those habitats. Within the water ponds one can find the vegetation of aquatic, marshland, meadow, shrubby and ruderal habitats [Pieńkowski et al. 2004, Gamrat 2006]. In open ponds, marshland and aquatic vegetation dominates. In overgrowing ponds, the species richness is greater, however, the vegetation of wet stands gives up. While in the post-water-ponds hollows there dominates ruderal vegetation, including nitrophilic vegetation, typical for agricultural landscape.

Phytocenosis, the most frequent of communities
Oenantho-Rorippetum, Phalaridetum arundinaceae community with: *Calamagrostis canescens, Deschampsia caespitosa, Elymus repens, Epilobium hirsutum, Galium aparine, Lemna minor, Phragmites australis, Rubus caesius, Typha latifolia, Urtica dioica*
Phytocenosis, moderately frequent communities
Calamagrostietum epigeji, Caricetum acutiformis, Epilobio-Juncetum effusi, Rumicerum maritimi, Salicetum pentandro-cinereae, Scirpetum sylvatici, Sparganietum erecti, Sparganio-Glycerietum fluitantis community with: *Agrostis stolonifera, Alisma plantago-aquatica, Alopecurus geniculatus, A. pratensis, Anthriscus sylvestris, Apera spica-venti, Artemisia vulgaris, Bidens tripartita, Cirsium arvense, Festuca pratensis, Glechoma hederacea, Glyceria maxima, Holcus lanatus, Iris pseudacorus, Poa pratensis-Festuca rubra*
Phytocenosis, rare communities
Acoretum calami, Caricetum elatae, Caricetum gracilis, Cicuto–Caricetum pseudocyperi, Hottonietum palustris, Spirodeletum polyrhizae, Leonuro-Arctietum tomentosi, Ranunculetum circinati community with: *Anthoxanthum odoratum, Arctium major, Arrhenatherum elatius, Bromus tectorum, Capsella bursa-pastoris, Carex nigra, C. rostrata, C. vulpina, Cerasium arvense, Cirsium palustre, Conium maculatum, Epilobium parviflorum, Equisetum arvense, Hydrocharis morsus-ranae, Lemna gibba, L. trisulca, Lychnis flos-cuculi, Lysimachia vulgaris, Polygonum amphibium, Rudbeckia hirta, Solanun dulcamara, Symphytum officinale, Typha angustifolia*

Table 1. The frequency of occurrence of the phytocenoses on the ponds (by Gamrat 2009)

The water ponds, on their edges, are often accompanied by woodlots lanes or patches. The vegetation acts as a biological filter protecting water from pollution with agrochemicals from arable fields. Ryszkowski and Bartoszewicz [1989] found that the concentration of nitrates in water flowing under woodlots can be even 30-times lower than in the environment without that vegetation.

Water ponds also occur among marshlands. Those are very important landscape elements playing hydrological and ecological functions and can affect the biodiversity of plants both on a local and regional scale [Thiere et al. 2009].

2. Biodiversity of plants in agroecosystems

The agroecosystems are an essential element of agricultural landscape. Agricultural ecosystems are in mutual ecological relationships with other ecosystems and elements of the environment. This can be illustrated schema (Fig. 2).

The biodiversity of agricultural ecosystems depends on their kind, method of use and management. The basic kind of agricultural land in the world are grasslands; meadows and pasture. Grasslands cover more than 10% of the land area of the Earth. About one third is taken by arable meadows and pasture and one fourth – by semi-natural and natural extensive pasture [Mooney 1993].

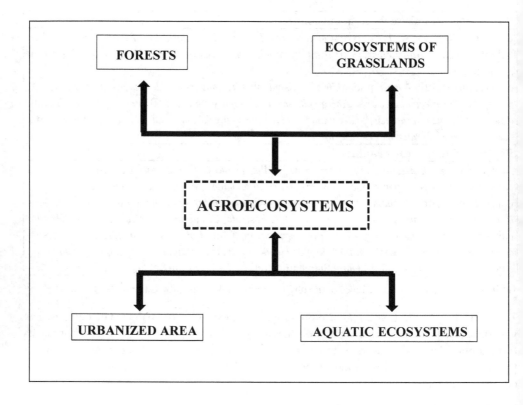

Figure 2. Depending agroecosystems in the environment

Grasslands play various non-production functions in the environment [Wasilewski 2009]:

- climatic, forming a mild microclimate, also covering adjacent areas,
- hydrological, with a large water-retention potential,
- protective; limiting the soil erosion and protecting the soil and water from pollution with agrochemicals and biogenes,
- phytosanitary, by stopping PMs and emissions of essential oils,
- health-enhancing, being the habitat of many herbs,
- landscape and aesthetic, due to the diversity of forms and colours of plant habitats.

The plant biodiversity of grasslands is, in general, greater than arable fields, which comes from the nature of meadow and pasture sward, made up of many species of grasses, papilionaceous plants, herbs and weeds. Many authors, cited by [Pärtel et al. 2005], show that per 100 cm² there can occur a few dozen or so plant species and per 1 m² – almost a hundred. The plant biodiversity of grasslands depends on the habitat conditions and on the method and the intensity of their use. The flora composition is greatly affected by soil properties; moisture, the rate of mineralization of organic nitrogen compounds, the kind of organic matter and the richness in nutrients [Pawluczuk and Alberski 2011]. In a moist habitat, frequently flooded or permeated there occurred, in general, the vegetation representing *Ranunculus, Equisetum, Carex* and *Rumex* genera and the grasses demonstrated a simplified flora composition. *Lotus uliginosus* Schkuhr, *Equisetum palustre* L., *Ranunculus acris* L. and *Ranunculus repens* L., *Lythrum salicaria* L., *Cirsium palustre* (L.) Scop., *Galium uliginosum* L. were abundant. In the habitat with a seasonally-changeable soil moisture, *Cirsium oleraceum* (L.) Scop., *Filipendula ulmaria* (L.) Maxim., *Geum rivale* L. were most abundant. When exposed to lower moisture and a greater organic matter mineralization dynamics, there were recorded numerous species of fodder grasses and other plants demonstrating high mineral nitrogen requirements: *Alopecurus pratensis* L., *Festuca pratensis* Huds., *Festuca rubra* L., *Poa pratensis* L., *Holcus lanatus* L., *Agropyron repens* (L.) P.Beauv., *Urtica dioica* L., *Agropyron repens* (L.) P.Beauv., *Cardaminopsis arenosa* (L.) Hayek. The plant biodiversity of grasslands also depends on their use: grazing, method and technique of cutting. The vegetation of grasslands is not permanent, climax in nature. Giving up the use leads to a secondary succession of those areas.

On a global scale, the arable land has a lower share in the total area than grasslands. However, in many countries it accounts for most agricultural land (Table 2). Agroecosystems are an area exposed to a strong anthropogenic impact on the environment. What is characteristic for those ecosystems is a low biodiversity, especially phytocenoses.

It covers a few crop species and a few, reduced by the farmer, non-crops. The anthropogenic impact on the environment concerns both the biotope and the biocenosis of those areas. The soil properties get changed according to the requirements of the crops. In the fields you will find mostly annual plants, shielding the soil only for some part of the year. Winter forms, e.g. wheat, rye, rape, occur in the field for about 300 days a year, spring crops with a long period of vegetation, including maize, beetroot, potato, for about 160 – 180 days. There exist, however, crops with a much shorter vegetation period. Spring barley stays in the field for about 100 days and some species – for a few weeks. Those crops have different ability to reduce soil erosion (Fig. 3). Most often, for a long period between the harvest and sowing of the successive crop, the soil remains with no vegetation. Only in some cases intercrops are grown or the mulch rests on the soil surface. The fields of crops are, in general, single-species. It is rarely the case that the mixtures of a few species or cultivars of the same crop are grown. Besides, non-crops; weeds and self-sown plants, are being removed.

Country	Agricultural land	of which	
		arable land	permanent pasture
Argentina	48,6	11,7	36,5
Australia	54,1	5,7	48,3
Belgium	48,8	27,9	17,4
Belarus	44,0	27,2	16,3
Brazil	31,2	7,2	23,1
Bulgaria	47,3	28,2	17,3
China	55,8	11,6	42,7
Denmark	63,7	56,6	7,1
Finland	7,4	7,4	0,0
France	53,1	33,3	18,0
Greece	35,7	16,3	10,9
Spain	55,8	25,0	21,2
India	60,5	53,2	3,5
Japan	12,6	11,8	0,0
Canada	7,5	5,0	1,7
Latvia	28,2	18,8	9,4
Mexico	52,9	12,8	38,7
Netherlands	54,6	31,6	23,0
Germany	48,6	34,2	13,8
Norway	3,5	2,8	0,7
New Zealand	38,8	1,7	37,1
Poland	49,9	38,7	10,2
Portugal	36,6	11,5	18,8
Czech Republic	54,9	39,2	13,1
Russian Federation	13,1	7,4	5,6
Romania	58,8	37,9	19,6
Slovakia	39,0	28,7	10,3
United States	44,9	18,6	26,0
Sweden	7,6	6,4	1,2
Turkey	50,7	28,0	18,9
Ukraine	71,3	56,1	13,6
Hungary	64,3	51,0	11,1
United Kingdom	73,2	24,8	47,9
Italy	45,7	24,2	12,3
WORLD	37,5	10,6	25,8

(based on Statistical Yearbook of Agriculture, CSO Warsaw 2010)

Table 2. The share (%) of agricultural land in the total area of some countries

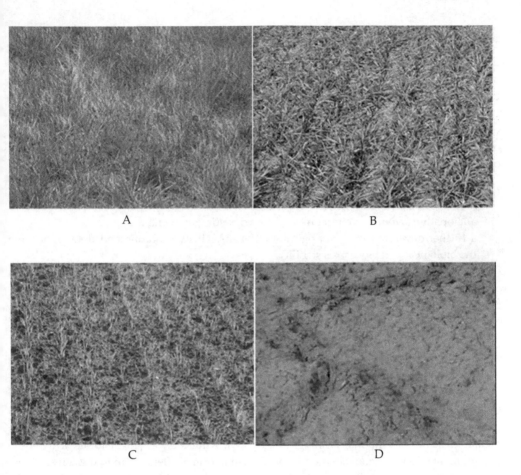

Figure 3. Covering the soil by vegetation in early spring, A - grassland (full coverage),
B - winter cereal (good coverage), C - spring cereal (poor coverage - the risk of erosion),
D - without plants (signs of erosion)

The impact on agrophytocenosis and its biodiversity depends e.g. on the farming system. On many conventional farms plant production dominates. Animal production, if any, often includes battery farming with the use of manufactured feedingstuffs or produced on arable land. The rotations of commodity and fodder crops get simplified even down to 2-3 species. Those are intensive single-species technologies with high inputs of mineral fertilizers and pesticides eliminating weeds and other agrophages. Intercrops are rarely grown; if so – to improve the stand value. In the sustainable farming system, especially in organic farming, the farm is perceived as an organism. Animal production should be its integral part, including ruminants, which require the animal feed base in a form of grasslands. Animal feed production on arable land involves perennials; e.g. *Fabaceae*. Crop rotations are multispecific, with legumes and intercrops being essential. The fields are quite frequently mixed and include species representing various genera. Weeds are their integral component. They are being limited in the fields of crops when they pose a threat to yields and their quality. To do so, there are applied various methods, also or only non-chemical. A greater biodiversity in organic than in conventional agriculture is mostly seen on a local scale, on the agricultural farms where there is a greater weed species richness; on a regional scale it can be similar in both farming systems and, to a greater extent, it depends on habitat conditions [Hawesa et al. 2010]. Irrespective of the farming system, weeds are an integral component of the agricultural landscape and agroecosystems. Especially on integrated and organic farms, their ecological role is noted, by incorporating e.g.:

- filling in the ecological niches and enhancing the diversity of flora and the quality of animal feeds from grasslands,
- allelopathic favourable effect on the crops coexisting in the field,
- a further development of ecological relationships between fauna and flora, enhancing the biological agrobiocenosis stability,
- soil protection from erosion and unproductive water evaporation, limiting non-point pollutions of soil and water,
- carbon sequestration in the environment,
- bioindication of the conditions and the state of the environment,
- application to the production of composts, biopreparations and herbal medicine.

A high biodiversity of agroecosystems in organic farming is not only due to the diversity of flora and fauna in arable fields but it also comes from the presence of a greater number of habitat components; e.g. woodlands, boundaries, hedgerows [Boutin et al. 2008]. Although the biodiversity of those components on organic and conventional farms can be similar, in organic farming the abundance of plants and their species in the fields is often many-times greater than in conventional farming [Hald 1999]. Krauss et al. [2011] found a five-time greater plant species richness in the triticale grown in organic fields than in the conventional ones, which, in turn, resulted in a greater richness and abundance of insects, including the pollinating ones. The greater number of predator insects resulted in a decrease in the number of aphids. It is especially precious that the biodiversity on organic farms is made up by the rare species of flora, broad-leaved weeds, pollinated by insects and legumes. Numerous research cited by Hole et al. [2005], and providing a comparison of the occurrence of non-crops in the fields in various farming systems, demonstrate that it is more diverse on organic than on the conventional farms. In the intensively-cultivated fields there decreases especially the number of broad-leaved weeds easily eliminated by the herbicides application, and to less extent – of grasses. A high diversity of flora in organic fields is found all over their area. On traditional farms it mostly focuses on the crop edges where the effect of herbicides is lower [Romero et al. 2008]. The farming system affects not only the plant abundance but also the abundance of their seeds. In organic fields a greater number of weed seeds is consumed by fauna, mostly birds [Navntoft et al. 2009].

The biodiversity of agrophytocenosis on arable land, especially in intensive farming, is determined by crops. The richness and diversity of crop species depend on habitat conditions and the plant production organization on the regional scale and on an agricultural farm. The diversity of crops defined by Jaskulski and Jaskulska [2011] in the Kujawy and Pomorze Province, in Poland, applying the algorithm of the Shannon-Weaver index depended on many features of the landscape, e.g. the share of components of high ecological value, including woodland, grasslands in the total area and the features of the agroecosystem and the farm; the soil quality and the crop structure. The number of crops and their diversity were an effect of the interaction between the habitat conditions and the farm organization. The number of crops in the arable fields in the region depended on the share of the woodland, woodlots and meadows in the total area and crops in the total acreage of arable land. The crop diversity index was an effect of the interaction between the soil quality index, the share of woodland in the total area, the share of pasture and set-aside land and crops in the total acreage of agricultural land or arable land.

To maintain the diversity of crops on arable land not only a high number of crop species is essential but also a lack of a strong domination of the crop structure by single crops. The analysis of changes in the crop diversity on arable land in Poland over 1960 – 2009 confirms that hypothesis. Despite the production intensification and an increased farm size (Fig. 4A), the crop diversity index H' value from 1960 to 1990 was increasing (Fig. 4B), which must have been due to the share of rye in crops getting strongly decreased and that of a few other crops getting increased, to include wheat, barley, triticale, cereal mixtures, and rape. At the beginning of the 21st century the diversity index value got slightly lower due to an increase in the domination of wheat in crops (Table 3).

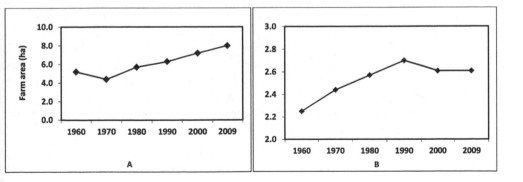

Figure 4. Changes in farm size - A and index of crop diversity by Shannon-Weaver H' - B

Crop	Year					
	1960	1970	1980	1990	2000	2009
Wheat	8,9	13,3	11,1	16,0	21,2	20,2
Rye	33,4	22,8	20,9	16,3	17,2	12,0
Barley	4,7	6,2	9,1	8,2	8,8	10,0
Oat	10,7	10,2	6,9	5,2	4,6	4,5
Triticale	-	-	-	5,3	5,6	12,6
Grain mixtures	1,7	2,7	5,1	8,2	11,9	11,5
Potato	18,8	18,2	16,2	12,9	10,1	4,4
Sugar beet	2,6	2,7	3,2	3,1	2,7	1,7
Oil plants	0,9	2,2	2,3	3,5	3,5	7,0
Fodder	4,7	4,9	5,5	4,9	2,4	3,6

Table 3. Share (%) of the main crops in the crop structure in Poland

The genetic variation pool within a given species is made up by cultivars. Creative breeding gives rise to new genotypes meeting the expectations of producers and consumers. Breeding work involves not only the plant yield-forming potential but also the physiological and morphological traits determining the reaction of the plants to habitat factors. The cultivars of a given species differ in their vegetation period length. The phenotype variation is seen from the morphology of the underground and above-ground parts. The size, the extent and the physiological activity of the root system differ. Cultivars vary in the habitat, height, foliage, and the colour of flowers. Breeding differentiates the resistance of the plants to abiotic stress habitat factors, including: low temperature, water deficit, soil reaction. The resistance to diseases, pests and weed infestation varies. The richness of cultivars of crop species demonstrating varied biological and functional traits facilitates the plant production compliant with the principles of various farming systems using the advantages of the agricultural production space. To maintain the biodiversity in agroecosystems, it is especially important to grow old traditional crop cultivars. They are adapted to local habitat conditions and extensive agrotechnical practises. At present there is a need to breed cultivars adapted to organic farming. They should differ in terms of physiology and morphology from the cultivars in conventional farming, which guarantees easily available nutrients and the protection from agrophages [Konvalina et al. 2009]. They should demonstrate a fast initial growth, a high foliage index and high stems. Such plants are competitive towards weeds, which allows for eliminating the application of herbicides from agrotechnical practises. Such cultivars should also show high resistance to diseases and pests.

Numerous breeding directions meeting the requirements of producers and consumers make, on the regional scale of respective countries, the cultivation of a few dozen or so and even over a hundred cultivars of some plant species possible. The real diversity of cultivars of a given species in field plant production is, in general, lower. It depends, on the one hand, on the desired quality and the methods of yield use as well as habitat-economic growing conditions and, on the other hand, on the available scope of cultivars with genetic-phenotypic traits allowing for such production.

In the Kujawy and Pomorze Province, in Poland, Jaskulska et al. [2012] found a variation in the richness of cultivars of crops grown on agricultural farms. They were determined as a ratio of the number of cultivars of a given species grown to the number of its plantations. The highest value of the cultivar richness index was recorded for potato plantation (0.71). It means that per 100 plantations there were 71 various cultivars. A high value of the richness index also concerned sugar beet (0.65) and maize (0.64). Lower cultivar richness was reported in cereals and in winter rape. The index value for rye cultivar richness was only 0.31 and it must have been due to a low number of cultivars and a high domination of crops by one of them. A strong domination by single cultivars was also reported for sugar beet and potato crops, which resulted in relatively low cultivar diversity. The diversity index was determined using the Shannon-Weaver algorithm; it ranged from 2.39 for rye to 3.98 for winter wheat. A high diversity was also found for spring barley, maize, winter rape, and winter triticale.

In contemporary agroecosystems, dominated by single-species crops, the cultivation of mixtures plays a very essential ecological role. On arable land it is possible to find the fields of genetically-diversified crops. They can be made up of the crops of various, even systematically distant, species or of the same species, however, of various cultivars. Not only the production but also ecological role of that kind of plant growing method are considered both in the agri-environmental research and the policy [Østergård and Fontaine 2006]. In the interspecific mixtures most often various species of cereals and cereals with papilionaceous plants are grown. The mixtures of single-species cultivars are usually arranged for cereal crops but also for others [Sobkowicz and Podgórska-Lesiak 2007]. Undersown crop is also a kind of mixed crop.

A positive ecological role of mixtures in agroecosystems comes from:

- complementary effect of various plant genotypes in the field,
- a better filling-in of the ecological niche by the stems of a few morphologically verified genotypes and their root systems, which facilitates the production of a greater amount of biomass,
- conditions facilitating the presence of a greater number of fauna,
- conditions for the self-control and maintaining the biocenotic equilibrium in the fields,
- a greater plant resistance to agrophages and a possibility to limit the application of pesticides,
- the possibility of restricting mineral fertilization, especially with nitrogen in the multi-specific fields with papilionaceous components.

According to FAOSTAT, the greatest share of mixtures in crops in Europe is found in Poland; in 2010 out of a total of about 1.54 million ha of grain mixtures, 1.33 million ha - in this country. Those are mostly the mixtures of spring cereals: hulled barley with hulled oats or naked oats, hulled barley + naked or hulled oats, oats with triticale, barley with triticale, wheat + oats or barley [Szempliński and Budzyński 2011]. In the mixtures also cereals with legumes or fodder grasses are grown. For the biodiversity of crops in arable fields, growing mixtures of a few cultivars of the same species is of similar importance.

The diversity of crops on arable land is supplemented by intercrops. In contemporary agriculture those are important components of field plant production essential for the environment and agroecology. They demonstrate a direct and indirect effect on the biodiversity of agroecosystems and agricultural landscape. They are an element of agri-environmental programs. In Poland in the mid of the first decade of the 21st century intercrops accounted for about 4.5% of arable land. Many plant species representing families *Fabaceae, Brassicaceae, Poace,* and others, are sown as intercrops. In crop rotation placed between two main yields, they increase the biodiversity of plants in rotation significantly. The effect of intercrops in the agroecosystem is comprehensive. It concerns both the period of their vegetation and the effect of the biomass remaining on the surface or introduced into soil. A short review of agricultural and environmental research [Jaskulska and Gałęzewski 2009] includes its numerous examples.

Intercrops limit non-point pollutions. They play the function of a biological filter. At present in the fields in the periods between successive production cycles they intake nutrients from soil protecting them from leaching to drainage and ground waters. The surface soil layer bound with the root system and covered with the stem biomass is secured from water and wind erosion, protecting not only directly arable fields but also indirectly landscape components limiting the eutrophization of reservoirs and watercourses, midfield ponds shallowing. The phytomass of vegetating plants, post-harvest residue and mulch stimulate the occurrence of other organisms in the habitat, which increases the agroecosytem stability. The biomass can increase the count, diversity and the activity of bacteria, fungi, protozoa and the nematodes. It also enhances the presence of parasitoids and pollinating insects.

Growing intercrops affects the carbon economy in agroecosystems and in the environment. The production of phytomass by those plants is an ecological method of carbon sequestration. Carbon dioxide bound in the biomass increases the content of organic carbon in soil. The plants and mulch decrease the amplitude of temperature of the soil surface layer, restrict its heating. It reduces the intensity of organic matter mineralization and the emissions of carbon dioxide to the atmosphere, which can decrease the contribution of agriculture to global climate warming.

3. Conclusions and recommendations

Agricultural landscape in many countries is the dominant landscape. As a result of human activity it has been transformed. Currently, it is primarily the production function. For the realization of social and cultural needs of the human need to preserve the natural values of those areas. It should cultivate and reclaim mosaic character of agricultural landscape and agroecosystems. They must be a lot of ecological islands and natural landscape components. The particular is the role of forest enclaves, midfield shelterbelts, avenues of trees, wetlands, swamps, bogs, ponds, streams, ditches, roads midfield, borders, etc. Their values are a large diversity of plants. Those components demonstrate a high flora diversity; aquatic plants and land plants of various stands. There exist clusters and single trees, shrubs, herbaceous communities of annual and perennial plants. High richness and diversity of plants determines the occurrence of many fauna. Those components play the role of microclimate and protection. They limit the effects of extreme weather events, soil degradation, pollution, greenhouse gas emissions. In agricultural ecosystems must be maintained semi-natural grasslands with their rich of flora and fauna and the preservation of environmental functions. In the field production should be limited assemblage of single crops in large fields. It should be kept of multispecies crop rotation in small fields with plants belonging to different botanical taxonomy, use groups, and cultivars. In addition to new varieties of crops should be present the old local genotypes. In crop canopies and their mixtures is also possible occurrence of non-cultivated plants. The interval between production cycles should be used for the cultivation of intercrops.

Author details

Dariusz Jaskulski and Iwona Jaskulska
Department of Plant Production and Experimenting,
University of Technology and Life Sciences, Bydgoszcz, Poland

4. References

Altieri M.A. 1999. The ecological role of biodiversity in agroecosystems. Agric. Ecosys. Environ. 74: 19–31.

Baudry J., Bunce R., Burel F. 2000. Hedgerows: An international perspective on their origin, function and management. J. Environ. Manag. 60: 7–22.

Benton T., Vickery J., Wilson J. 2003. Farmland biodiversity: is habitat heterogeneity the key? Trends Ecol. Evol. 18:182–188.

Billeter R., Liira J., Bailey D., Bugter R., Arens P., Augenstein I., Aviron S., Baudry J., Bukacek R., Burel F., Cerny M., De Blust G., De Cock R., Diekötter T., Dietz H., Dirksen J., Dormann C., Durka W., Frenzel M., Hamersky R., Hendrickx F., Herzog F., Klotz S., Koolstra B., Lausch A., Le Coeur D., Maelfait J.P., Opdam P., Roubalova M., Schermann A., Schermann N., Schmidt T., Schweiger O., Smulders M.J.M., Speelmans M., Simova P., Verboom J., van Wingerden W.K.R.E., Zobel M., Edwards P.J. 2008. Indicators for biodiversity in agricultural landscapes: a pan-European study. J. Appl. Ecol. 45: 141–150.

Boutin C., Baril A., Martin P. 2008. Plant diversity in crop fields and woody hedgerows of organic and conventional farms in contrasting landscapes. Agric. Ecosys. Environ. 123: 185–193.

Boutin C., Jobin B., Belanger L. 2003. Importance of riparian habitats to flora conservation in farming landscapes of southern Quebec, Canada. Agric. Ecosys. Environ. 94: 73–87.

Brandle J.R., Hodges L., Zhou X.H. 2004. Windbreaks in North American agricultural systems. Agrofor. Sys. 61: 65–78.

Central Statistical Office Warsaw. Statistical Yearbook of Agriculture 2010. http://www.stat.gov.pl/gus

Czarnecka J. 2011. Baulks of Western Wolhynia as the habitats of rare calciphilous plant species. Water-Environment-Rural Area 11, 2(34): 43–52. (in Polish)

FAOSTAT, http://faostat.fao.org/site/567/default.aspx#ancor

Fisher J., Lindenmayer D.B. 2007. Landscape modification and habitat fragmentation: a synthesis. Glob. Ecol. Biogeogr. 16: 265–280.

Gamrat R. 2006. Threat of small midfield ponds on Wełtyń Plain. Int. Agrophys. 20(2): 97–100.

Gamrat R. 2009. Vegetation in small water bodies in the young glacial landscape of West Pomerania. Monograph Ed. Łachacz A. Wetlands – their functions and protection. Department of Land Reclamation and Environmental Management, University of Warmia and Mazury Olsztyn

Hald A.B. 1999. Weed vegetation (wild flora) of long established organic versus conventional cereal fields in Denmark. Ann. Appl. Biol. 134: 307–314.

Hawesa C., Squirea G.R., Halletta P.D., Watsonb C.A., Young M. 2010. Arable plant communities as indicators of farming practice. Agric. Ecosys. Environ. 138, 1–2: 17–26.

Hole D.G., Perkins A.J., Wilson J.D., Alexander I.H., Grice P.V., Evans A.D. 2005. Does organic farming benefit biodiversity? Biol. Conserv. 122: 113–130.

Jakubowski W. 2007. Evaluation of biological diversity in agricultural landscapes in Poland. Water-Environment-Rural Area 7, 1: 79–90. (in Polish)

Jaskulska I., Gałęzewski L. 2009. Role of catch crops in plant production and in the environment. Fragm. Agron. 26(3): 48–57. (in Polish)

Jaskulska I., Osiński G., Jaskulski D., Mądry A. 2012. Diversity of crop cultivars in the farm group covered by the survey in the kujawy and pomorze region. Fragm. Agron. – in press (in Polish)

Jaskulski D, Jaskulska I. 2011. Diversity and dominance of crop plantations in the agroecosystems of the Kujawy and Pomorze region in Poland. Acta Agric. Scand. Sect. B. Soil Plant Sci. 61: 633–640.

Jongman R.H.G. 2002. Homogenisation and fragmentation of European landscape: ecological consequences and solutions. Landscape and Urban Planning 58, 2–4: 211–221.

Kizos T., Koulouri M. 2005. Economy, demographic changes and morphological transformation of the agri-cultural landscape of Lesvos, Greece. Human Ecol. Rev. 12: 183–192.

Konvalina P., Stehno Z., Moudrý J. 2009. The critical point of conventionally bred soft wheat varieties in organic farming systems. Agron. Res. 7(2): 801–810.

Kovalev N.G., Ivanov D.A., Kashtanov A.N. 2004. Optimisation of the proportion of meadows, woods and arable lands in humid zone of the European part of Russia. Journal of Water and Land Development 8: 117-126.

Krauss J., Gallenberger I., Steffan-Dewenter I. 2011. Decreased functional diversity and biological pest control in conventional compared to organic crop fields. PLoS ONE 6, 9

Lazarev M.M. 2006. Transformation of the annual water budget of soils under shelterbelts. Euras. Soil Sci. 39, 12: 1318–1322.

Le Coeur D., Baudry J., Burel F., Thenail C. 2002. Why and how we should study field boundary biodiversity in an agrarian landscape context. Agric. Ecosys. Environ. 89: 23–40.

Marshall E.J.P., Moone A.C. 2002. Field margins in northern Europe: their functions and interactions with agriculture. Agric. Ecosys. Environ. 89: 5–21.

Millennium Ecosystem Assessment 2005. Ecosystems and Human Well-being: Synthesis, Island Press, Washington

Mize C.W., Brandle J.R., Schoneberger M.M., Bentrup G. 2008. Ecological development and function of shelterbelts in temperate North America. USDA Forest Service / UNL Faculty Publications. Paper 40. http://digitalcommons.unl.edu/usdafsfacpub/40

Mooney H. A. 1993. Human impact on terriestral ecosystems-what we know and what we are doing it. Proceedings of the XVII International Grassland Congress, New Zealand: 11–14.

Navntoft S., Wratten S.D., Kristensen K., Esbjerg P. 2009. Weed seed predation in organic and conventional fields. Biol. Contr. 49: 11–16.

Onyewotu L.O.Z., Stigter C.J., Oladipo E.O., Owonubi J.J. 2004. Air movement and its consequences around a multiple shelterbelt system under advective conditions in semi-arid Northern Nigeria. Theor. Appl. Climatol. 79: 255–262.

Orłowski G., Nowak L. 2005. Species composition of woody vegetation of three types of mid-field woodlots in intensively managed farmland (Wrocław Plain, south-western Poland). Pol. J. Ecol. 53, 1: 23–34.

Østergård H., Fontaine L. (Eds) 2006. Proceedings of the COST SUSVAR workshop on cereal crop diversity: Implications for production and products, held in Domaine de La Besse (Camon, Ariège), France, 13-14 June

Pawluczuk J., Alberski J. 2011. Habitat conditions and grassland vegetation on peat-moorsh soils in the Olsztyn lakeland. Water-Environment-Rural Area 11, 3 (35): 183–195. (in Polish)

Pärtel M., Bruun H.H., Sammul M. 2005. Biodiversity in temperate European grasslands: origin and conservation. In: Lillak R., Viiralt R., Linke A. and Geherman V. (eds.), Integrating efficient grassland farming and biodiversity. Proceedings of the 13th International Occasional Symposium of the European Grassland Federation. Estonian Grassland Society: 1–14.

Pieńkowski P., Gamrat R., Kupiec M. 2004. Evaluation of transformations of midfield ponds in an agrosystem on Wełtyń Plain. Water-Environment-Rural Area 4, 2a (11): 351–362. (in Polish)

Romero A., Chamorro L., Sans F. 2008. Weed diversity in crop edges and inner fields of organic and conventional dryland winter cereal crops in NE Spain. Agric. Ecosys. Environ. 124: 97–104.

Ryszkowski L., Bartoszewicz A. 1989. Impact of agricultural landscape structure on cycling of inorganic nutrients. In: Ecology of arable land. Pr. zbior. Red. M.L. Clarholm, L. Bergstrom. Dordrecht: Kluwer Acad. Publ.: 241–246.

Ryszkowski L., Karg J., Bernacki Z. 2003. Biocenotic function of the midfield woodlots in west Poland: Study area and research assumptions – Pol. J. Ecol. 51: 269–281.

Sobkowicz P., Podgórska-Lesiak M. 2007. Experiments with crop mixtures: interactions, designs and interpretation. EJPAU 10, 2, #22. http://www.ejpau.media.pl/volume10/issue2/art-22.html

Symonides E. 2010. The role of ecological interactions in the agricultural landscape. Water-Environment-Rural Area 10, 4(32): 249–263. (in Polish)

Szempliński W., Budzyński W. 2011. Cereal mixtures in polish scientific literature in the period 2003-2007. Review article. Acta Sci. Pol., Agricultura 10(2): 127–140.

Thiere G., Milenkovski S., Lindgren P.-E., Sahlén G., Berglund O., Weisner S.E.B. 2009. Wetland creation in agricultural landscapes: Biodiversity benefits on local and regional scales. Biol. Conserv. 142: 964–973.

Tsitsilas A., Stuckey S., Hoffmann A.A., Weeks A.R., Thomson L. J. 2006. Shelterbelts in agricultural landscapes suppress invertebrate pests. Aust. J Exp. Agr. 46(10): 1379–1388.

Walker M.P., Dover J.W., Sparks T.H., Hinsley S.A. 2006. Hedges and green lanes: vegetation composition and structure. Biodivers. Conserv. 15, 8: 2595–2610.

Wasilewski Z. 2009. Present statut and directions of grassland management according to the requirements of the common agricultural policy. Water-Environment-Rural Area 9, 2(26): 169–184. (in Polish)

Chapter 3

Soil Fauna Diversity – Function, Soil Degradation, Biological Indices, Soil Restoration

Cristina Menta

Additional information is available at the end of the chapter

1. Introduction

Soil represents one of the most important reservoirs of biodiversity. It reflects ecosystem metabolism since all the bio-geo-chemical processes of the different ecosystem components are combined within it; therefore soil quality fluctuations are considered to be a suitable criterion for evaluating the long-term sustainability of ecosystems. Within the complex structure of soil, biotic and abiotic components interact closely in controlling the organic degradation of matter and the nutrient recycling processes. Soil fauna is an important reservoir of biodiversity and plays an essential role in several soil ecosystem functions; furthermore, it is often used to provide soil quality indicators. Although biodiversity was one of the focal points of the Rio conference, in the 1990s virtually no attention was paid to activities for the conservation of soil communities. However, with the new millennium, the conservation of soil biodiversity has become an important aim in international environmental policies, as highlighted in the EU Soil Thematic Strategy (2006), the Biodiversity Action Plan for Agriculture (EU 2001), the Kiev Resolution on Biodiversity (EU/ECE 2003) and afterwards in the Message from Malahide (EU 2004), that lay down the goals of the 2010 Countdown.

Human activities frequently cause a degradation of soil environmental conditions which leads to a reduction in the abundance and to a simplification of animal and plant communities, where species able to bear stress predominate and rare taxa decrease in abundance or disappear. The result of this biodiversity reduction is an artificial ecosystem that requires constant human intervention and extra running costs, whereas natural ecosystems are regulated by plant and animal communities through flows of energy and nutrients, a form of control progressively being lost with agricultural intensification. For these reasons the identification of agricultural systems which allow the combination of production targets and environmentally friendly management practices, protecting both soil and biodiversity, is essential in order to prevent the decline of soil fauna communities in agricultural landscapes.

2. Biodiversity of soil fauna

Soil biota are thought to harbour a large part of the world's biodiversity and to govern processes that are regarded as globally important components in the recycling of organic matter, energy and nutrients. Moreover, they are also key players in several supporting and regulating ecosystem services [1]. Furthermore, they are key components of soil food webs. Rough estimates of soil biodiversity indicate several thousand invertebrate species per site, as well as the relatively unknown levels of microbial and protozoan diversity. Soil ecosystems generally contain a large variety of animals, such as nematodes, microarthropods (Figure 1) such as mites and Collembola, Symphyla, Chilopoda, Pauropoda, enchytraeids and earthworms (Figure 2). In addition, a large number of meso- and macrofauna species (mainly arthropods such as beetles, spiders, diplopods, chilopods and pseudoscorpion (Figure 3) , as well as snails) live in the uppermost soil layers, the soil surface and the litter layer.

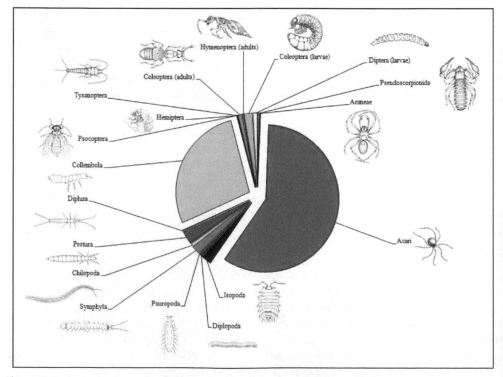

Figure 1. Soil microarthropod community in a beech forest of Northern of Italy

Figure 2. Earthworm belonging to megafauna

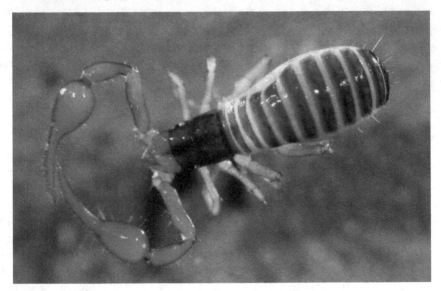

Figure 3. Pseudoscorpion, a typical organism that live in the wood litter

In general, soil invertebrates are classified according to their size in microfauna, mesofauna, macrofauna and megafauna [2] .

Microfauna: organisms whose body size is between 20 μm and 200 μm. Just one group, protozoa, is found wholly within this category; among the others, small mites, nematodes, rotifers, tardigrades and copepod crustaceans all fall within the upper limit.

Mesofauna: organisms whose body size is between 200 μm and 2 mm. Microarthropods such as mites and springtails, are the main representatives of this group, which also includes nematodes, rotifers, tardigrades, small araneidae, pseudoscorpions, opiliones, enchytraeids, insect larvae, small isopods and myriapods.

Macrofauna: organisms whose size is between 2 mm and 20 mm. This category includes certain earthworms, gastropods, isopods, myriapods, some araneidae and the majority of insects.

Megafauna: organisms whose size exceeds 20 mm. The members of this category include large size invertebrates (earthworms, snails, myriapods) and vertebrates (insectivores, small rodents, reptiles and amphibians).

Despite several decades of soil biological studies it is still very difficult to provide average abundance and biomass values for soil invertebrates. This is partly caused by their high variability in both time and space, as well as by differences in the sampling methods used [1]. In addition, most work has been performed in the forest soils of temperate regions, while other ecoregions such as the tropics, or other land uses such as agriculture, have been seriously neglected [1].

Soil fauna are highly variable and the majority are also highly adaptable with regard to their feeding strategies, ranging from herbivores to omnivores and including carnivores. Depending on the available food sources many soil fauna are able to change their feeding strategies to a greater or lesser extent with many carnivorous species able to feed on dead organic matter in times of low food availability. The interactions between soil fauna are numerous, complex and varied. As well as the predator / prey relationships and in some instances parasitism, commensalism also occurs.

The degree of interaction between soil organisms and the soil itself can be highly variable among taxa and dependent on the part of the life cycle that is spent in the soil [2]. In particular, in this regard, combined with the morphological adaptations and the ecological functions of organisms, it is possible to classify soil fauna into four main groups: temporarily inactive geophiles, temporarily active geophiles, periodical geophiles, and geobionts (Figure 4). It should be noted that these groupings do not have any taxonomical significance but rather are useful when studying the life strategies of soil invertebrates. Temporarily inactive geophiles are organisms that live in the soil for only certain phases of their life, such as to overwinter or to undergo metamorphosis, when protection from climatic instability is more necessary. Due to their relative inactivity, the organisms belonging to this group have a weak influence on the ecological function of soil, although they can be important as prey for other organisms. Temporarily active geophiles live in the soil in a stable manner for a large part of their life (i.e. for one or more development stages, and emerge from the soil as adults). Most of these organisms are insects, such as Neuroptera, Diptera, Coleoptera and Lepidoptera. Organisms having a "pupae" stage in their life cycle, play a minor role in the soil during these phases, while the "larvae" stage is much more important for the ecology of soil, especially when the population density is high [1,2]. Most larvae can act as both detritivores and predators. Periodical geophiles spend a part of their life cycle in the soil, generally as larvae, but throughout their lives they occasionally go back to the soil to perform various activities, such as hunting, laying of eggs or to escape dangers. Several Coleoptera groups (e.g. carabides, scarabeids, cicindelids) spend their larval stage in the litter or in the upper layers of mineral soil, and when adults, use soil as a food source, a refuge and for other purposes [1,2]. Geobionts are organisms that are very well adapted to life in soil and cannot leave this environment, even temporarily, having characteristics that prevent survival outside of the soil environment due to their lacking protection from desiccation and temperature fluctuations, as well as the sensory organs necessary to survive above ground by finding food and avoiding predators. Several species of myriapods, isopods, Acari, molluscs and the majority of Collembola, Diplura and Protura, belong to this group [1]. These different types of relationships between soil organisms and the soil environment determine a differentiated level of vulnerability among various groups, as a consequence of any possible impact on soil environment. For instance, if soil contamination occurs, any impact will be highest on geobyonts (because they cannot leave the soil and must spend all their life there) and lowest on temporarily inactive geophiles.

Figure 4. The main four grouping that can be individualized between soil invertebrates, depending on their life strategies and how closely they are linked with soil

There are many extremely old groups of microarthropods in soils, such as collembola and mites, dating from the Devonian period (more than 350 million years ago). In relation to the origin of soil microarthropods, it is possible to form two hypothetical groups: the first group originated in epigeous (above the soil) habitats and only subsequently adapted to soil (e.g. Coleoptera, Chilopoda (Figure 5), Diplopoda (Figures 6 and 7) and Diptera. The second group possibly originated directly in the soil. This group contains organisms such as Protura, Diplura, Symphyla, Pauropoda (Figure 8), and Palpigrada which do not have forms in epigeous, or aquatic habitats (some exceptions are found in caves, where the environmental conditions are very similar to that of soils).

Figure 5. Centipede (Chilopoda)

Figure 6. Millipede (Diplopoda)

Figure 7. Polyxena, a curious group of Diplopoda

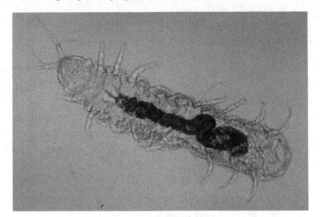

Figure 8. Pauropoda, a microarthropod well adapted to soil

Over the very long period of adjustment to life below ground, the bodies of euedaphic microarthropods became adapted with characteristics that allow them to survive within the soil habitat. During this process of adaptation, impressive levels of convergence have occurred with many of the adaptation characteristics being morphological, and easily explainable and understood. For example, the reduction of the visual apparatus, loss of pigmentation or cryptal coloration (camouflage), reduction of appendages and the acquiring of special structures essential for life below ground. Some of these characteristics, such as the reduction of body length (miniaturisation), loss of the appendages (legs, antenna, etc.) and the loss of eye functionality, which in some cases leads to the complete disappearance of eyes, are direct consequences of degenerative processes in structures which are very important in above ground habitats but useless in the soil. Conversely, soil microarthropods have developed characteristics that permit them to live in the particular conditions present in the soil, such as chemico- and hydroreceptors, often distributed not only in the oral region, as they are for most above ground organisms, but also on other structures of the body. The confinement of these groups to soils, i.e. the groups' incapacity to leave them, is due to the relative stability of these habitats. In fact, diverse factors such as water, temperature and organic matter vary only slightly over the short- and medium-term, as compared to large variations in above ground environments. In addition, there is obviously no light in soil at depths greater than a few millimetres. As a result of all of these factors combined, euedaphic microarthropods are sensitive and unable to survive abrupt variations in environmental factors. They are particularly sensitive to soil degradation and to the disturbances caused, for example, by agricultural cultivation and trampling. Collembola (springtails) represent one of the most important groups of soil microarthropods, both because of the number of species, and the number of individuals, generally present in soils. They have some characteristics that make this group very interesting and useful for studying soil evolution convergence phenomena. Furthermore, they are very useful as indicators of soil quality as their biodiversity and density are influenced by numerous soil factors (in particular organic matter and water content but also other factors such as contamination).

3. The role of soil fauna in soil ecosystem processes

Some researchers have defined edaphic fauna as a "super organism" that assumes a crucial significance due to the chemico-physical and biological processes that are rooted in the soil. Soil biota play an essential role in soil functions as they are involved in processes such as the decomposition of organic matter, the formation of humus and the nutrient cycling of many elements (nitrogen, sulphur, carbon). Moreover, edaphic fauna affect the porosity and aeration of, as well as the infiltration and distribution of organic matter within soil horizons. The ecosystem services provided by soil fauna are one of the most powerful arguments for the conservation of edaphic biodiversity. Decomposition of organic matter by soil organisms is crucial for the functioning of an ecosystem because of its substantial role in providing ecosystem services for plant growth and primary productivity [3].

Due to the absence of light, which makes photosynthesis unfeasible, among the organisms populating the soil we find very few real phytophages, unless we extend the definition of soil animals to surface organisms, or if we consider that pests also include fungivorous microarthropods [2]. The activity of animals, among them typically protozoa, nematodes, rotifers, certain springtails and mites, which feed on microflora, consisting of bacteria, actinomycetes and fungi (both hyphae and spores) is of crucial importance both for regulating the density and for diffusing these micro-organisms. For example, through their faeces, springtails, which feed on fungi, can spread fungal spores that are still viable to areas as far as a few metres away from their point of origin.

The detritus food chain takes on an essential role within the soil, as it becomes the basis of the hypogean food web; in fact, many organisms such as isopods, certain myriapods, earthworms, springtails, many species of mites, and the larvae and adults of many insects feed on the vegetable and animal detritus that is deposited on the soil. For example in the soil of a temperate forest, in which the contribution of litter each year can amount to 400 g/m^2, about 250 g/m^2 are ingested by earthworms and enchytraeids, 30-40 g/m^2 by mites and 50-60 g/m^2 by springtails. Soil fauna performs a mainly mechanical action, whereas chemical degradation is essentially performed by fungi and bacteria, both free and intestinal symbionts of other organisms; furthermore during digestion, organic substances are enriched by enzymes that are dispersed in the soil along with the faeces, contributing to humification.

Earthworms are among the most important organisms in many of the soils of the world. According to an ecological and functional classification [4] it is possible to identify three groups of earthworms that differ in size, burrowing capability, type of food and habitat: epigeic, endogeic and anecic. The epigeic, whose sizes ranges from 0.5 to 5 cm, have a red coloured body, are poorly adapted to burrowing and have a good tolerance to low pH values. They inhabit the superficial organic layers where they feed on litter. The endogeic, whose sizes range from 1 to 8 cm, have a non-pigmented body; exposure to soil with pH values below 5 restricts their activity. They live in the first few centimetres of mineral soil, feeding on humic compost and dead roots, they are capable burrowers and they make tunnels that extend mainly horizontally. Earthworms defined as anecic, whose sizes generally are in excess of 5 cm and have a reddish-brown body, are excellent burrowers, making vertical tunnels that can reach several metres in depth; they tend to avoid soil that is asphyxiated or lacking in moisture. They live in the mineral layer of the soil but rise to the surface, mainly at night, to feed on litter. The activity of earthworms produces a significant effect, not just on the structure, but also on the chemical composition of the soil, since a large part of the organic matter ingested by earthworms is returned to the soil in a form easily used by plants. While they are feeding, earthworms also ingest large quantities of mineral substances (minimally so in the case of the epigeic), that are then mixed with the organic matter ingested and, after having been cemented with a little mucous protein, are expelled in piles called worm casts. In addition to being rich in nitrogen and other nutritive substances such as calcium, magnesium and potassium, worm casts also contain a large quantity of non-digested bacteria which proliferate easily in this sub-stratum and contribute to the humification and mineralisation of organic matter [5]. Vermicasts have more favourable physico-chemical properties, increased microbial population, enzyme activities and nutrient mineralization that support plant growth and yeld [6]. Many studies have reported increased microbial activities during the passage of food through the gut in earthworms [6,7] and higher numbers of fungi, bacteria and actinomycetes in vermicasts [7]. [6] showed enhanced microbial populations and activity in the freshly deposited pressmud vermicasts of *Lampito mauritii* and *Eudrilus eugeniae* in relation to nutrient rich substrate concentrations, multiplication of microbes after passing through the gut, optimal moisture level and large surface areas of vermicasts ideally suited for better feeding and multiplication of microbes. It is important to note that often worm casts are not released in the same layer in which the earthworm fed; in effect the components of the anecic group live at depth and release their mineral-rich faeces on the surface. In contrast the endogeic live at the surface and release their faeces, once again rich in organic matter, at lower depths. These forms of behaviour together with the direct action due to the burrowing, ensure that the soil is mixed, thereby increasing its fertility. The burrowing of the earthworms is also essential for increasing the aeration of the soil and for improving the circulation of water since their tunnels increase the porosity of the ground by 20-30% [5] , enabling those organisms that are not good burrowers to move around easily even at lower depths of the soil. Other than with their worm casts, earthworms contribute to the increase in the amount of nitrogen present in the ground through the excretion of ammonia and urea, forms that are directly useable by plants; furthermore a sizeable quantity of nitrogen is returned to the soil on the death of animals, which have a 72% protein content [8].

Within the context of edaphic fauna it is possible to identify, based on the type of locomotion, swimming organisms, capable of moving around in capillary or gravitational water, reptants that move by taking advantage of natural porosity or cavities produced by other organisms and burrowing organisms. The last-mentioned can open up cavities in the ground in various ways; for example, earthworms compress the ground outside their bodies, diplopods use their legs and backs to push the sediment upwards, the larvae of click beetles use their mandibles, while moles, scarabs and mole-crickets (Figure 9) have specialised legs for burrowing. This continual burrowing activity contributes to the creation of spaces within the soil with the resultant increase in its porosity; the increase of the pores between the particles in turn increases the aerobic bacterial activity and the consequent speed of demolition of organic substances. This bioturbation also has positive effects on water retention, percolation processes and the development of the rhizosphere. The burrowing activity also enables the soil to be mixed and organic matter from the surface layers to be incorporated into the lower layers, while mineral substances are brought towards the surface. This process is carried out in a very evident way by the anecic earthworms discussed a little earlier, which move vertically in the ground even reaching depths of several metres.

Figure 9. Mole cricket

The anthill is one of the most interesting and elaborate examples of the modification of the soil by cunicular burrowing organisms. It consists of a complex of chambers generally constructed on several levels and linked together by tunnels and corridors. Some nests can reach a depth of more than 5 metres and contain over 2,000 chambers, some of them set aside for the cultivation of fungi. The tunnels that connect the chambers contribute to the circulation of air and water within the anthill. Each chamber is inhabited by numerous individuals, some are reserved for incubating eggs, others for rearing the larvae, and yet others for the development of the nymphs which are moved to chambers where the humidity and temperature are more conducive to their development. An anthill can rise above ground level (Figure 10) and can have one or more entrances, or be completely underground and communicate with the surface through one or more exits that are constantly guarded by sentries. Special devices prevent the water that penetrates the soil from flooding the chambers and in this way ensure the survival of the eggs and the nymphs, which are incapable of leaving the nest. As far as the soil is concerned, the presence of channels and tunnels increases the porosity, assisting the penetration of air and water. In addition, a consequence of the movement of the fine material towards the surface by ants during the course of the construction and maintenance of the anthill, is the creation at the surface of a layer with a fine particle size, which is more mineral than organic in nature [9].

Figure 10. A typical anthill of *Formica rufa*

The soil fauna, in particularly molluscs (Figure 11) and earthworms, also has an effect on the soil through the secretion of cutaneous mucous, that have a cementing effect on the particles in the ground, assisting the stability and structure of the soil and making it less vulnerable to processes of erosion. The mucous secretions, the faeces (especially those of earthworms) and the bodies themselves of the animals (when they die) influence in large measure the concentration of nutrients present in the soil particularly potassium, phosphorous and nitrogen, reducing the C/N ratio of the litter and facilitating decomposition.

Figure 11. A slug (a) and a snail of the species *Helix pomatia* (b)

The presence of roots is generally associated with a greater density of micro-organisms in the nearby soil compared with soil devoid of roots; the term rhizosphere is used in a broader sense to refer to the portion of soil surrounding roots in which the micro-organisms are influenced by their presence; its extension is very variable but in general it is considered to be the cylinder of soil used by the root hairs and in which they emit exudates [10]. The rhizosphere can be distinguished from the majority of the soil on the basis of its chemical, physical and biological characteristics. Penetrating the ground, the roots act on the clay minerals and the particles of soil surrounding them; this leads to the formation of an area around them in which the water pathway, and the movement of nutrients and microflora is more heavily channelled than in the rest of the soil. For the same reasons, the organic matter released by the roots accumulates close to them. The chemical nature of the rhizosphere is significantly different from the rest of the soil; this results in large part from the release of carbon and the selective capturing of ions in solution in the groundwater by the roots. The plants act as carbon pumps fixing what is available in the atmosphere in the root exudates that are quickly captured by bacteria; for this reason the level of carbon available around the roots is never very high. The selective absorption of ions instead, causes the depletion of some of them in the rhizosphere, while others, not absorbed by the roots, tend to accumulate. The relationship between the roots and the microflora can be very close and lead to bacteria or fungi becoming an integral part of the roots as in mycorrhizal symbiosis and in the association of bacteria and legumes. The peculiar characteristics of the rhizosphere are also reflected in a selectivity of the animal element. The interaction between soil animals and the roots of the plant can take a variety of forms that lead to benefits or repress the growth of the plant, and often involve interactions with the microbial population of the soil. The dispersion of the inocula of mycorrhizal fungi by soil animals can have beneficial effects for the plants; this dispersion is particularly favoured by burrowing organisms belonging to the mega- and mesofauna. The hyphae of mycorrhizal fungi may make up a significant proportion of the total microbial biomass in some soils and can become one of the most important sources of food for fungus grazing animals such as springtails. Numerous soil animals feed directly on the roots of plants; among them are a large number of species of springtails and myriapods. It is still not entirely clear how much of the damage inflicted on plants can be attributed to the direct action caused by the grazing of the roots or from the subsequent vulnerability of the roots to pathogens in the soil, especially fungi.

It is extremely rare for the biological relationships between soil organisms to consist of a simple and clear interaction. The actual conditions are the result of many complex interactions that typically involve multiple participants in life in the soil such as the plants, the microbes, the fungi and the animals, the last mentioned at different levels of organization. For example, examination of the faeces of earthworms reveals a mixture of fragments of plants, microorganisms, fungi, and fragments of encysted animals and protozoa capable of surviving the unfavourable conditions in the gut of the earthworm [10] .

4. Diversity of soil fauna in different ecosystems and soil managements

A rapid survey of invertebrate and vertebrate groups reveals that at least ¼ of described living species are strictly soil or litter dwelling. The greatest diversity is observed in systems where equilibrium exists between the productivity level and the perturbation rate, whereas local species extinctions may occur through lack of demographic recuperation when perturbations increase or through competitive exclusion when productivity increases. Land use changes and the intensification of agriculture also generate severe habitat degradation or destruction of soil biota. The environment of soil organisms in managed ecosystems can be influenced by any combination of land use factors, such as tillage, pesticide and fertilizer application, soil compaction during harvest, and removal of plant biomass.

The profile of a mature and undisturbed soil can be divided into a number of spheres, whose chemico-physical properties are determined by vegetation cover, by geographical location and by climate; this stratification is reflected in the edaphic population and as a result in its vertical distribution. Usually two groups of organisms are recognised in the soil: the euèdaphons, which include organisms that live in the strip of mineral soil, and the hemiedaphons, found in the strip of organic soil; to these can be added the epiedaphons (or epigeons), consisting of organisms that live in the surface of the soil, and hyperedaphons that extend to the herbaceous layer. The moisture content and the pH of the soil have the greatest influence on the distribution of the hemi- and euedaphon fauna, even if the characteristics of the litter, the porosity of the soil and numerous other factors are also important in determining the vertical distributions of edaphons. Humidity has often been used to establish further subdivisions within the hemiedaphons; one example is the classification in which hemiedaphon springtails were subdivided into three categories: hydrophilic (living on the surface of free water), mesophilic (living in damp organic litter) and xerophilic (living in more exposed dry areas, such as in lichens, mosses and tree bark). In reality this type of stratigraphical classification is difficult to apply since soil organisms migrate on both a daily and a seasonal basis; in fact, many species, for example mites, springtails and isopods, can move towards the surface over distances ranging from a few millimetres to a few centimetres. An interesting case is that of sinfili that can go down as much as 40-50 cm. Nematodes are usually most abundant in the first 10 cm of soil, but their distribution can vary greatly depending on the type of vegetation cover and conditions of dampness. For example, in soil with abundant vegetation cover and a high presence of roots, these organisms concentrate in the upper part of the ground which, being rich in exudates and decomposing organic matter, maintain a numerous population of their potential prey. In soils with sparse vegetation cover, the high frequency of dry conditions on the surface cause nematodes to be more common at a depth of 5-10 cm whereas, in cultivated land a uniform distribution of the population can be observed down to a depth of about 20 cm, attributable to the mixing effects caused by ploughing. The density of these organisms is particularly high in grasslands, where they can be present with a density in the order of 20 million/m², whereas in moorland the density varies greatly and usually is less than one million/m², even though in some cases it can exceed 5 milioni/m².

The abundance, diversity, composition and activity of species of the soil community can be affected by plant species, plant diversity and composition, as well as by animal grazing [3]. The two factors that most likely influence the soil community are the nutrient resources available, and the diversity of microhabitats [3]. Resource type is determined by the tissue chemistry of plant species and the nutrient content of excreted wastes from grazing by consumers [3]. The microhabitat is directly affected by the physical and chemical properties of the system as well as the amount of nutrient input [11].

Various studies have described the structure of soil invertebrate communities in relation to forest diversity, dynamics and management [12-18]. In central Italy silvicultural practices and the composition of deciduous forests do not seem to have any important effect on the structure of microarthropod communities [19]. The absence of a change in soil community structure could be linked to the litter layer that in these hardwood stands is thick enough to maintain a high level of organic matter and a favourable microclimate in every season. Soil mesofauna seem to recover quickly after disturbances such as tree cutting [12] indicating a good level of ecosystem integrity (community resilience). The same aspect also emerged in conventional tillage conditions where soil arthropod abundance was significantly higher in autumn compared with summer [20]. The authors suggested that, given sufficient time without soil disturbance, soil arthropods are able to recover within the growing season. These results agree with previous studies conducted on a temperate cool rain forest in west Canada where there were no significant differences in the population density of arthropods between undisturbed forests on harvested plots and those on unharvested patches [21]. Additionally, in another recent study involving a beech forest, the hypothesis of [22], which predicts community changes during forest rotations, was refuted from a functional view point [14]. However several studies considering the effects of silvicultural practices on soil fauna found important impacts on soil forest fertility/productivity and in the terrestrial food chain [23]. It is generally accepted that the removal of trees by clear-cutting, or other methods, has a significant effect on the invertebrate fauna of the forest floor [24,25]. The effects on arthropod communities are complex and difficult to analyze since various taxonomic groups are affected and they react to impacts differently [12,26-29].

A study that compared the soil community in different grasslands with a semi-natural woodland area and an arable land site showed that microarthropod communities of the three land use typologies differed in terms of both observed groups and their abundance [30]. Typical steady soil taxa characterised woodland and grassland soils, whereas their abundances were significantly higher in the former. The mean highest abundances of Acari and Collembola were observed in grassland, confirming the suitable trophic conditions of this habitat, whose taxa diversity and soil biological quality did not significantly differ from woodland samples. On the contrary in the arable land, the microarthropod community showed a reduction both in taxa number and soil biological quality compared with the other sites. Besides, mite and springtail abundances were significantly lower than in grassland. The authors concluded that soil biological quality and edaphic community composition highlighted the importance of grassland habitat in the protection of soil biodiversity, especially because it combines fauna conservation with the production of resources for human needs.

5. Principal factors affecting the loss of soil fauna diversity

Around the world there are numerous soils that have lost their fertility or their capacity to carry out their function due to the impact of man. The causes are mainly related to processes that are accelerated or triggered directly by human activities and that often act in synergy with each other, amplifying the effect. Among them, the most widespread at a worldwide level are erosion, the loss of fertility and a decline in organic matter, compaction, salinisation, phenomena of flooding and landslides, contamination and the reduction in biodiversity. Reduction of soil biodiversity as a result of urbanization can be even more severe. The urbanization process leads to the conversion of indigenous habitat to various forms of anthropogenic land use, the fragmentation and isolation of areas of indigenous habitat, and an increase in local human population density. The urbanization process has been identified as one of the leading causes of declines in arthropod diversity and abundance.

Soil properties determine ecosystem function and vegetation composition/structure, serve as a medium for root development, and provide moisture and nutrients for plant growth [31]. Disturbances linked to natural forces and to human activities can alter physical, chemical and biological properties of soils, which can, in turn, impact long-term productivity [32,33]. Humans have extensively altered the global environment and caused a reduction of biodiversity. These change in biodiversity alter ecosystem processes and change the resilience of ecosystems to environmental change. It is estimated that human activities increased the rates of extinction 100-1000 times [34]. In the absence of major changes in policy and human behaviour our effect on the environment will continue to alter biodiversity. Land use change is projected to have the largest impact on biodiversity by the year 2100 [35]. Within agricultural land use, that covers 10.6 % of land surface, the intensity of agronomic practices and crop management can also affect biodiversity (Figure 12). Land use is considered to be the main element of global change for the near future. In a review on changing biodiversity [35] consider that land use will be the main cause of change in biodiversity for tropical, mediterranean and grassland ecosystems. Forests, tropical or temperate, generally represent the biomes with the largest soil biodiversity. Consequently any land use change resulting in the removal of perennial tree vegetation will produce a reduction of soil biodiversity. In some cases forests are succeeded by pasture or perennial grasslands, while in others arable land replaces formerly wooded areas. The change in soil biodiversity will therefore be influenced by the subsequent use of land following the forest.

Figure 12. Image of Val D'Adige (Italy) where it's possible to note the intense agricultural utilization

The abundance, biomass and diversity of soil and litter animals are influenced by a wide range of management practices which are used in agriculture. These management practices include variations in tillage, treatment of pasture and crop residues, crop rotation, applications of pesticides, fertilisers, manure, sewage and ameliorants such as clay and lime, drainage and irrigation, and vehicle traffic [36]. Furthermore differences in agricultural production systems, such as integrated, organic or conventional systems, have been demonstrated to affect soil fauna in terms of numbers and composition [37,38]. The impact of soil tillage operations on soil organisms is highly variable, depending on the tillage system adopted and on soil characteristics. Conventional tillage by ploughing inverts and breaks up the soil (Figure 13), destroys soil structure and buries crop residues [39] determining the highest impact on soil fauna; the intensity of these impacts is generally correlated to soil tillage depth. Minimum tillage systems can be characterized by a reduced tillage area (i.e. strip tillage) and/or reduced depth (i.e. rotary tiller, harrow, hoe); crop residues are generally incorporated into the soil instead of being buried. The negative impact of these conservation practices on soil fauna is reduced compared with conventional tillage. Under no-tillage crop production, the soil remains relatively undisturbed and plant litter decomposes at the soil surface, much like in natural soil ecosystems. The influence on soil organism populations is expected to be most evident when conservation practices such as no-till are implemented on previously conventionally tilled areas because the relocation of crop residues to the surface in no-till systems will affect the soil decomposer communities [40]. No-till [41] and minimum tillage generally determine an increase in microarthropod numbers.

Figure 13. Conventional tillage by ploughing inverts and breaks up the soil, destroy soil structure and buries crop residues

A multidisciplinary study was carried out over four years in Northern Italy on a silt loam under continuous maize [42] to evaluate two factors: the soil management system (conventional tillage and no-tillage) and N fertilisation. This study showed that total microarthropod abundances were higher in NT compared with CT (+29%). Acari showed higher sensitivity to tillage compared with collembola. Moreover, N fertilisation with 300 kg N ha^{-1} had a negative effect on the total microarthropod abundance. [43] also observed that higher values of mite density were associated with a decrease in tillage impact. [38] similarly reported that the mite community, in particular oribatid, was more abundant in no-tillage compared with conventional tillage, but the differences were found only in some periods of the year. Conventional tillage caused a reduction of microarthropod numbers as a result of exposure to desiccation, destruction of habitat and disruption of access to food sources [44]. The infuence of these impacts on the abundance of soil organisms will be either moderated or intensified depending on their spatial location; that is, in-row where plants are growing, near the row where residues accumulate or between rows being subjected to possible compaction from mechanized traffic [45].

Observations on the impacts of different forms of agricultural management on communities of microarthropods showed that the high input of intensively managed systems tends to promote low diversity while lower input systems conserve diversity [11,46]. It is also evident that high input systems favour bacterial-pathways of decomposition, dominated by labile substrates and opportunistic, bacterial-feeding fauna. In contrast, low input systems favour fungal pathways with a more heterogeneous habitat and resource leading to domination by more persistent fungal feeding fauna [11].

The effects of fertilizers on soil invertebrates are a consequence of their effects on the vegetation and, directly on the organisms. Increases in quantity and quality of food supplied by vegetation are frequently reflected in greater fecundity, faster development and increased production and turnover of invertebrate herbivores [47]. The effects of organic and inorganic fertilizers in terms of nutrient enrichment may be comparable, but these two types of fertilizers differ in that organic forms provide additional food material for the decomposer community. [48] concluded that the total soil microbial biomass and the biomass of many specific groups of soil organisms will reflect the level of soil organic matter inputs. Hence, organic or traditional farming practices, that include regular inputs of organic matter in their rotation, determine larger soil communities than conventional farming practices [48]. Also [49] reported that the soil microbial and faunal feeding activity responded to the application of compost with higher activity rates than with mineral fertilization. Generally the responses of soil fauna to organic manure will depend on the manure characteristics, and the rates and frequency of application. Herbivore dung, a rich source of energy and nutrients, is exploited initially by a few species of coprophagous dung flies and beetles and, later, by an increasingly complex community comprising many general litter-dwelling species [47].

A study related to sewage sludge application on agricultural soils showed an increase in the abundance of Collembola [50], Carabidae [51], Oligochaeta [52], soil nematodes [53] and Arachnida [53]. In some cases, the application of sewage sludge to agricultural land can bring toxic substances that, accumulating in the soil, reach potentially toxic levels for soil fauna [53]. Field studies have suggested that metals contained in sewage sludge don't reduce the abundance of euedaphic [50] and epigeic collembola [54] but may alter their population structure. [53] reported negative effects on collembola communities in soil treated with sewage sludge and this effect may be attributed to anaerobic conditions and high ammoniacal level. In effect, the knowledge gained in relation to the effects of sewage sludges showed that species more sensitive to the toxic substances contained in sewage sludge can disappear, while others which are more tollerant, can dramatically increase.

Organic wastes are usually stored in dump areas with associated high management costs and problems of environmental impact. On the other hand organic wastes could become an easily available and cheap source of organic matter after composting processes. The use of compost obtained by organic waste in agricultural activity enables the converting of waste materials into a useful resource. Therefore, the national authorities have over recent years stimulated both the use of compost to reduce the soil fertility loss and research aimed at assessing its effects on both agricultural production and soil environment [55-57]. But negative effects on soil fauna could be related to the use of organic waste, such as the accumulation of trace metals in soil. In fact many trace elements contained in organic wastes can reduce the abundance and diversity of soil microarthropod communities, or can influence the survival potential and the rate of growth of more sensitive species [58]. [59] concluded that negative effects of compost use on soil fauna abundance or biodiversity were not observed in two Italian sites studied that were treated with compost, supporting the use of this product derived from waste in order to add organic matter into the soil.

Pesticide application to the soil can affect the soil fauna influencing the performance of individuals and modifyng ecological interactions between species. When one or more ecosystem component is impacted by a pesticides, this will affect the microarthropod communities in terms of number and composition. Pesticide toxicity on soil fauna is determined by different factors, such as the pesticide's chemical and physical characteristics, the species' sensitivity and the soil type. In fact among soil microarthropods, different taxa showed a variety of responses. The physical and chemical characteristics of the soil, such as its texture, structure, pH, organic matter content and quality, and nature of clay minerals, are important factors determining the toxic effects of pesticides or other xenobiotics. A study carried out by [60] showed reduced toxic effects, as a function of soil type, in the following order: sand>sandy-loam>clay>organic soil. Often, the toxicity of pesticides can be related directly to soil organic matter content [61]. However pesticide application does not always cause negative impacts on the entire soil microarthropod community. For example for certain types of soil there is evidence that some taxa can obtain a competitive advantage from the application of some specific pesticides.

6. Use of soil invertebrates in soil biodiversity assessment and as soil biological quality indices

As previously illustrated, the increasing anthropic pressure on the environment is leading, in most parts of the world, to a rapid change in land use and an intensification of agricultural activities. These processes often result in soil degradation and consequently loss of soil quality. Soil quality could be defined as the capacity of a specific kind of soil to function, within natural or managed ecosystem boundaries, to sustain plant and animal productivity, to maintain or enhance water and air quality, and to support human health and habitation [62,63]. A common criterion for evaluating the long-term sustainability of ecosystems is to assess the fluctuations of soil quality [64]. Soil reflects ecosystem metabolism; within soils, all bio-geo-chemical processes of the different ecosystem components are combined [65]. Monitoring ecosystem components plays a key role in acquiring basic data to assess the impact of land management systems and to plan resource conservation. Maintaining soil quality is of the utmost importance to preserving biodiversity and to the sustainable management of renewable resources.

Soil quality can be evaluated through its chemical-physical properties and biological indicators and indices. The importance of some of these parameters is generally accepted. Soil organic matter among the chemical indicators, bulk density [66-69,42] and aggregate stability [70,71] among the physical indicators, were the most often used but there were few examples of biological indicators of soil quality [72,66,68]. However, biological monitoring is required to correctly assess soil degradation and correlated risks [73,74] . Indicators of soil health or quality should fulfil the following criteria [75]: 1) sensitivity to variations of soil management; 2) good correlation with the beneficial soil functions; 3) helpfulness in revealing ecosystem processes; 4) comprehensibility and utility for land managers; 5) cheap and easy to measure. The growing interest in the use of living organisms for the evaluation of soil conditions is justified by the great potential of these techniques, that allow the measurement of factors difficult to detect with physical-chemical methods and give more easily interpretable information [76]. Biotic indices, based on invertebrate community studies, were recently developed as a promising tool in soil quality monitoring. These organisms are highly sensitive to natural and human disturbances and are increasingly being recognized as a useful tool for assessing soil quality. The complex relationships of soil fauna with their ecological niches in the soil, their limited mobility and their lack of capacity to leave the soil environment, make some taxa (e.g. Collebola, Protura, Pauropoda) particularly vulnerable to soil impact [77]. For these reasons soil fauna communities represent an excellent candidate for soil bioindication and for evaluating soil impact. The basic idea of bio-indications is that the relationship between soil factors and soil communities can be tight [76]. When soil factors influence community structure, the structure of a community must contain information on the soil factor [78]. To retrieve information about soil quality, different properties of community structures, such as the richness and diversity of species, the distribution of numbers over species, the distribution of body-size over species, the classification of species according to life-history attributes or ecophysiological preferences and the structure of the food-web can be used [78]. The number of bio-indicator systems using soil invertebrates is relatively high; some approaches use Nematode, Enchytraeid, mites, Collembola, Diptera, Coleoptera or all microarthropod communities [79,78,77,80,81]. Moreover the use of bioindicators makes it possible to highlight the interactions among the different pollutants and between them and the soil. Often, bio-monitoring techniques are not very specific in identifying the pollutant or environmental variable that creates stress in the organisms. For this reason bio-monitoring must not be considered a substitute for physical-chemical analysis, but a complementary methodology which allows a broader outlook on the case study.

Most edaphic animals have life cycles that are highly dependent on their immediate environment, interacting with soil in several different ways. To be able to evaluate their role and function, it is important to use methodologies that highlight either the number of species present or the processes and roles that they play in the soil environment. In particular mesofauna groups are a key component of soil biota. They are very abundant, their role in soil formation and transformation is well-recognized, the area covered during their life cycle is representative of the site under examination, their life histories permit insights into soil ecological conditions and, several species have already been recognized as useful biological indicators of soil quality. In general, soil invertebrate-based indices consider the consistency and richness of populations [82]. Some species in a single taxon may be specified as indicators of soil quality or as test organisms and used in toxicology tests. In the collembolan taxon, *Folsomia candida* (Figure 14) is the most frequently used species in both sub-lethal and lethal testing [83-86]. *Onychiurus armatus* [87,58], *Orchesella cincta* [88-90], *Isotoma notabilis* [58], *Tetrodontophora bielanensis* [91] and other collembolan species [92] have been used in laboratory tests but have not reached the same level of routine use as has *F. candida*. Because of the species-specific differences in responses to contaminants, the tests conducted on *F. candida* provide partial indications as to the effects provoked by these substances on the collembolans; this information has also been useful in calibrating experiments on other species. Some collembolan species like *Folsomia quadrioculata*, *Folsomia fimetariodes*, *Isotoma minor* and others species have been used to evaluate the effects of chemicals on collembola in the field [85].

Figure 14. *Folsomia candida*, a collembola used in toxicology tests

The Maturity Index (MI) [93] is a bio-indicative method based on the soil nematological community composition (Figures 15 and 16) that sorts the families into five categories according to the reproductive characteristics which define them as colonizer or persistent organisms. Each family is assigned a score ranging between 1 and 5 passing from the colonizer to persistent forms. These values, called c-p (v), are then multiplied by the organisms' frequency (f) and finally inserted in a summation.

$$MI = \sum_{i=1}^{n} v(i) f(i)$$

Index values close to 1 indicate the predominance of colonizer forms and therefore an environmental situation that presents no great stability. On the other hand, values ranging between 2 and 4 highlight the presence of a situation with persistent forms and more stable conditions.

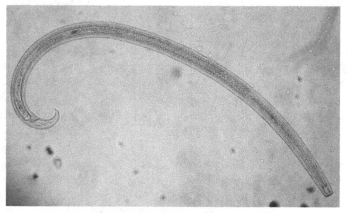

Figure 15. A nematode belonging to Mononchidae family

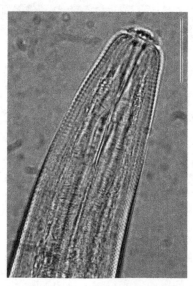

Figure 16. A nematode belonging to Plectidae family (*Chiloplectus andrassyi*). In the picture is showed a particular of the head

Moreover, soil biological quality could be expressed using an Acari/Collembola ratio (A/C) and the QBS-ar index. The first indicator is based on the densities of Acari and Collembola communities where in natural conditions the ratio of the number of mites to the number of collembola is greater than one. On the other hand, in the case of soil degradation, that ratio shifts towards collembola and its value diminishes [94]. The QBS-ar index [77] is based on the following concept: the higher the soil quality is, the higher the number of microarthropod groups morphologically well adapted to soil habitat will be. Soil organisms are separated into biological forms according to their morphological adaptation to the soil environment; each of these forms is associated with a score named EMI (eco-morphological index), which ranges from 1 to 20 in proportion to the degree of adaptation. The QBS-ar index value is obtained by summing the EMI of all the collected groups. If biological forms with different EMI scores are present in a group, the higher value (more adapted to soil form) is selected to represent the group in the QBS-ar calculation. QBS-ar was applied in several agricultural ecosystems, grasslands, urban soil and woods at different levels of naturality and anthropic impact [95,42,30,96,19]. This index reached higher values in grassland and woods and lower values in agricultural ecosystems [30]. Moreover, [96] demonstrated that QBS-ar was highest in the soil with the lowest metal content and the highest density and taxa richness of the invertebrate community. The authors suggested that this index seems to be appropriate in defining the quality of the investigated soils. In Figure 17 are showed some QBS-ar values detected in different soils.

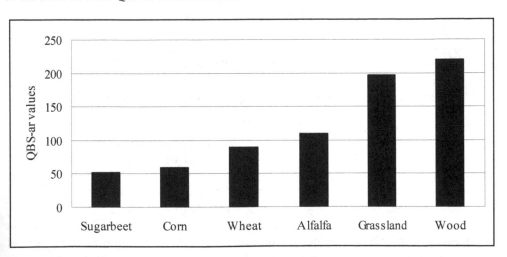

Figure 17. QBS-ar values detected in different soils

[19] concluded that in forest ecosystem management, QBS-ar could be an efficient index for evaluating the impacts on soil of forest harvesting (i.e. soil compaction due to logging) and so for determining the sustainable use of renewable resources. At the same time, QBS-ar can be a valuable tool in ecosystem restoration programmes for monitoring the development of soil functions and biodiversity and for preventing the negative effects of soil compaction when mechanization is used (e.g. in Europe many LIFE projects include mechanized

operations). Furthermore this index could be implemented in environmental management programmes of urban forestry and protected areas in relation to recreational use to prevent the negative effects of trampling.

Another index that could be applied to soil fauna communities is the V index [97], which expresses the magnitude of the response to tillage.

The V index was calculated as:

$$V = \frac{2M_{CT}}{M_{CT} + M_{NT}} - 1$$

where M_{CT} and M_{NT} could be the abundance of taxa under conventional tillage and no-tillage, respectively.

Six magnitude categories were provided for the V index [97]:

Extreme inhibition by tillage (or treatment): $V < -0.67$

Moderate inhibition by tillage (or treatment): $-0.33 > V > -0.67$

Mild inhibition by tillage (or treatment): $0 > V > -0.33$

Mild stimulation by tillage (or treatment): $0 < V < 0.33$

Moderate stimulation by tillage (or treatment): $0.33 < V < 0.67$

Extreme stimulation by tillage (or treatment): $V > 0.67$

Wardle V index proved to be a good indicator of the response to tillage [42].

7. Effect of actions for restoration and conservation of soil fauna diversity

The realisation that degradation of the soil is an environmental problem of global significance, with immediate consequences at an economic and social level, and the recognition of the importance of protecting it, have led to an increase in international initiatives. The Convention on Biological Diversity is the first global agreement aimed at conservation and the sustainable use of biological diversity (Secretariat of the Convention on Biological Diversity 2000). The CBD lies at the heart of biodiversity conservation initiatives. It offers opportunities to address global issues at a national level through locally grown solutions and measures. One important requirement is the development of National Biodiversity Strategies and Action Plans mainstreaming them into relevant sectors and programmes, a principal means for implementation of the Convention at the national level (United Nations 1992). The recent Conference of the Parties of the Convention on Biological Diversity (May 2008, Bonn) demonstrated that the need for action to protect biodiversity is unanimously acknowledged. Biodiversity conservation is essential both for ethical reasons and especially for the ecosystem services that the complex of living organisms provides for current and future generations. These ecosystem services are essential for the functioning of our planet. A necessary starting point for achieving the objective of preserving soil biodiversity is to reach an adequate level of knowledge on its extent and on its spatial and temporal distribution. Among the most important tasks that man should set himself for safeguarding the assets of the soil is to rectify the damage caused to ecosystems, where it is still possible, through works of environmental recovery. There are a number of actions that could reduce the risk of damage to ecosystems, ecological receptors and to humans, that include land reclamation, environmental restoration and halting the exposure to sources of pollution (either by physical means or through communication such as information and education) [98]. These initiatives are often problematical and carry a heavy price tag. It is well known that the reclamation of waste disposal sites is usually characterized by soil quality problems; soil used to cover the dump is generally affected by physical, biological and sometimes chemical degradation. These conditions affect both the plant and the animal communities and, more generally, the effectiveness of the restoration processes. In this habitat the soil fauna can be inherited or can have established itself on the reclaimed waste disposal site; generally the more structured the soil is the more complex is the soil fauna community. Management of waste disposal is an extremely emotive issue and the level of acceptance of this kind of facility by the local community is often dependent on the mitigation of its impact and on a good restoration of the interested sites. Beyond the well-known hygienic-health risks and its strong impact on the landscape, the construction of a dump requires the removal of a large quantity of soil that cannot be considered an unlimited resource. In order to prevent the risks of groundwater contamination, the bottom of waste disposal sites is sealed using layers of waterproof materials and the top of the dump is sealed to prevent rain seepage and the leakage of biogas produced by decomposition. The surface layer of a dump is usually covered by filling soil that is suited to being re-colonized by plants and animals, even though this soil does not represent the same physical, chemical and biological characteristics of the removed soil. Thinking about the complexity of the issues relating to dumps, it is important to obtain exhaustive and multi-disciplinary information on the environment and it is necessary to support traditional physical-chemical analyses with bio-monitoring techniques. As previously stated, the growing interest in the employment of living organisms for the evaluation of soil conditions is justified by the great potential of these techniques, that allow the measurement of factors difficult to detect with physical-chemical methods and that give more easily interpretable information [76]. The great differences in abundance and maturity shown by nematode communities (Maturity Index MI 2.76) in the top soil of a reclaimed waste disposal site compared with permanent grassland and wood could have been caused by many factors [99]. In this study disturbance in the soil from a dump is reflected by the lower maturity [93] due to the absence of omnivorous nematodes like Thornenematidae [100]. The greater abundance and maturity in the nematode communities from woods and grasslands may be due to the fact that the soil is less disturbed. The differences in soil litter composition and in root distribution had probably affected the predominance of plant feeding nematodes in grasslands and the predominance of dorylaimids (Qudsianematidae, Leptonchidae) in the woods [101]. [102], in a study related to nematode communities in ash dumps covered with turf and reclaimed from different times reported MI values ranging from 2.0 and 2.3. [102] reported that in the ash dump reclaimed over a longer period the total abundance of nematodes was higher than those reclaimed over a shorter time and in some samples it was similar to the lowest abundances observed in grasslands in Poland. The author suggested that the species with high ability to colonize new habitats had the best chance of survival in these conditions. Probably, lack of soil structure, high salt content and low organic matter content may be responsible for low MI values and the low density observed in the dump that was the subject of the [99] study. Moreover, the poor and little structured covering of vegetation in the dump, that consequently did not create homogeneity in soil structure and organic matter content, may be a very important reason also limiting the microarthropod community. [103] showed the vulnerability of springtails and pauropods. The authors observed that the reduction of collembolan and pauropod densities in high-input management systems is largely explained by the mechanical and chemical perturbations produced by conventional agricultural management practices and by particular abiotic soil conditions present in the intensively managed sites that are unfavourable for these organisms. The authors reported that symphylans were more abundant in the mixed management site. Extraction activities have a significant impact on the community, affecting both vegetation and soil microbes and animals. The studies related soil community changes during ecological succession in degraded soils are still scarce. After the extraction, the ecosystem would be able to recover spontaneously if the mineral substratum and the environmental conditions were right, but in many cases the physical, chemical and biological conditions of the soil are too disturbed (e.g. unbalanced particle sizes, low organic matter content, inadequate biological component of the soil) or the start of a secondary succession is impeded due to isolation from the colonisation resources [104]. In the process of open-cast mining, the vegetation is completely removed and this causes major changes in the physical, chemical and microbiological properties of the soil [105]. Topsoil is an essential component in abandoned quarries for the growth of vegetation and must be preserved for the restoration of the ground once the extraction work has been completed [106]. Generally a significant period of time passes between the initial removal of the topsoil and the final distribution of the same over the restored area. Because of this, the properties of the stored soil can deteriorate and it can become biologically sterile. In [108] it was demonstrated that the microbial population in the accumulated stockpile falls dramatically in comparison with a control sample of soil that had not been removed. In the same study results were compared of samples taken in the quarry and those taken from a control area, and the particle sizes of the mineral components were analysed. It was found that the proportion of sand particles in the quarry site had risen, while the particles of lime and clay had fallen in comparison with the control soil, phenomena probably due to the process of erosion. This is a consequence of a low stability of the aggregates and, consequently, a high rate of infiltration [107]. In the accumulated stockpile instead, it was observed that there was an increase in density and a reduction of porosity, caused by compaction by machinery during the excavation. These changes make the diffusion of gases more difficult and they restrict the growth of the deep roots of the plants, thus representing one of the reasons why in the shrub stage they cease to grow. To this must be added a change in the pH of the stockpile, with an increase in its acidity due to the separation of the base cations and the scarcity of nutrients, probably caused by the reduction of soil microbes induced by the accumulation of the soil stockpile. If it is not possible to deposit the stored soil in the quarry site within the maximum period for preserving its fertility, it becomes necessary to initiate a biological restoration in order to preserve the topsoil, but it must still be carried out within the conservation period, that is before the cessation of microbiological activity and the breakdown of the nutrient cycle, in order to prevent the soil from becoming completely unproductive [108]. The motor for the succession is the interaction between the trophic levels of the ecosystem. Given that plants are at the bottom of the food chain and that they play an important role in the formation of the physical structure of the soil, the changes in the vegetation during the succession are crucial to the successional development of the other organisms, including soil animals. At the same time, the succession of plants depends on the abiotic conditions of the site, on the pool of species and the intra-species competition, but it is also influenced by other trophic levels, among them the herbivores and the soil invertebrates. The latter can influence the successional changes of the plants through soil phytophagous, the effects on the availability of nutrients, and by influencing the formation and the modification of the soil as a habitat for plants. Soil communities are important in the processes of soil formation because they influence the distribution of the organic matter and as a consequence, the rate of decomposition [109]. The study by [109] on the restoration of an extraction site demonstrated that there are strict timing synchronisations between the changes in the vegetation, the soil and its being populated. This indicated that the interaction between all these components can play an important role in successional changes in the ecosystem. The study of these components of fauna are therefore important for monitoring environmental recovery processes, given the links between edaphic fauna, soil and vegetation. As with vegetation, the post-restoration recovery of the invertebrate community is slow and not less than 15 years [110,111], with 80-102 years estimated for the recovery of springtail communities in forests [112]. The maximum density of the mesofauna is generally reached during the 2-3 years of the "pioneering" phase, which is followed by a drastic reduction in the density to levels of less than 20% within the following 10 years [113]. Successive changes in the taxonomical composition and relative abundance may be correlated to successional changes in factors such as vegetation cover, the pH of the soil and the content of organic matter, etc. [114]. The richness of the animal taxa is indicative of the maturity of the community of vegetation in the recovered area. After recovery of the soil, the process of secondary succession involves an increase in the diversity of the structure and in the available ecosystem energy, that facilitate the development of high trophic levels. A study conducted in northern Italy in an open-cast quarry after the restoration phase showed mature microarthropod communities and higher abundances in the sites where the extraction activity had finished earlier. The presence of edaphic organisms generally associated with stable soil conditions, such as pauropods, symphylans, proturans and diplurans, was found only in these sites (personal data unpublished). Succession to a naturalized grassland from former agricultural land and pasture is accompanied by changes in plant biodiversity and in the soil community [3]. These change are the result of a reduction or elimination of management, fertilizer applications and of grazing by large herbivores. The response of the soil faunal community and diversity might not be in-step with plant succession. Species of soil biota found in early successional stages persist in later stages, although with changes in dominance and species frequency [3]. Species replacement is either less pronounced or it occurs on a different time scale. In a study of grassland succession from 7 to 29 years into restoration, the changes in soil faunal species in Isopoda, Chilopoda and Diplopoda did not correspond with plant successional changes, although the macro-invertebrate diversity and density increased with field age, but decreased in the oldest field [115]. Environmental changes during succession increased the amount of basal resources that provided various micro-habitat and nutrient resources for macro-invertebrates that could lead to the establishment of a diverse community [3]. An increase in the amount of habitable space created by increasing pore surface area would increase the abundance of the macro-invertebrates [115]. It is still not clear to what extent different groups of organisms, such as nematodes, microarthropods or bacteria, respond separately or as an integrated food web community to plant succession. Mechanisms of feedback interactions of soil organisms among themselves and with roots are complex, and not well understood at a molecular level [3].

8. Conclusion

Too rarely do we pause to reflect on the fact that soil is the foundation upon which society is sustained and evolves, that it is a vital component of ecological processes and cycles, as well as the basis on which our infrastructure rests. Often not enough importance is given to the fact that soil quality and its protection contribute significantly to preserving the quality of life, and that the nutrition and health of humans and animals cannot be separated from the quality of the soil. Growing pressures from an ever increasing global population, as well as threats such as climate change and soil erosion, are placing increasing stresses on the ability of soil to sustain its important role in the planet's survival. Evidence suggests that while increased use of mono-cultures and intensive agriculture has led to a decline in soil biodiversity in some areas, the precise consequences of this loss are not always clear [1]. Soil is one of the fundamental components for supporting life on Earth. It is the processes that occur within soil, most of which are driven by the life that is found there, which drive ecosystem and global functions and thus help maintain life above ground. Soil performs numerous ecosystem functions and services, ranging from providing the food that we eat to filtering and cleaning the water that we drink. It is used as a platform for building and provides vital products such as antibiotics, as well as containing an archive of our cultural heritage in the form of archeological sites. Life within the soil is hidden and so often suffers from being 'out of sight and out of mind' [1]. A more complete knowledge of soil fauna is needed for biodiversity conservation.

Only by knowing soil in all its complexity, while maintaining its functionality and quality through actions aimed at protecting its properties, and acknowledging the importance it assumes in the quality of life worldwide, can we embark on a truly sustainable use of soil perceived as a resource and build a proper Man / Soil relationship to be left to future generations.

Author details

Cristina Menta
Department of Evolutionary and Functional Biology, University of Parma, Parma, Italy

9. References

[1] Jeffery S, Gardi C, Jones A, Montanarella L, Marmo L, Miko L, Ritz K, Peres G, Römbke J, van der Putten WH, editors (2010). European Atlas of Soil Biodiversity. Publications Office of the European Union, Luxembourg.

[2] Wallwork JA (1970) Ecology of soil animals. McGraw-Hill, London.

[3] Maharning AR, Mills AA, Adl SM (2008) Soil community changes during secondary succession to naturalized grasslands. Appl. Soil Ecol. 41: 137-147.

[4] Bouché MB (1972) Lombriciens de france. Ecologie et systematique. INRA 72-2 Institut National Des Recherches Agriculturelles, Paris.

[5] Zanella A, Tomasi M, De Siena C, Frizzera L, Jabiol B, Nicolini G (2001) Humus forestali. Centro Ecologia Alpina.

[6] Parthasarathi K, Ranganathan L S (1999) Longevity of microbial and enzyme activity and their influence on NPK content in pressmud vermicasts. Eur. J. Soil Biol. 35 (3): 107-113.

[7] Burges A, Raw F (1967) Soil Biology. Academic Press.

[8] Dindal DL (1990) Soil biology guide. Wiley.

[9] Bachelier G, Vannier G, Pussard M, Bouché MB, Jeanson C, Boyer P, Massoud Z, Revière J, Chalvignac MA, Keilling J, Dommergues Y (1971) La vie dans les sols, Gauthier-Villars.

[10] Killham K (1994) Soil Ecology. Cambridge University Press.

[11] Bardgett RD, Cook R (1998) Functional aspects of soil animal diversity in agricultural grasslands. Appl. Soil Ecol. 10: 263–276.

[12] Bird S, Robert N C, Crossley D A (2000) Impacts of silvicultural practices on soil and litter arthropod diversity in a Texas pine plantation. Forest Ecol. Manag. 131: 65-80.

[13] Doblas-Miranda E, Wardle DA, Peltzer DA, Yeates GW (2007) Changes in the community structure and diversity of soil invertebrate across the Franz Josef Glacier chronosequence. Soil Biol. Biochem. 40: 1069-1081.

[14] Hedde M, Aubert M, Bureau F, Margerie P, Decaens T (2007) Soil detritivore macro-invertebrate assemblages throughout a managed beech rotation. Annals of Forest Science 64: 219-228.

[15] Jabin M, Mohr D, Kappes H, Topp W (2004) Influence of deadwood on density of soil macro-arthropods in a managed oak-beech forest. Forest Ecol. Manag. 194: 61-69.

[16] Kaneko N, Salamanca E (1999) Mixed leaf litter effects on decomposition rates and soil microarthropod communities in an oak-pine stand in Japan. Ecol. Res. 14: 131-138.

[17] Paquin P, Coderre D (1997) Changes in soil macroarthropod communities in relation to forest maturation through three successional stages in the Canadian boreal forest. Oecologia 112 (1): 104-111.

[18] Theenhaus A, Schaefer M (1995) The effects of clear-cutting and liming on the soil macrofauna of a beech forest. Forest Ecol. Manag. 77: 35-51.

[19] Blasi S, Menta C, Balducci L, Conti FD, Petrini E , Piovesan G (in press) Soil microarthropod communities from mediterranean forest ecosystems in Central Italy under different disturbances. Env. Monit. Asses.

[20] Neave P, Fox CA (1998) Response of soil invertebrates to reduce tillage systems established on a clay loam soil. Appl. Soil Ecol. 9: 423-428.

[21] Addison J (2007) Green tree retention: a tool to maintain ecosystem health and function in second-growth coastal forests. In: D W Langor editor. Arthropods of Canadian Forest. Natural Resources Canada, Canadian Forest Service.

[22] Ponge JF, André J, Zackrisson O, Bernier N, Nilsson M-C, Gallet C (1998) The forest regeneration puzzle. BioScience 48: 523-528.

[23] Moore JD, Ouimet R, Camiré C, Houle D (2002) Effects of two silvicultural practices on soil fauna abundance in a northern hardwood forest, Québec, Canada. Canadian Journal of Soil Science 82: 105-113.

[24] Heliovaara K, Vaisanen R (1984) Effects of modern forestry on northwestern European forest invertebrates-a synthesis. Acta Forestalia Fennica 83: 1-96.

[25] Hoekstra JM, Bell RT, Launer AE, Murphy DD (1995) Soil arthropod abundance in coastal redwood forest: effect of selective timber harvest. Env. Ent. 24: 246-252.

[26] Hill SB, Metz LJ, Farrier MH (1975) Soil mesofauna and silvicultural practices. In: Bernier B, Winget CH editors. Forest soil and Forest Management. Laval: Les Presses de l'Université Laval, France, 119-135.

[27] Huhta V, Karppingen E, Nurminen M, Valpas A (1967) Effect of silvicultural practices upon arthropod, annelid and nematode populations in coniferous forest soil. Annales Zoologici Fennici 4: 87-143.

[28] Lasebikan BA (1975) The effect of clearing on the soil arthropods of a Nigerian rain forest. Biotropica 7: 84-89.

[29] Vlug H, Borden JH (1973) Acari and Collembola populations affected by logging and slash burning in a coastal British Columbia coniferous forest. Env. Ent. 2: 1016-1023.

[30] Menta C, Leoni A, Gardi C, Conti F (2011) Are grasslands important habitats for soil microarthropod conservation? Biodiv. Conserv. 20: 1073-1087.

[31] Minnesota Forest Resources Council (1999) Sustaining Minnesota Forest Resources: Voluntary site-level forest management guidelines for landowners, laggers, and resources managers. St. Paul: Minnesota forest resources council, 473.

[32] Buger JA, Zedaker SM (1993) Drainage effects on plant diversity and productivity in lobolly pine (Pinus taeda L.) plantations on wet flats. Forest Ecol. Manag. 61: 109-126.

[33] Gupta S R, Malik V (1996) Soil ecology and sustainability. Tropical Ecology 37(1): 43-55.

[34] Lawton JH, May RM (1995) Extinction Rates. Oxford Univ. Press, Oxford.

[35] Chapin FS III, Zavaleta ES, Eviner VT, Naylor RL, Vitousek PM, Reynolds HL, Hooper DU, Lavorel S, Sala OE, Hobbie SE, Mack MC, Diaz S (2000). Consequences of changing biodiversity. Nature 405: 234-242.

[36] Baker GH (1998). Recognising and responding to the influences of agriculture and other land-use practices on soil fauna in Australia. Appl. Soil Ecol. 9: 303-310.

[37] Hansen B, Alroe HF, Kristensen ES (2001) Approaches to assess the environmental impact of organic farming with particular regard to Denmark. Agric. Ecosys. Environ. 83: 11-26.

[38] Cortet J, Gillon D, Joffre R, Ourcival J-M, Poinsot-Balanguer N (2002) Effects of pesticides on organic matter recycling and microarthropods in a maize field: use and discussion of the litterbag methodology. Eur. J. Soil Biol. 38 : 261-265.

[39] Dittmer S, Schrader S (2000) Longterm effects of soil compaction and tillage on Collembola and straw decomposition in arable soil. Pedobiologia 44: 527-538.

[40] Beare MH, Parmelee RW, Hendrix PF, Cheng W, Coleman DC, Crossley Jr DA (1992) Microbial and faunal interactions and effects on litter nitrogen and decomposition in agroecosystems. Ecological Monographs 62 (4): 569-591.

[41] Hendrix PF, Parmelee RW, Crossley Jr DA, Coleman DC, Odum EP, Groffman PM (1986) Detritus food webs in conventional and no-tillage agroecosystems. Bioscience 36: 374-380.

[42] Tabaglio V, Gavazzi C, Menta C (2009) Physico-chemical indicators and microarthropod communities as influenced by no-till, conventional tillage and nitrogen fertilisation after four years of continuous maize. Soil Till. Res. 105:135-242.

[43] Ferraro DO, Ghersa CM (2007) Exploring the natural and human-induced effects on the assemblage of soil microarthropod communities in Argentina. Eur. J. Soil Biol 43:109-119.

[44] House GJ, Del Rosario Alzugaray M (1989) Influence of cover cropping and no-tillage practices on community composition of soil arthropods in a North Carolina agroecosystem. Environ. Entomol. 18: 302-307.

[45] Fox CA, Fonseca EJA, Miller JJ, Tomlin AD (1999) The influence of row position and selected soil attributes on Acarina and Collembola in no-till and conventional continuous corn on a clay loam soil. Appl. Soil Ecol. 13: 1-8.

[46] Siepel H, van de Bund C (1988) The influence of management practices on the microarthropod community of grassland. Pedobiologia 31: 339-354,

[47] Curry JP (1994) Grassland Invertebrates. Ecology, influence on soil fertility and effects on plant growth. Chapman & Hall.

[48] Ryan M (1999) Is an enhanced soil biological community, relative to convenctional neighbours, a consistent feature of alternative (organic and biodynamic) agricultural systems? Biol. Agr. Hort. 17 (2): 131-144.

[49] Pfotzer GH, Schuler C (1997) Effects of different compost amendments on soil biotic and faunal feeding activity in an organic farming system. Biol. Agr. Hort. 15 (1-4): 177-183.

[50] Lübben B (1989) Influence of sewage sludge and heavy metals on the abundance of Collembola on two agricultural soils. In: Dallai R editor. Third International Seminar on Apterygota. Università di Siena, Siena, 419-428.

[51] Larsen KJ, Purrington FF, Brewer SR, Taylor DH (1986) Influence of sewage sludge and fertilizer on the ground beetle (Coleoptera: Carabidae) fauna of an old-field community. Env. Ent. 25: 452-459.

[52] Cuendet G, Ducommun A (1990) Peuplements lombriciens et activité de surface en relation avec les boues d'epuration et autres fumures. Revue Suisse die Zoologie 97 : 851-869.

[53] Bruce LJ, McCracken DL, Foster G, Aitken M (1999) The effects of sewage sludge on grassland euedaphic and hemiedaphic collembolan populations. Pedobiologia 43: 209-220.

[54] Bruce LJ, McCracken DL, Foster G, Aitken M (1997) The effects of cadmium and zinc-rich sewage sludge on epigeic Collembola populations. Pedobiologia 41: 167-172.

[55] Pinamonti F, Stringari G, Gasperi F, Zorzi G (1997) The use of compost: its on heavy metal level in soil and plant. Resource, Conservation and Recycling 21: 129-143.

[56] Bazzoffi P, Pellegrini S, Rocchini A, Morandi M, Grasselli O (1998) The effect of urban refuse compost and different tractors tyres on soil physical properties, soil erosion and maize yield. Soil Till. Res. 48: 275-286.

[57] Allievi L, Marchesini A, Salardi C, Piano V, Ferrari A (1993) Plant quality and soil residual fertility six years after a compost treatment. Bioresource Technology 43: 85-89.

[58] Tranvik L, Bengtsson G, Rundgren S (1993) Relative abundance and resistance traits of two Collembola species under metal stress, J. Appl. Ecol. 30: 43-52.

[59] Menta C, Leoni A, Tarasconi K, Affanni P (2010) Does compost use affect microarthropod soil communities? Fres. Env. Bull. 19: 2303-2311.

[60] Joy VC, Chakravorty PP (1991) Impact of insecticides on nontarget microarthropods fauna in agricultural soil. Ecotox. Envir. Saf. 22 (1): 8-16.

[61] Van Gestel CAM, van Straalen NM (1994) Ecotoxicological Test Systems for Terrestrial Invertebrates. In: Donker MH, Eijsackers H, Heimbach F editors. Ecotoxicology of Soil Organisms. SETAC Lewis Publishers, 205-228.

[62] Doran JW, Parkin TB (1994) Defining and assessing soil quality. SSSA special publication 35: 3-21.

[63] Karlen DL, Mausbach MJ, Doran JW, Cline RG, Harris RF, Schuman GE (1997) Soil quality: a concept, definition, and framework for evaluation. Soil Science Society of American Journal 61 (1): 4–10.

[64] Schoenholtz SH, Van Miegroet H, Burger JA (2000) A review of chemical and physical properties as indicators of forest soil quality: challenges and opportunities. Forest Ecol. Manag. 138: 335-356.

[65] Dylis NV (1964) Principles of construction of a classification of forest biogeocoenoses. In: Sukachev VN, Dylis NV editors. Fundamentals of Forest Biogeocoenology. Edinburgh and London, 572-589.

[66] Liebig MA, Doran JW (1999) Impact of organic production practices on soil quality indicators. J. Env. Qual. 28: 1601-1609.

[67] Kettler TA, Lyon DJ, Doran JW, Powers WL, Stroup WW (2000) Soil quality assessment after weed-control tillage in a no-till wheat-fallow cropping system. Soil Sci. Soc. of Am. J. 64: 339-346.

[68] Gilley JE, Doran JW, Eghball B (2001) Tillage and fallow effects on selected soil quality characteristics of former conservation reserve program sites. J. of Soil and Water Cons. 56: 126-132.

[69] Li Y, Lindstrom MJ, Zhang J, Yang J (2001) Spatial variability of soil erosion and soil quality on hillslopes in the Chinese Loess Plateau. Acta Geologica Hispanica 35: 261-270.

[70] Bowman RA, Nielsen DC, Vigil MF, Aiken RM (2000) Effects of sunflower on soil quality indicators and subsequent wheat yield. Soil Sci. 165: 516-522.

[71] Six J, Elliott ET, Paustian K (2000) Soil structure and soil organic matter: II. A normalized stability index and the effect of mineralogy. Soil Sci. Soc. of Am. J. 64: 1042-1049.

[72] Pankhurst CE (1997) Biodiversity of Soil Organisms as an Indicator of Soil Health. In: Pankhurst CE, Doube BM, Gupta VVSR editors. Biological Indicators of Soil Health. CAB International, 297-324.

[73] Eijsackers H (1983) Soil fauna and soil microflora as possible indicators of soil pollution. In: Best EPH, Haeck J editors. Ecological Indicators for the Assessment of the Quality of Air, Water, Soil, and Ecosystems. Reidel Publishing Company, Dordrecht, 307-316.

[74] Turco RF, Kennedy AC, Jawson MD (1994) Microbial indicators of soil quality. In: Doran JW, Coleman DC, Bezdicek DF, Stewart BA editors, Defining Soil Quality for a Sustainable Environment. SSSA, Madison, WI, 73-90.

[75] Doran JW, Zeiss MR (2000) Soil health and sustainability: managing the biotic component of soil quality. Appl. Soil Ecol. 15: 3-11.

[76] Van Straalen NM (1997) Community Structure of Soil Arthropods. In: Pankhurst CE, Doube BM, Gupta VVSR editors, Biological Indicators of Soil Health. CAB International, 235-264.

[77] Parisi V, Menta C, Gardi C, Jacomini C, Mozzanica E (2005) Microarthropod community as a tool to asses soil quality and biodiversity: a new approach in Italy. Agr. Ecos. Env. 105: 323-333.

[78] Van Straalen NM, 2004. The use of soil invertebrates in ecological survey of contaminated soils. In: Doelman P, Eijsackers HJP editors, Vital Soil Function, Value and Properties. Elsevier, 159-194.

[79] Cortet J, Gomot de Vauflery A, Poinsot-Balaguer N, Gomot L, Texier C, Cluzeau D (1999) The use of invertebrate soil fauna in monitoring pollutant effects. Eur. J. Soil Biol. 35:115-134.

[80] Cluzeau D, Guernion M, Chaussod R, Martin-Laurent F, Villenave C, Cortet J, Ruiz-Camacho N, Pernin C, Mateille T, Philippot L, Bellido A, Rougé L, Arrouays D, Bispo A,

Pérès G (2012) Integration of biodiversity in soil quality monitoring: baselines for microbial and soil fauna parameters for different land-use types. Eur. J. Soil Biol. 49: 63-72.

[81] Cameron KH, Leather SR (2012) How good are carabid beetles (Coleoptera, Carabidae) as indicators of invertebrate abundance and order richness? Biodivers. Conserv. 21: 763-779.

[82] Van Straalen NM (1998) Evaluation of bioindicator systems derived from soil arthropod communities. Appl. Soil Ecol. 9: 429-437.

[83] Cortet J, Gomot-De Vauflery A, Poinsot-Balaguer N, Gomot L, Texier C, Cluzeu D (2000) The use of invertebrate soil fauna in monitoring pollutant effects. Eur. J. Soil Biol. 35 : 115-134.

[84] Crommentuijn T, Stab JA, Doornekamp A, Estoppey O, van Gestel CAM (1995) Comparative ecotoxicity of cadmium, chlorpyrifos and triphenyltin hydroxide for four clones of the parthenogenetic collembolan *Folsomia candida* in an artificial soil. Funct. Ecol. 9: 734-742.

[85] Hopkin SP (1997) Biology of the Springtails (Insecta: Collembola). Oxford University Press.

[86] Menta C, Maggiani A, Vattuone Z (2006) Effects of Cd and Pb on the survival and juvenile production of *Sinella coeca* and *Folsomia candida*. Eur. J. Soil Biol. 42: 181-189.

[87] Bengtsson G, Gunnarsson T, Rundgren S (1985) Influence of metals on reproduction, mortality and population growth in *Onychiurus armatus* (Collembola). J. Appl. Ecol. 22: 967-978.

[88] Nottrot F, Joosse ENG, van Straalen NM (1987) Sublethal effects of iron and manganese soil pollution on *Orchesella cincta* (Collembola). Pedobiologia 30: 45-53.

[89] Posthuma L, Hogervorst RF, van Straalen NM (1992) Adaptation to soil pollution by cadmium excretion in natural population of *Orchesella cincta* (L.) (Collembola). Arch. Environ. Cont. Tox. 22: 146-156.

[90] Van Straalen NM, Schobben JHM, de Goede RGM (1989) Population consequences of cadmium toxicity in soil microarthropods. Ecotox. Environ. Safe. 17: 190-204.

[91] Gräff S, Berkus M, Alberti G, Köhler HR (1997) Metal accumulation strategies in saprophagous and phytophagous soil invertebrates: a quantitative comparison. Biometals 10: 45-53.

[92] Chauvat M, Ponge JF (2002) Colonization of heavy metal-polluted soils by collembola : preliminary experiments in compartmented boxes. Appl. Soil Ecol. 21: 91-106.

[93] Bongers T (1990) The Maturity Index: An Ecological Measure of Environmental Disturbance Based on Nematode Species Composition. Oecologia 83: 14-19.

[94] Bachelier G (1986) La vie animale dans le sol. ORSTOM, Paris.

[95] Gardi C, Menta C, Leoni A (2008) Evaluation of environmental impact of agricultural management practices using soil microarthropods. Fresen. Environ. Bull. 17 8(b): 1165-1169.

[96] Santorufo L, Van Gestel CAM, Rocco A, Maisto G (2012) Soil invertebrates as bioindicators of urban soil quality. Environmental Pollution 161: 57-63.

[97] Wardle DA (1995) Impacts of disturbance on detritus food webs in agro-ecosystems of contrasting tillage and weed management practices. Adv. Ecol. Res. 26: 105–185.

[98] Burger J (2008) Environmental management: Integrating ecological evaluation, remediation, restoration, natural resource damage assessment and long-term stewardship on contaminated lands. Science of The Total Environment 400: 6-19.

[99] Menta C, Leoni A, Bardini M, Gardi C, Gatti F (2008) Nematode and microarthropod communities: comparative use of soil quality bioindicators in covered dump and natural soils. Envi. Bioind. 3 (1): 35-46.

[100] Bongers T (1999) The Maturity Index, the evolution of nematode life history traits, adaptive radiation and c-p scaling. Plant and Soil 212:13-22.

[101] Yeates G W (1999) Effects of plants on nematode community structure. Annu Rev Phytopatol 37:127-49.

[102] Dmowska E (2005) Nematodes colonizing power plant ash dumps. II. Nematode communities in ash dumps covered with turf – effect of reclamation period and soil type. Pol. J. Ecol. 53(1): 37-51.

[103] Bedano JC, Cantú MP, Doucet ME (2006) Soil springtails (Hexapoda: Collembola), symphylans and pauropods (Artropoda: Myriapoda) under different management systems in agroecosystems of the subhumid Pampa (Argentina). Eur. J. Soil Biol. 42: 107-119.

[104] Pilar A, Eduardo M (2005) Soil mesofaunal responses to post-mining restoration treatments. Appl. Soil Ecol. 33: 67-78.

[105] Sendlein VA, Lyle Y H, Carison L C (1983) Surface mining reclamation handbook. Elsevier Science Publishing Co. Inc 290.

[106] Kundu N K, Ghose M K (1994) Studies on the topsoil of an opencast coal mine. Environ. Conserv. 21:126-132.

[107] Donhuer R L, Miller R W, Shickleena J G (1990) Soils – An introduction to soils and plant growth. Prentice – Hall.

[108] Ghosemrinal K (2004) Effect of opencast mining on soil fertility. J. Scient. Indust. Res. 63:1006-1009.

[109] Frouz J, Prack K, Pizl V, Háněl L, Starý J, Tajovký K, Materna J, Balík V, Kalčík J, Řehounkova K (2008) Interactions between soil development, vegetation and soil fauna during spontaneous succession in post mining sites. Eur. J. Soil Biol. 44: 109-121.

[110] Neuman FG (1991) Responses of litter arthropods to major natural or artificial ecological disturbances in mountain ash forests. Aust. J. Ecol. 1:19–32.

[111] Webb NR (1994) Postfire succession of Cryptostigmatic mites (Acari, Cryptostigmata) in a calluna-heathland soil. Pedobiologia 38 (2): 138–145.

[112] Addison JA, Trofymow JA, Marshall VG (2003) Abundances, species diversity and community structure of collembola in successional coastal temperate forests on Vancouver island. Can. Appl. Soil Ecol. 24: 233–246.

[113] Koehler H (1998) Secondary succession of soil mesofauna: A 13 year study. Appl. Soil Ecol. 9: 81–86.

[114] Black HIJ, Parekh NR, Chaplow JS, Monson F, Watkins J, Creamer R, Potter ED, Poskitt JM, Rowland P, Ainsworth G, Hornung M (2003) Assessing soil biodiversity across Great Britain: national trends in the occurrence of heterotrophic bacteria and invertebrates in soil. J. Environ. Manag. 67: 255–266.

[115] Berg MP, Hemerik L (2004) Secondary succession of terrestrial isopod, centipede, and millipede in grasslands under restoration. Biol. Fertil. Soils 40: 163-170.

Genetics and Life Sciences

Species Distribution Patterns, Species-Area and Species-Temperature Relationships in Eastern Asian Plants

Jianming Deng and Qiang Zhang

Additional information is available at the end of the chapter

1. Introduction

It is a fascinating issue for ecologists to develop a general theory or principle to interpret the mechanisms of global gradients and stabilization of biodiversity. This question has perplexed biogeographers and ecologists for about 100 years, and the diverse theories and hypotheses have been put forward to account for latitudinal gradients in biodiversity (Wright 1983; Rohde, 1992; Waide *et al.*, 1999; Colwell & Lees, 2000; Gaston, 2000; Allen *et al.* 2002; Hawkins *et al.*, 2003; Willig *et al.* 2003; Ricklefs, 2004; Evans & Gaston, 2005; Evans *et al.*, 2005; Mittelbach *et al.* 2007; Gillooly& Allen, 2007; Storch *et al.* 2007; Cardinale, *et al.*, 2009), Recent decade, the metabolic theory of biodiversity (MTB) is developed and attracting a lot of attentions of ecologists as a novel hypothesis based on metabolic theory of ecology (MTE) and the energetic-equivalence rule (West *et al.* 1997, 1999; Enquist *et al.* 1998; Allen *et al.* 2002, 2007; Brown *et al.* 2004; Deng *et al.* 2006, 2008). The MTB is recognized as a general principle that can quantify relationships between the dynamic processes of population and biodiversity patterns in ecosystem, and between species richness and environmental factors (see also Allen *et al.* 2003, 2006; Allen & Gillooly 2006; Gillooly & Allen 2007). The metabolic eco-evolutionary model of biodiversity, the most recent extension of the MTB, has been developed by Stegen *et al.*, (2009).

The MTB considered that species richness, S, in plots of fixed area, A, should be described by a form of equation as following $S = (J/A)(B_0/B_T)e^{-E/kT}$. In this expression, E is the activation energy of metabolism, -0.6 to -0.7 eV, k is Boltzmann's constant (8.62×10^{-5} eV K^{-1}, where K is degrees Kelvin), T is environmental temperature in degrees kelvin, $B_0 = b_0 M^{3/4}$, where b_0 is a normalization constant that varies by taxonomic group, M is individual body size, B_T is the total energy flux of a population per unit area, varying with taxonomic group and plot area A, and J/A is the total density of individual per unit area. Apparently, the species richness should vary as a function of abundance, body size and environmental temperature. So, when abundance and size are both presumed constants across geographical space, the relationship between the natural logarithm of species richness (lnS) and the inverse temperature (1/kT) will be re-expressed by a linear equation: $\ln S = -E(kT)^{-1} + C$. In this equation, the intercept term ($C = \ln[(B_0/\overline{B_T})(J/A)]$) incorporates the effect of mean body size of the study taxon, area and total community abundance (Allen *et al.*, 2002; Gillooly & Allen, 2007).

The intense and continuous controversies for the MTB have been focusing on two primary predictions: 1) whether ln-transformed species richness is linearly associated with an inverse rescaling of ambient temperature or not, and 2) if so, whether the slope of the relationship is encompassed in the theoretical value range of -0.6 to -0.7. The proponents argue that this theory accounts for diversity gradients over a range of spatial scales from mountain slopes to continental and global gradients, and for many groups of plants and ectothermic animals (Allen *et al.* 2002, 2003; Brown *et al.*, 2003, 2004; Gillooly & Allen, 2007). Kaspari *et al.*'s (2004) and Hunt *et al.*'s (2005) analyses using respectively ant communities and deep-sea communities datasets were also in favor of the Allen *et al.*'s (2002) predictions. Concomitantly, the disagreements for MTB emerged in some literature. Hawkins *et al.* (2007a) tested the predictions of this theory with 46 different data sets compiled from a variety of terrestrial plants, invertebrates, and ectothermic vertebrates, and found that the results were partly deviated from the predictions of the MTB (Allen *et al.* 2002; Brown *et al.* 2004), Accordingly, they considered that MTB was a poor predictor for the observed diversity gradients in most terrestrial system. Latimer (2007) subsequently reanalyzed some Hawkins *et al.*'s (2007a) data sets using a Bayesian approach and supported their conclusions. Algar *et al.* (2007) recently showed that the relationship between richness and temperature was actually curvilinear for several data sets in North America, and slopes varied systematically in geographical space, which were strongly consistent with Cassemiro *et al* (2007) analyzing for New World amphibians. As a consequence, they claimed that Allen *et al.*'s (2002) model did not give an adequate fit to the data.

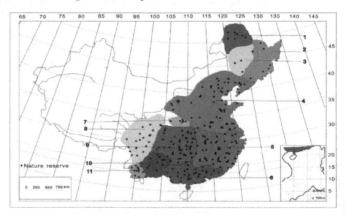

Figure 1. Location of the eleven floristic regions and 270 nature reserves of China used in this study. Floristic regions was marked with Arabic numerals as following (1, Daxinganling; 2, NE China; 3, NE China plain; 4, North China region; 5, East China; 6, Lingnan region China; 7, Tsinling Mountains; 8, The region of Hengduan Mountain; 9, Central China; 10, Dian-qian-gui region; 11, The region of Yunnan Plateau)(also see Zhang et al. 2011).

Sample type	Variable	N	Minimm	Maximum	Mean	SD	Skewness	Kurtosis
Floristic region	**Species richness**							
	Compositae	9	104	655	293.78	163.66	1.34	2.50
	Poaceae	9	86	597	254.44	155.37	1.29	2.52
	Rosaceae	9	49	406	197.33	122.44	0.45	-0.64
	Liliaceae	9	35	164	93.22	41.32	0.11	-0.32
	Labiatae	9	28	253	130.33	79.02	0.07	-1.10
	Angiosperm	11	1009	7891	4363.45	2513.16	-0.02	-1.47
	Gymnosperm	10	4	63	31.90	19.53	0.27	-0.73
	Seed plant	11	1019	7954	4392.45	2524.83	-0.03	-1.47
	Climatic variables							
	MAT (°C)	11	-2.80	20.88	12.34	7.48	-0.98	0.002
	Location and area							
	Mean latitude (°)	11	23	50	35.33	9.54	0.35	-1.49
	Mean longitude (°)	11	100	132	115.1	9.94	0.25	-0.44
	Area (km²)	11	52000	960000	401660	273070	0.64	0.73
Nature reserve	**Species richness**							
	Compositae	71	11	324	78.66	44.51	2.55	12.34
	Poaceae	70	8	131	63.13	28.75	0.40	-0.16
	Rosaceae	71	2	185	54.93	33.76	1.27	2.24
	Liliaceae	49	1	85	31.7	19.52	1.08	0.93
	Labiatae	56	3	96	30.41	17.51	1.31	2.99
	Angiosperm	255	79	3893	1186.44	705.08	0.84	0.91
	Gymnosperm	234	1	110	14.21	12.33	3.32	19.32
	Pteridophyte	189	1	594	108.86	92.75	1.41	3.51
	Vascular plant	193	138	4543	1388.18	809.92	0.86	0.99
	Climatic variables							
	MAT (°C)	270	-2.8	29	13.93	5.71	-0.61	0.44
	Location and area							
	Mean latitude (°)	270	18.4	51.6	30.74	6.55	0.71	0.37
	Mean longitude (°)	270	95	130.6	111.88	7.04	0.12	-0.37
	Area (km²)	270	0.64	6698	490.89	922.28	3.94	18.69

Table 1. Summary statistics of species richness in different plant groups, climate variables, and areas of 11 floristic regions and 270 nature reserve used in this paper.

Empirical evaluations of how well observed richness patterns fit the central predictions of the MTB are now appearing in several literature (Allen et al., 2002; Kaspari et al., 2004; Hunt et al., 2005; Algar et al., 2007; Cassemiro et al., 2007; Latimer, 2007; Hawkins et al., 2007a, b; Sanders et al., 2007). Although to date the observed patterns in biodiversity have been taxonomically and geographically limited (Ellison, 2007), the data sets for the detailed plant groups are relatively absent. Wang et al. (2009) showed that magnitude of temperature dependence (i.e. E) of tree species richness in both eastern Asia and North America increases with spatial scale at the large scale of 50×50km to 400×400km. Therefore, we conjectured that the species richness inherent dependence of spatial scale may influence on the successful tests for the predictions of the MTB (also see Zhang et al. 2011). However, it is unclear (i) how the species richness responds to temperature at the variant spatial scale, especially at the small scale level. (ii) what spatial scale range is appropriate to the MTB.

Here we aimed to evaluate how the relationship between species richness and temperature predicted by MTB varied with respect to sampling scales, as well as with respect to different plant taxonomic group using an extensive plant data sets including three divisions in vascular plant at two different sample scales including nature reserve grain and floristic grain.

2. Methods

We compiled species richness and other basic characteristics of 11 floristic regions and 270 natural reserves. All of the plant species richness data sets used in our analysis were collected from the previous reports involving eleven floristic regions and 270 nature reserves across the eastern China (Zhao & Fang, 2006, many others; for details see Zhang et al.2011). All species were compiled and classified into three groups (pteridophyte, gymnosperm, and angiosperm) at both floristic and reserve scales (the details see Zhang et al.2011). Here the alien species were excluded from our data analyses and only the native species retained. The areas of nature reserves and floristic regions were respectively range from 0.64 to 6689 and from 52000 to 960000 square kilometers (km^2) between 18.4° N and 51.6° N latitude and between 95° E and 130.6° E longitude covering a total terrestrial area of 132,540 km^2 (See Fig1 and Table1). The temperature and the size distribution of the 270 nature reserves also were showed in Table 1.

The mean annual temperature (MAT), assigned to each nature reserve based on its location and used to analyze the relationship between temperature and species richness, was compiled from a 1971-2000 temperature database of China generated from 722 climate stations across China. Flora's MAT was an average value of all the covered climate stations within each floristic region. Other environmental variables such as geographical range and area were also documented.

Descriptive statistics of plant species richness and environmental variables were produced to interpret the information on the data distributions (Table1). The observed slopes of ln-transformed richness versus $1/kT$ relationships and at the floristic and reserve grains across the taxonomic group (three divisions and five families of the angiosperms) were estimated by reduced major axis (RMA) regression (Brown et al., 2004; Hawkins et al., 2007). The species richness-area relationships were also analyzed to evaluate the effect of the region size on the species richness (See Fig 4 and 5).

3. Results

The natural logarithm of species richness was significantly linear with $(kT)^{-1}$ at the floristic grain. The 95% CI values of all slopes estimated by RMA did not exclude the second primary prediction of MTB, implying that no significant heterogeneity occurred in slopes among taxonomic groups (Fig2; Table2). However, at the nature reserve grain, only two of three large plant groups （angiosperm, Fig3-f and pteridophyte, Fig3-h）showed significantly linear relationship and met the first prediction of MTB, and gymnosperm were not linear and rejected the entire MTB (Fig3-g; Table3). The slope values estimated by RMA regression for all taxonomic groups were significantly exclusive from the second prediction of MTB (Table2).

The species-area relationships for all taxonomic divisions at both the floristic and nature reserve special scales indicated that the area size of community have more impact on the species richness for subdivision (e.g. family) than for division (Fig. 4 and 5). Moreover, the observed slope values were close to or encompass (95% CI) the theoretical values predicted by MBT at the spatial scale range of 50- 6698 km², excluding the size of area class less than 50 km² (Fig. 6; Table 4).

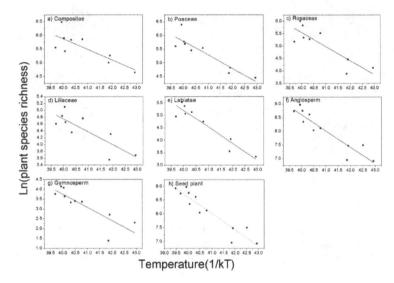

Figure 2. The relationship between natural logarithm of species richness (lnS) and inverse temperature (1/kT) for seven groups in 11 floristic regions: two divisions (gymnosperm and angiosperm) and five families of angiosperm (Compositae, Poaceae, Rosaceae, Liliaceae and Labiatae).

Group	Figure	N	R²	P	RMA slope(95%CI)
Compositae	2-a	9	0.64	0.009	-0.48(-0.74- -0.23)
Poaceae	2-b	9	0.82	<0.001	-0.55(-0.76- -0.34)
Rosaceae	2-c	9	0.73	0.003	-0.66(-0.97- -0.35)
Liliaceae	2-d	9	0.67	0.007	-0.46(-0.69- -0.22)
Labiatae	2-e	9	0.89	<0.001	-0.70(-0.92- -0.50)
Angiosperm	2-f	11	0.86	<0.001	-0.64(-0.83- -0.46)
Gymnosperm	2-g	10	0.71	0.002	-0.80(-1.15- -0.45)
Seed plant	2-h	11	0.89	<0.001	-0.64(-0.80- -0.48)

Table 2. Summary of regressions testing Model II (RMA) slopes of richness-temperature relationships for cases with linear relationship between inverse scaled temperature and ln-transformed richness in 11 floristic regions.

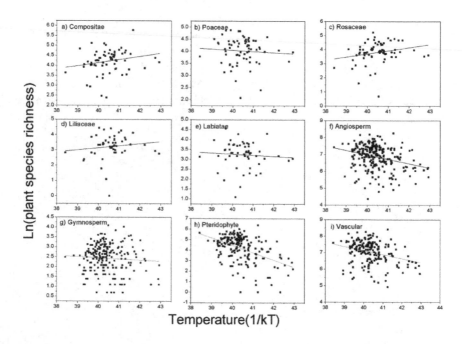

Figure 3. The relationship between natural logarithm of species richness (lnS) and inverse temperature (1/kT) for eight plant groups in 270 nature reserve: three divisions (Pteidophyte, Gymnosperm and Angiosperm) and five families of angiosperm (Compositae, Poaceae, Rosaceae, Liliaceae and Labiatae).

Group	Figure	N	R^2	P	RMA slope(95%CI)
Compositae	3-a	71	0.05	0.08	0.73(0.56–0.91)
Poaceae	3-b	70	0.00	0.58	-0.71(-0.89–-0.54)
Rosaceae	3-c	71	0.05	0.05	0.96(0.73–1.18)
Liliaceae	3-d	49	0.02	0.36	0.81(0.58–1.05)
Labiatae	3-e	56	0.01	0.60	-0.81(-1.03–-0.59)
Angiosperm	3-f	255	0.05	0.09	-0.90(-1.01–-0.79)
Gymnosperm	3-g	234	0.00	0.326	-0.91(-1.03–-0.79)
Pteridophyte	3-h	189	0.21	<0.001	**-1.55(-1.75–-1.35)**
Vascular plant	3-i	193	0.09	<0.001	**-0.82(-0.93–-0.71)**

Table 3. Summary of regressions testing Model II (RMA) slopes of richness-temperature relationships for cases with linear relationship between inverse rescaled temperature and ln-transformed richness in nature reserve.

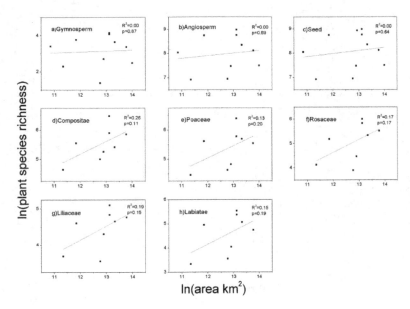

Figure 4. Species richness-area relationships for the seven plant groups in floristic regions: two divisions (gymnosperm and angiosperm) and five families of angiosperm (Compositae, Poaceae, Rosaceae, Liliaceae and Labiatae).

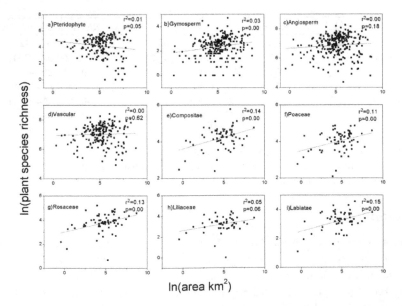

Figure 5. Species richness-area relationships for the eight plant groups in 270 nature reserves: three divisions (Pteidophyte, Gymnosperm and Angiosperm) and five families of angiosperm (Compositae, Poaceae, Rosaceae, Liliaceae and Labiatae).

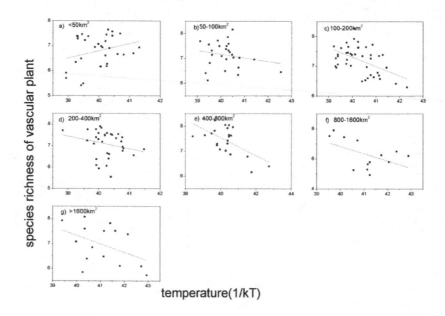

Figure 6. Temperature–richness relationships for the vascular plant group along an area classes. Note that the classification of area is following: a)<50, b)50-100, c)100-200, d)200-400, e)400-800, f)800-1600, g)>1600.

Area classes(km^2)	Figure	N	R^2	P	RMA slope(95%CI)
<50	7-a	32	0.08	0.12	1.04(0.67-1.42)
50-100	7-b	25	0.04	0.32	-0.72(-1.02– -0.41)
100-200	7-c	44	0.25	<0.001	**-0.63(-0.80– -0.46)**
200-400	7-d	33	0.06	0.19	-1.24(-1.69– -0.80)
400-800	7-e	22	0.36	0.003	**-0.60(-0.82– -0.37)**
800-1600	7-f	12	0.25	0.08	-0.96(-1.55– -0.38)
>1600	7-g	16	0.18	0.10	-0.84(-1.27– -0.40)

Table 4. Summary of regressions testing Model II (RMA) slopes of richness-temperature relationships for vascular plant group with linear relationship between inverse rescaled temperature and ln-transformed richness along an area classes.

4. Discussion

Hawkins' *et al.* (2007a) suggested that the relationship of logarithm transformed species richness and inverse temperature was nonlinear through analyzing the datasets of Chinese angiosperm taken from nature reserves with a range of area from 100 km^2 to 247 km^2. Here we similarly failed to observe significantly linear relationships between them at the nature reserve grain with the regions ranging from 0.64 km^2 to 6689 km^2, excepting for two large groups (angiosperm and pteridophyte). Moreover, almost all slope values were exclusive from the predictive range of MTB (Table 2) as the pattern of tree species distribution in eastern Asia (Wang *et al.*, 2009). However, when we analyzed these data sets at the floristic regions ranging from 52000 km² to 960000 km², not only this linear relationship was observed, but also the slopes is highly in agreement with the theoretical values of MTB (Allen *et al.* 2002; Brown *et al.* 2004). Therefore, the plant species richness patterns predicted by MTB apparently depended on the grain size (Ellison, 2007). This scenario may be due to the fact that the number of species at the large scale overwhelmed the number of species at the relative small sample scale (e.g. nature reserve). However our analysis of species richness-area relationships showed no significant relations at floristic grain (Fig 4). The adjacent nature reserves frequently have the similar annual temperature, but the other environmental factors (i.e. water, elevation and nutrition) may exhibit a lot of variations between them that can also strongly influence the local plant species richness (Storch *et al.*, 2007). The large-scale (floristic region) patterns are not simply explicable in terms of knowledge of small-scale (nature reserve) processes (Storch and Gaston, 2004). On the contrary, despite the habitat heterogeneity including annual temperature is large between plant flora, it is usually overwhelmed within plant flora because of the enormous sample scale (Field *et al.*, 2009).

For the purpose of evaluating the MBT's robustness, Hawkins *et al.* (2007a) show the relationship between the inverse of temperature and the natural log of richness in terrestrial ectotherms (including amphibians, reptiles), invertebrates, mammals and plant around the world. However, in their plant data sets, detailed taxonomic unit (e g, pteridophyte, gymnosperm and family unit) were not contained. In their 46 data sets, 14 had no significant relationship; 9 of the remaining 32 were linear, meeting the first prediction of the MBT, but the slope values against its second prediction. So, they contended that it was important to use appropriate taxonomic ranges for accepting or refusing the prediction of MBT (see also, Ellison, 2007).

Our results clearly showed that the significant taxonomic dependence in the nature reserve data sets. Pteridophyte unit which potentially supported the first prediction of MTB dominantly differs from the other groups in particular. Pteridophytes have a reproductive strategy based on the high dispersibility of spores, and have a strong moisture dependence of the sexual reproduction (Pausas & Sáez, 2000; Lehmann, *et al.*, 2002; Castán & Vetaas, 2005). Thus, the life history and growth cycle for pteridophytes are probably more directly and tightly linked to abiotic factors than many other groups of plants because of the lack of co-evolved relationships with animal vectors (Barrington, 1993; Lwanga *et al.*, 1998; Pausas & Sáez, 2000; Castán & Vetaas, 2005). So our results possibly support the perspective that the ability of MBT to predict richness patterns will also depend on dispersal ability (Latimer, 2007).

The plant species were not subdivided into division group to test the slopes converge around the predicted value -0.65 by MTB (Allen *et al.* 2002; Brown *et al.* 2004). Whereas the significant heterogeneity of slopes were observed at both floristic region and reserve scale among the different taxonomic groups as the most recently reported by Hawkins *et al.* (2007a,b) and Wang *et al.*(2009), indicating that the plant groups may hold variable activation energies rather than an invariant value. Our more recently research showed that validity of the MTB lies on if the area size of the community has no significant effect on species richness (Zhang et al. 2011). Therefore we believe that the slope value for each taxonomic group should be co-influenced by the restriction of distribution range, the area size of sampling community and other abiotic factors, as well as the inherent activation energy differences.

5. Conclusion

Our results suggested that the relationship predicted by MTB between the plant richness and temperature can be tested at the larger regional scale (e.g. floristic region) well. However, at the small scale (e.g. nature reserve), the predicted relationships were easily influenced by the many other factors such as area size of community, taxonomic divisions, seed dispersal and so on. Allen *et al.* (2003) claimed that the theory of biodiversity proposed by themselves is not complete and comprehensive. Here we consider that the theory must integrate the fundamental influences of multifactor involving temperature, area size, water, elevation and nutrition on the species richness patterns in small scale regions where the disturbance of environmental factors easily result in change of the species diversity. At the same time, we should also seek the more biological interpretation for the noticeable differences among taxonomic groups in the future.

Author details

Jianming Deng and Qiang Zhang
Key Laboratory of Grass and Agriculture Ecosystem,
School of Life Science, Lan Zhou University, Lanzhou, China

Acknowledgement

This study were supported by the Natural Science Foundation of China (31000286), the Program for New Century Excellent Talents in University to J.M.D. and Key Project of Ministry of Education of China (no. 109152).

6. References

Algar, A.C., Kerr J.T., Currie, D. J. 2007 A test of metabolic theory as the mechanism underlying broad-scale species-richness gradients. *Global Ecol. Biogeogr* 16, 170–178.

Allen, A. P., Brown, J. H., Gillooly, J. F. 2002 Global biodiversity, biochemical kinetics, and the energetic–equivalence rule. *Science* 297, 1545–1548.

Allen, A. P. & Gillooly, J. F. 2006 Assessing latitudinal gradients in speciation rates and biodiversity at the global scale. *Ecol Lett* 9, 947–954.

Allen, A. P., Gillooly, J. F., Brown, J. H. 2003 Response to comment on "Global biodiversity, biochemical kinetics and the energetic–equivalence rule". *Science* 299, 346c.

Allen, A. P., Gillooly, J. F., Savage, V. M., Brown, J. H. 2006 Kinetic effects of temperature on rates of genetic divergence and speciation. *Proc. Natl Acad. Sci. USA* 103, 9130–9135.

Brown, J. H., Gillooly, J. F., Allen, A. P., Savage, V. M., West, G. B. 2004 Toward a metabolic theory of ecology. *Ecology* 85, 1771–1789.

Cardinale, B. J., Hillebrand, H., Harpole, W. S., Gross, K., Ptacnik, R. (2009) Separating the influence of resource 'availability' from resource 'imbalance' on productivity–diversity relationships. *Ecology letters*, 12, 475–487.

Cassemiro, F. A. S., Barreto, B. S., Rangel, T. F. L. V. B., Diniz-Filho, J. A. F. 2007 Non-stationarity, diversity gradients and the metabolic theory of ecology. *Global Ecol. Biogeogr* 16, 820–822.

Colwell, R. K., Lees, D. C. (2000) The mid-domain effect: geometric constraints on the geography of species richness. *Trends in Ecology and Evolution*, 15, 70–76.

Deng, J.M., Li, T., Wang, G.X., Liu, J., Zhao, C.M., Ji, M.F., Zhang, Q., Liu, J.Q. 2008 Trade-offs between the metabolic rate and population density of plants. *Plos One*, 3 (3), 1799.

Deng, J. M., Wang, G. X., Morris, E. C., Wei, X. P., Li, D. X., Chen, B. M., Zhao, C. M., Liu, J., Wang, Y. 2006 Plant mass–density relationship along a moisture gradient in north-west China. *Journal of Ecology* 94, 953–958.

Ellison, A. M. (2007) Metabolic theory and patterns of species richness. *Ecology*, 88, 1889.

Enquist, B. J., Brown, J. H., West, G. B. 1998 Allometric scaling of plant energetics and population density. *Nature* 395,163–165.

Evans, K. L., Gaston, K. J. (2005) Can the evolutionary–rates hypothesis explain species–energy relationships? *Functional Ecology*, 19, 899–915.

Evans, K. L., Warren, P. H., Gaston, K. J. (2005) Species–energy relationships at the macroecological scale: a review of the mechanisms. *Biological Reviews*, 80, 1–25.

Field, R., Hawkins, A. B., Cornell, H. V., Currie, D. J., Diniz-Filho, J. A. F., Guégan, J. F., Kaufman, D. M., Kerr, J. T., Mittelbach, G. C., Oberdorff, T., O'Brien, E. M. and Turner, J. R. G. (2009) Spatial species-richness gradients across scales: a meta-analysis. *Journal of Biogeography*, 36, 132–147.

Gaston, K. J. (2000) Global patterns in biodiversity. *Nature*, 405, 220–227.

Gillooly, J. F. & Allen, A. P. 2007 Linking global patterns in biodiversity to evolutionary dynamics using metabolic theory. *Ecology* 88, 1890–1894.

Hawkins, B. A., Albuquerque, F. S., Araújo, M. B., Beck, J., Bini, L. M., Cabrero-Sañudo, F. J., Castro-Parga, I., Diniz-Filho, J. A. F., Ferrer-Castán, D., Field, R., *et al.* 2007a A global evaluation of metabolic theory as an explanation for terrestrial species richness gradients. *Ecology* 88, 1877–1888.

Hawkins, B.A., Diniz-Filho, J. A. F., Bini, L. M., Araújo, M. B., Field, R., Horta,l J., Kerr, J. T., Rahbek, C., Rodríguez, M. Á., Sanders, N. J. (2007b) Metabolic theory and diversity gradients: where do we go from here? *Ecology*, 88, 1898–1902.

Hawkins, B. A., Field, R., Cornell, H. V., Currie, D. J., Guegan, J. F., Kaufman, D. M., Kerr, J. T., Mittelbach, G. G., Oberdorff, T., O'Brien, E. M., Porter, E. E., Turner, J. R. G. (2003) Energy, water, and broad-scale geographic patterns of species richness. *Ecology*, 84, 3105–3117.

Hunt, G., Cronin, T. M., Roy, K. 2005 Species–energy relationship in the deep see: a test using the Quaternary fossil record. *Ecol Lett* 8, 739–747.

Kaspari, M., Ward, P. S., Yuan, M. 2004 Engery gradients and the geographic distribution of local ant diversity. *Oecologia* 140, 407–413.

Latimer, A. M. 2007 Geography and resource limitation complicate metabolism-based predictions of species richness. *Ecology* 88, 1895–1898.

Mittelbach, G. G,, Schemske, D. W., Cornell, H. V., Allen, A. P., Brown, J. M., Bush, M. B., Harrison, S. P., Hurlbert, A. H., Knowlton, N., Lessio ,H.A., *et al*. 2007 Evolution and the latitudinal diversity gradient: speciation, extinction and biogeography. *Ecol Lett*, 10, 315–331.

Ricklefs, R. E. 2004 A comprehensive framework for global patterns in biodiversity. *Ecol Lett* 7, 1–15.

Rohde, K. 1992 Latitudinal gradients in species diversity: the search for the primary cause. *Oikos* 65, 514–527.

Rosenzweig, M. L. 1995 *Species Diversity in Space and Time*. Cambridge University Press, Cambridge, UK.

Sanders, N. S., Lessard, J. P., Fitzpatrick, M. C., Dunn, R. 2007 Temperature, but not productivity or geometry, predicts elevational diversity gradients in ants across spatial grains. *Global Ecol. Biogeogr* 16, 640–649.

Stegen, J. C., Enquist, B. J. and Ferriere, 2009 R. Advancing the matabolic theory of biodiversity. *Ecology Letters* 12, 1001–1015.

Storch, D. 2003 Comment on "Global biodiversity, biochemical kinetics, and the energetic-equivalence rule". *Science* 299, 346.

Storch, D., Gaston, K. J. (2004) Untangling ecological complexity on different scales of space and time. *Basic and Applied Ecology*, 5, 389–400.

Storch, D., Marquet, A., Brown, J. H. (2007) *Scaling Biodiversity*. Cambridge University Press, Cambridge, UK.

Waide, R. B., Willig, M. R., Steiner, C. F., Mittelbach, G., Gough, L., Dodson, S. I., Juday, G. P., Parmenter, R. (1999) The relationship between productivity and species richness. *Annual Review of Ecology and Systematics*, 30, 257–300.

Wang, Z., Brown, J. H., Tang, Z., Fang, J. 2009 Temperature dependence, apatial scale, and tree species diversity in eastern Asia and North America. *Proc. Natl Acad. Sci. USA* 106, 13388–13392.

West, G. B., Brown, J. H., Enquist, B. J. 1997 A general model for the origin of allometric scaling laws in biology. *Science* 276, 122–126.

West, G. B., Brown, J. H., Enquist, B. J. 1999 A general model for the structure, function, and allometry of plant vascular systems. *Nature* 400, 664–667.

Willig, M. R., Kaufman, D. M., Stevens, R. D. 2003 Latitudinal gradients of biodiversity: pattern, process, scale, and synthesis. *Annu Rev Ecol Evol Syst* 34, 273–309. Wright, D. H. 1983 Species-energy theory: an extension of species-area theory. *Oikos* 41, 496–506.

Zhang, Q. Wang, Z. Q., Ji, M. F., Fan, Z. X., Deng, J. M. 2011. Patterns of species richness in relation to temperature, taxonomy and spatial scale in eastern China. *Acta Oecologica* 37 (4), 307-313.

Zhao, S. & Fang, J. 2006 Patterns of species richness for vascular plants in China's nature reserves. *Diversity Distrib* 12, 364–372.

Chapter 5

Biological Identifications Through DNA Barcodes

Hassan A. I. Ramadan and Nabih A. Baeshen

Additional information is available at the end of the chapter

1. Introduction

Although much biological research depends upon species diagnoses, taxonomic expertise is collapsing. We are convinced that the sole prospect for a sustainable identification capability lies in the construction of systems that employ DNA sequences as taxon 'barcodes'. It was established previously that the mitochondrial gene cytochrome *c* oxidase I (COI) can serve as the core of a global bio- identification system for animals. A new tools were developed recently to be complementary markers for (COI) DNA barcoding.

Species identification is essential in food quality control procedures or for the detection and identification of animal material in food samples. Recent food scares e.g. avian flu and swine flu, malpractices of some food producers and religious reasons have tremendously reinforced public awareness regarding the composition of food products. However, because labels do not provide sufficient guarantee about the true contents of a product, it is necessary to identify and/or authenticate the components of processed food, thus protecting both consumers and producers from illegal substitutions [1]. In addition, trade of endangered species has contributed to severe depletion of biodiversity.

Numerous analytical methods that rely on protein analysis have been developed for species identification, such as electrophoresis techniques [2], immunoassays [3] and liquid chromatography [4]. However, these methods are of limited use in species identification. The progress of molecular biology introduced a new approach, which is based on nucleotide sequence diversities among species in particular regions of DNA [5–7]. The nucleotide regions chosen for species identification were varied by researchers. Within vertebrates, a cytochrome b (cyt b) gene in the mitochondrial DNA has been studied from multiple viewpoints including the nucleotide diversity among species [6] and the availability of nucleotide sequence data for references [5]. Many of the other regions studied are also located in the mtDNA. The coding regions for 12S and 16S ribosomal RNA [8–10], and the noncoding D-loop region [7, 11, 12] have shown their potential to be the targets for the species test.

Although central to much biological research, the identification of species is often difficult. DNA sequencing, with key sequences serving as a pattern "barcode", has therefore been proposed as a technology that might expedite species identification [13].

DNA barcoding promises fast, accurate species identifications by focusing analysis on a short standardized segment of the genome [14]. Several studies have now established that sequence diversity in a 650-bp fragment of the mitochondrial gene cytochrome c oxidase I (cox1; also referred to as COI) provides strong species-level resolution for varied animal groups including birds [15], fishes [16] and Lepidoptera [17].

Besides the cox1 gene, other mitochondrial markers also have been widely sequenced across vertebrates for their utility in phylogenetic or to complement cox1 in DNA barcoding.

In amphibians the 16S ribosomal RNA gene (16S) has been suggested as a complementary DNA barcoding marker [18]. Another protein coding gene, cytochrome b, has also been suggested as a marker to determine species boundaries [19, 20].

An attempt was made to present a phylogenetic systematic framework for an improved barcoder as well as a taxonomic framework for interweaving classical taxonomy with the goals of 'DNA barcoding' [21]. Another study showed that DNA arrays and DNA barcodes are valuable molecular methods for biodiversity monitoring programs [22]. In this chapter we introduce the use of specific fragments of mitochondrial ribosomal RNA from Egyptian buffalo to be used as a perfect barcode for identification of closely related species. Also, we will extend this study to include distantly species identification [23-24]. Our studies were also extended for chickens and small organisms like mites to be studied by both nuclear and mitochondrial markers. Identification of these mites is very important for biological control programs.

All these methods could be used for global bio-identification system or forensic science development.

2. Materials and methods

2.1. DNA purification

Genomic DNA was extracted from peripheral blood of Egyptian buffalo's and chickens by using standard commercial Kit (Pure-gene Genomic DNA purification Kit) as recommended by the manufacturer (www.gentra.com). In case of mites, Genomic DNA was extracted using Capture Column kit method, total DNA was purified using generation DNA purification system.

2.2. Primers used for amplifications of PCR specific fragments

2.2.1. D-loop primers

These primers yielded a PCR product of 1142 base pairs. This encompasses the whole of the D-loop and includes flanking sequence at both ends [12].

IL0500: 5'AGGCATTTTCAGTGCCTTGC-3'
IL0501: 5'TAGTGCTAATACCAACGGCC-3'

Two additional new forward primers (SH-1 and SH-2) specific for buffalo were designed inside the D-loop sequence to facilitate sequencing and correction processes.

SH-1: 5' CCT CGC ATG TAC GGC ATA CA-3'
SH-2: 5'CAA CCC TTC AGG CAA GGA TC-3'

2.2.2. Primers used for amplification of specific fragments from mites

Two target DNA fragments of the predatory mite, *A. swirskii* were PCR amplified and sequenced: a fragment in the central part of the mitochondrial cytochrome oxidase subunit I gene (COI) and the fragment of the nuclear ribosomal transcribed spacers (ITS) [25-26]. The COI primers were designed specifically for tetranychid mites. They were:

5'TGATTTTTTGGTCACCCAGAAG3' and 5'TACAGCTCCTATAGATAAAAC 3'.

The ITS region was amplified using the primers 5'AGAGGAAGTAAAAGTCGTAACAAG 3' for the 3' end of 18S rDNA and 5' ATATGCTTAAATTCAGGGGG 3' for the 5' end of the 28S.

2.2.3. Primers used for amplification of the first 539 base fragment of the D-loop region of the birds

The conserved primer pair, L16750 (forward; 5'-AGG ACT ACG GCT TGA AAA GC-3') and H 547 (reverse; 5'- ATG TGC CTG ACC GAG GAA CAA G-3') were used to amplify the first 539 base fragment of the D-loop region of the birds. The primer number refers to the positions of the 3' end of the primer in the reference sequence [27].

2.2.4. 12S primers

Primers specific for mitochondrial 12S rRNA gene were synthesized [23]:

5'-CAAACTGGGATTAGATACCCCACTAT-3'; 5'-AGGGTGACGGGCGGTGTGT-3' and directed towards the two conserved regions of the gene. The primers were synthesized by Amersham Pharmacia Biotech (U.K.).

2.2.5. 16S primers

PCR amplification and direct sequencing With two universal primers (sense, 5'-GTGCAAAGGTAGCATAATCA-3' and antisense, 5'-TGTCCTGATCCAACATCGAG-3') directed toward conserved regions [24], the polymerase chain reaction was used to amplify homologous segments of mitochondrial 16S rRNA from four animal species belonging to family Bovidae, including river buffalo, cattle, sheep and goat.

2.3. The amplification reaction

The amplification reaction used for amplification of the D-loop fragment was also used (with little modifications in temperature cycling) in the other experiments according to the conditions of each experiment.

The amplification reaction was carried out in a 25 μl reaction mixture consisting of 1.25 unit Taq polymerase (DyNAzyme), 1X enzyme buffer (1X is 10 mM Tris-HCl, pH 8.8 at 25 0C, 1.5 mM MgCl2, 50 mM KCl and 0.1% Triton X-100) supplied by the manufacture, 1 μM of each forward and reverse primer, 0.2 mM dNTPs and 100 ng of DNA. The reaction mixture was overlaid with sterile mineral oil and was run in an MJ research PTC-100 Thermocycler. The temperature cycling was as follows: 30 cycles of 45 seconds at 94°C; 1 minute at 58°C and 1 minute at 71°C, followed by a final extension at 71°C for 5 minutes. All PCR amplifications included a negative control reaction which lacked template DNA. No product was seen in any negative control. Small quantities of the reaction products (5 μl each) were used for electrophoresis with an appropriate size marker on 1.5% agarose in 1X-Tris acetate buffer (TAE).

After electrophoresis the gels were stained with ethidium bromide and were examined with UV lamp at a wave length 312 nm to verify amplification of the chosen specific fragment. The PCR products were purified using QIAquick PCR purification kit (Qiagen, Inc.) and the resulting purified products were used in the subsequent sequencing reactions. Sequencing was performed on an Applied Biosystems 310 genetic analyzer (Applied Biosystem) using Big Dye terminator cycle sequencing ready reaction mixture according to manufacturer's instructions (Applied Biosystems).

2.4. Sequence analysis and multiple sequence alignment

Pairwise sequence alignments were carried out using NCBI-BLASTN 2.2.5 version & PSI BLAST. Multiple sequence alignments were done using the MUSCLE 3.6 software and CLUSTALW (1.82). Analysis, manipulation, conservation plots, positional entropy plot and conserved region analysis was done using the BIOEDIT package. Variable sites were extracted from the multiple sequence alignment using the MEGA 3.1 package [12].

2.5. Phylogenetic analysis

Phylogenetic model selection was done using the FINDMODEL server available from the HCV LANL database at (http://hcv.lanl.gov/ /content/hcv-db/findmodel/). A Bayesian phylogenetic tree was constructed by Markov chain Monte Carlo (MCMC) method as implemented in the MR BAYES 3.1 package using the Hasegawa-Kishino-Yano plus Gamma model HKY+G substitution model with an invariant four category gamma distribution among sites. A 50% consensus tree was generated and the analysis was repeated two times. Maximum parsimony tree was conducted using MEGA version 4, with 1000 bootstraps for reliability.

The mean overall, within group and between groups genetic distances were done using the MEGA 4.0 software [12].

3. Results

Our experience in the field of molecular identification or DNA barcoding through a series of published research papers are represented in this section Results with some illustrated figures and tables are represented here but the complete information could be obtained through obtaining the complete published papers from the publication section.

Shows the Positional entropy plot of the D-loop for the buffalo, and cow sequences The Bayesian phylogenetic trees of cow and buffalo sequences were constructed using MRBAYES software (Figure 2) and Maximum parsimony tree using the Kimura two-parameter model and the closest neighbor interchange method of the MEGA 3.1 software package (Figure 3). Table 1. Shows the Substitution events detected in complete D-loop sequences from multiple sequence alignments between cows and buffaloes.

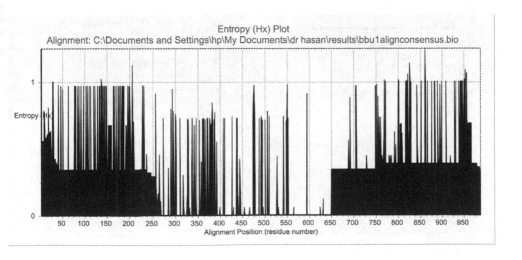

Figure 1. Positional Entropy Plot of the D loop for the Buffalo, and cow sequences. The X axis is the nucleotide position, while the Y axis is the entropy (lack of information) at that position. The central region is mainly conserved, and the beginning and end regions are highly variable (shaded areas).

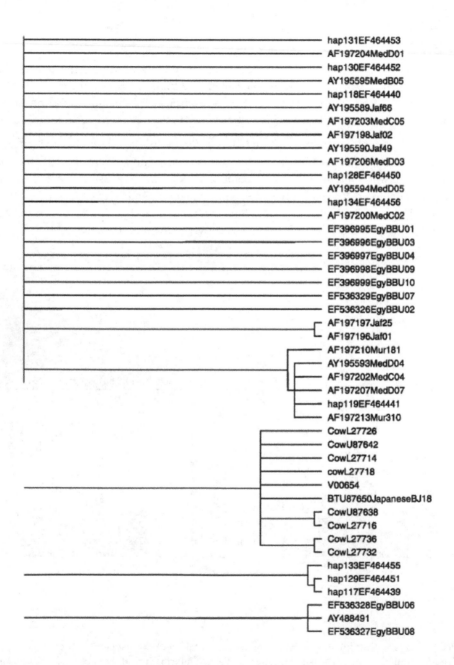

Figure 2. Bayesian phylogenetic tree of the cow and buffalo sequences using the MRBAYES software.

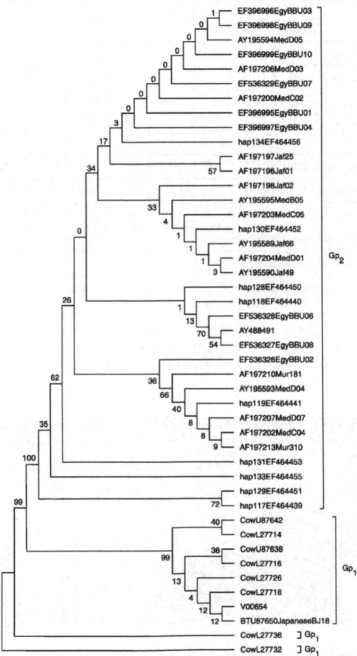

Figure 3. Maximum parsimony tree constructed using the Kimura two-parameter model and the closest neighbor interchange method of the MEGA 3.1 software package. The numbers show the percentage of bootstrap confidence.

Animals	Cows (35)		Buffaloes (53)		Cows & buffaloes (88)	
Substitutions and genetic distances	Value	S E	Value	S E	Value	S E
Total numbers of transitions	44		20		100	
Total numbers of transversions	4		1		61	
Total number of indels (insertions/deletions)	1		7		15	
Total number of substitutions	49		28		176	
R ratio (transversions/transitions)	0.09		0.05		0.61	
Genetic distance within group	0.023	0.003	0.007	0.002		
Genetic distance between the two groups					0.156	0.016
Overall (all animals) distance					0.06	0.006

Table 1. Substitution events detected in complete D-loop sequences from multiple sequence alignment between Cows (35 animals) and buffaloes (53 animals).

Shows the PCR amplification of chicken mitochondrial D loop fragments while the phylogenetic tree constructed between the Egyptian and GenBank database chicken samples is represented in Figure 5. The Polymorphic sites and their positions are shown in Table 2.

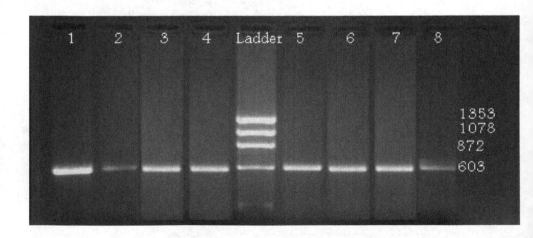

Figure 4. PCR amplification of chicken mitochondrial D loop fragment. The PCR reactions were run on 1% agarose gel, stained with ethidium bromide and examined with UV. Samples from 1 to 4 for Dandarawi breed and samples from 5 to 8 for Fayoumi breed. The *Hae* III digest of Φ X174 DNA was used as ladder (1353, 1078, 872, and 603 base pairs).

output2.ph Sat Dec 29 00:51:38 2007 Page 1 of 1

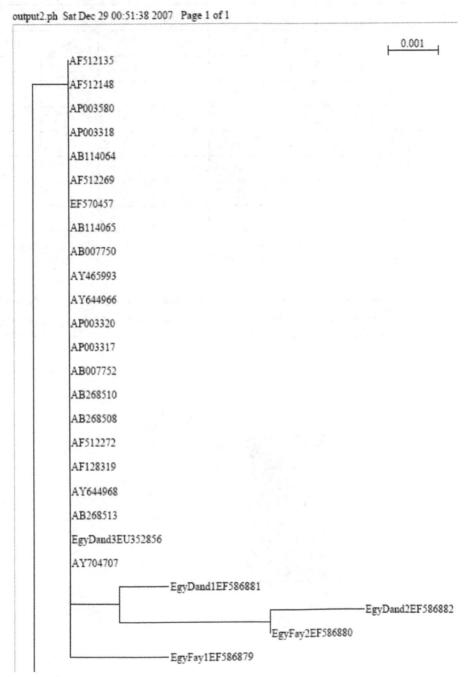

Figure 5. Phylogenetic tree constructed between the Egyptian and GenBank database chicken samples. Sample DQ629875 was used as out-group for its high diversity.

Breed & Accession number	Variable sites and their positions					
	23	35	276	457	464	483
EgyDand1 EF586881	T	A	C	A	G	T
EgyDand2 EF586882	T	A	C	C	T	T
EgyDand3 EU352856	T	A	C	A	G	A
EgyFay1 EF586879	T	A	A	A	G	A
EgyFay2 EF586880	T	A	C	C	T	A
DQ629875	A	*	C	A	G	A
Database public sequence	T	A	C	A	G	A

Table 2. The Polymorphic sites and their positions. The nucleotide positions were given with respect to the Egyptian nucleotide numbers in GenBank database. The left column shows the breeds with their accession numbers.

The following Figures: show polyacrylamide gel representing the PCR-amplified fragment of CO1 and its sequence (Figures 6 & 8) while the PCR-amplified fragment of ITS region and its sequence were presented in Figures 7 & 9.

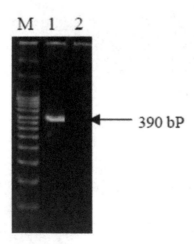

Figure 6. 10% polyacrylamide gel representing the PCR-amplified product of CO1. M: 50 bp DNA size marker, lane 1: CO1 fragment and lane 2: blank (PCR cocktail without DNA).

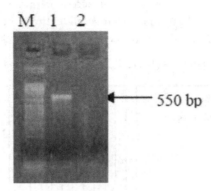

Figure 7. Agarose gel representing the PCR-amplified product of ITS. M: 50 bp DNA size marker, lane 1: ITS fragment and lane 2: blank.

5'CAATCTAATAAGTTTCAGACTTTCGGGGCTTTGGGTATAATTTATGCTATATTGTCTATTGGG
ATTTTAGGGTTTATTGTATGAGCTCATCATATATTTACAGTAGGAATAGATGTTGATTCTCGAG
CTTATTTTACCGCTGCTACGATAATTATTGCGGTACCTACAGGTATTAAGGTTTTTTCTTGATT
ATCTACAATGTATGGGTCAGTAGTAAAGTGAGGAGTAGTTGTTTTGTGAAATTTTGGTTTTATT
TTTTTATTCAGGTTAGGTGGTTTAACCGGGGTGGTTTTGTCAAA3'

Figure 8. Sequence of the mitochondrial gene fragment coding for cytochrome oxidase subunit I(COI) in the predacious mite, *Amblyseius swirskii.*

5'GNCACCTGATTTAGATCACGTCATAATGGTTTGTTTTGGTTACTCTTGTGACCAAGAGAATT
TATACCATTGCATGGCGATACCGACGAGGGCTGCAGCAAAGGTAAATACGTATGATACGGTT
TATATACTCGAAACAAATGTAGCACAGATACAGTAAGTGTCCGTGCTGAAATTTTCATTCAA
AAACACAATGCTCACAAAATTTCACAACTCACATCAATTTCCACAAATTACTATGTTTTTCAT
CGATTTAAGGACTAAGTGATCCCCCATATTGAGTCTTGGTTTTTTTCTTCTCGATAGCACAACT
TACTTCCAAAGGAAGTGAAAAGGTTTGTCGGAATGGTTACCGACTTACTGTCGCATACGCCTT
CCCCGTAACCAAAAGGGACCGGTTAAACACCCACCAATTCGAAGCGGTTGTCGCCCAAAAA
GGGACCGGTTTGGAAGACCGTTATCCACCGTTAAAAACCTAAATTACAGGTTTGTTTTTTTCAT
TGGTCCTTTTTCGAAAAAACACAN3

Figure 9. Sequence of the internal transcribed spacers (ITS) region in the predacious mite, *Amblyseius swirskii.* N: unknown base.

Sequencing of a fragment of ITS region (ITS1, ITS2, 5.8S) indicated almost complete identity of the Egyptian samples with *Neoseiulus swirskii*, accession number EU 310505 (= *A. swirskii*). Regardless the locality within Egypt, taxa were less identical when compared to the related species *N. Cucumeris* (Oud.) *N. andersoni* (Chant), and *N. fallacis* (Gar-man).

According to the molecular analysis (Table 3), samples are grouped into three groups.

Nucleotide number (Accession No. EU924213) — variable sites in ITS1, 5.8S rRNA and ITS2 regions. Nucleotide positions (read vertically): 23, 33, 35, 88, 145, 146, 147, 148, 149, 163, 182, 216, 223, 249, 262, 271, 274, 287, 300, 304, 3118, 313, 314, 3119, 345, 332.

Sample	Accession Number	Phylogenetic Group	23	33	35	88	145	146	147	148	149	163	182	216	223	249	262	271	274	287	300	304	3118	313	314	3119	345	332
			ITS1			5.8S rRNA															ITS2							
Sample 3	EU924213		c	c	c	a	c	c	a	t	a	c	c	a	g	t	c	a	t	a	c	c	-	a	g	-	a	g
Sample 4	EU924214	Group 1	c	c	c	a	c	c	a	t	a	c	c	a	g	t	c	a	t	a	c	c	-	a	g	-	a	g
Sample 2	EU924215		a	a	t	a	t	c	a	t	a	-	c	a	g	t	c	a	t	a	c	c	-	a	g	-	a	g
Sample 1	EU924212		a	a	t	-	t	g	t	-	-	-	t	-	t	t	t	t	-	t	-	t	t	t	t	t	a	-
N. swirskii	EU310505	Group 3	a	a	t	-	t	g	t	-	-	-	t	-	t	t	t	t	-	t	-	t	t	t	t	t	a	t
Sample 5	EU924216		n	a	t	-	t	g	t	-	-	-	t	a	g	a	t	t	-	c	c	t	t	a	t	g	g	t
Sample 6	EU924217	Group 2	n	n	n	-	t	g	t	-	-	-	t	a	g	a	t	a	-	c	-	t	t	a	t	-	g	t

Table 3. The variable sites (a = Adenine, c = Cytosine, g = Guanine, t = Thymine, - = deletion and n = Not detected) detected in a fragment of nuclear ITS region of six samples of *A. swirskii* collected from citrus and grapes in the Nile delta of Egypt.

The results of 12S rRNA showed that, two haplotypes of 12S rRNA sequences were identified from the multiple alignment results between the nine tested Egyptian buffalo sequences and other examples of homologous buffalo sequences selected from GenBank database. Two buffalo haplotypes were revealed, of which haplotype 1 which include Egyptian buffaloes and haplotype 2 which include Chinese swamp buffalo; breed: Haikou (accession AY702618), Mediterranean (accession AY488491) and *Bubalus bubalis* (accession AF231028). The detected SNPs can be classified as shown in table 4.

Eleven SNPs were detected which can be used to discriminate between subfamily Bovinae, represented by buffalo and cattle and the subfamily Caprinae represented by sheep and goat.

No. of SNPs	Specificity	Representing nucleotide	Base position
8	all buffaloes (haplotypes 1 & 2)	Guanine	at positions 110, 132 and 196
		Thymine	position 172
		Cytosine	positions 71, 269, 271 and 348.
3	buffalo haplotype 1 only	Guanine	position 158
		Cytosine	position 267
		Adenine	position 293
4	buffalo haplotype 2 only	Thymine	positions 32 and 267
		Guanine	position 293
		Cytosine	position 72
12	cattle (Bos taurus)	Guanine	positions 193, 266 and 273
		Adenine	27 and 174
		Thymine	positions 26, 36, 186 & 190
		Cytosine	positions 253, 294 and 295.
6 + 2 indels	sheep (Ovis aries)	Guanine	positions 172, 231 and 300
		Adenine	position 190
		Thymine	position 271
		Cytosine	position 349
		insertion of Adenine	position 164-165
		deletion of Adenine	position 259
6 + 1 deletion	goat (Capra hircus)	Guanine	position 299
		Thymine	positions 260, 261 and 317
		Cytosine	positions 225 and 259
		deletion of Adenine	deletion at position 273

Table 4. Nucleotide variation in a specific 12S rRNA gene fragment of four studied Bovidae species. Nucleotide positions correspond to Egyptian buffaloes GenBank accession numbers (FJ828575-FJ828583).

Considering multiple alignment results between homologous 16S rRNA sequences obtained from GenBank database with the reference sequence, it was shown that, the entire 16S rRNA fragment (422 bp. in size) contains more than 57 variable sites (from base no. 21 to base no. 323) inside the two conserved regions. The bases outside this variable region are completely conserved in the four species (Figure 10 and Table 5). From these variable sites, 25 specific nucleotides were chosen (which gave clear significant results in both types of alignment comparisons (two and multiple alignment sequences programs) as a reference for identification of unknown species (from base no. 21 to base no. 308). It was also shown that the size of the amplified fragments were less by one nucleotide (421 bp) in case of goat and two nucleotides (420 bp) in case of both cattle and sheep.

Detection of specific variable sites between Egyptian buffalo 16S rRNA gene fragment and the other studied three species is shown to be a good marker for identification of the four studied species. The detected variable sites can be classified as represented in both Fig. 10 and Table 5.

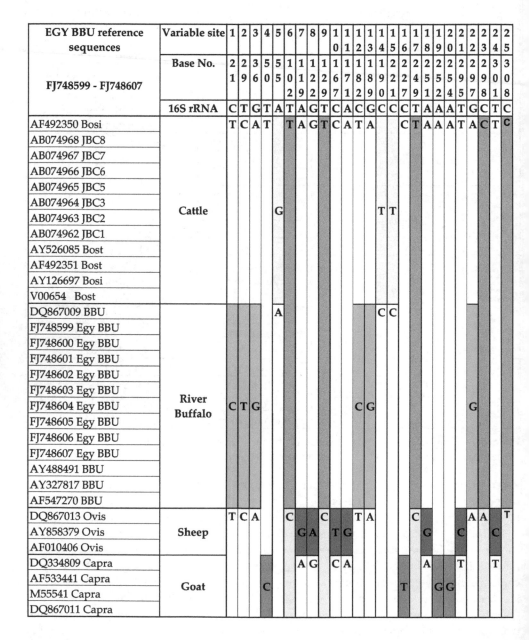

Figure 10. Multiple sequence alignments result showing the total variable sites between river buffalo (BBU), Cattle (Bost, Bosi and JBC), Sheep (Ovis) and goat (Capra) in the specific 16S rRNA fragment. Sequences of the 16S rRNA fragment of Egyptian buffaloes (FJ748599–FJ748607). Differing nucleotides are noted (T, A, G and C)

No. of Variable sites (SNPs)	Specificity	Representing nucleotides	Base position
Six	River buffaloes	Guanine	(36, 189, and 297)
		Thymine	(29)
		Cytosine	(21 and 182)
Three	Cattle (*Bos taurus*)	Guanine	(55)
		Thymine	(190 and 221)
Seven	Sheep (*Ovis aries*)	Guanine	(119, 171, and 251)
		Adenine	(122)
		Thymine	(167)
		Cytosine	(295 and 301)
Four	Goat (*Capra hircus*)	Guanine	(252 and 254)
		Thymine	(227)
		Cytosine	(50).
Five	Group one (river buffaloes and cattle, Subfamily Bovinae)	Thymine	(102, 129 and 249)
		Cytosine	(298 and 308)
Five	Group two (sheep and goat, Subfamily Caprinae).	Adenine	(298)
		Cytosine	(102, 129 and 249)
		Thymine	(308)

Table 5. Nucleotide variation in a specific 16S rRNA gene fragment of four studied Bovidae species. Nucleotide Positions correspond to Egyptian buffaloes GenBank accession numbers (FJ748599–FJ748607)

4. Discussion

4.1. DNA barcoding, genome evolution & phylogenetic trees

The ability of molecular trees to encompass both short and long periods of time is based on the observation that different genes evolve at different rates. The DNA specifying ribosomal RNA (rRNA) changes relatively slowly, so comparisons of DNA sequences in these genes are useful for investigating relationships between taxa that diverged hundreds of millions of years ago. Studies of the genes for rRNA have shown, for example, that fungi are more closely related to animals than to green plants—something that certainly could not have been deduced from morphological comparisons alone.

In contrast, the DNA in mitochondria (mtDNA) evolves relatively rapidly and can be used to investigate more recent evolutionary events.

The methodology used in DNA barcoding has been straightforward. Sequences of the barcoding region are obtained from various individuals. The resulting sequence data are then used to construct a phylogenetic tree using a distance-based 'neighbour-joining' method. In such a tree, similar, putatively related individuals are clustered together. The term 'DNA barcode' seems to imply that each species is characterized by a unique sequence, but there is of course considerable genetic variation within each species as well as between species. However, genetic distances between species are usually greater than those within species, so the phylogenetic tree is characterized by clusters of closely related individuals, and each cluster is assumed to represent a separate species.

An evolutionary tree (or *phylogenetic tree*) is a branching diagram that represents the evolutionary history of a group of organisms. For example, we might use morphological and genetic data to figure out a phylogenetic tree of animals. Such a tree can provide a huge amount of information. For any particular group of animals our tree could identify the ancestors and closest relatives of the group. If we traced the history of animals all the way back, we could use the tree to help us answer questions such as, What did the earliest animals look like? What features did they pass on to all their descendants?. Phylogenetic trees also have great practical value. The same techniques we use to reconstruct evolutionary history have been used in forensics, where phylogenetic trees have helped solve criminal cases, and epidemiology, where trees have been used to estimate when and where diseases such as AIDS originated.

Now that we can compare entire genomes, including our own, some interesting facts have emerged. As you may have heard, the genomes of humans and chimpanzees are strikingly similar. An even more remarkable fact is that homologous genes are widespread and can extend over huge evolutionary distances. While the genes of humans and mice are certainly not identical, 99% of them are detectably homologous. And 50% of human genes are homologous with those of yeast.

It is not a coincidence that DNA barcoding has developed in concert with genomics-based investigations.

DNA barcoding (a tool for rapid species identification based on DNA sequences) and genomics (which compares entire genome structure and expression) share an emphasis on large scale genetic data acquisition that offers new answers to questions previously beyond the reach of traditional disciplines. DNA barcodes consist of a standardized short sequence of DNA (400–800 bp) that in principle should be easily generated and characterized for all species on the planet (1). A massive on-line digital library of barcodes will serve as a standard to which the DNA barcode sequence of an unidentified sample from the forest, garden, or market can be matched. Similar to genomics, which has accelerated the process of recognizing novel genes and comparing gene function, DNA barcoding will allow users to efficiently recognize known species and speed the discovery of species yet to be found in nature. DNA barcoding aims to use the information of one or a few gene regions to identify all species of life, whereas genomics, the inverse of barcoding, describes in one (e.g., humans) or a few selected species the function and interactions across all genes.

To be practical as a DNA barcode a gene region must satisfy three criteria: (*i*) contain significant species-level genetic variability and divergence, (*ii*) possess conserved flanking sites for developing universal PCR primers for wide taxonomic application, and (*iii*) have a short sequence length so as to facilitate current capabilities of DNA extraction and amplification. A short DNA sequence of 600 bp in the mitochondrial gene for cytochrome *c* oxidase subunit 1 (CO1) has been accepted as a practical, standardized species-level barcode for animals (see www.barcoding.si.edu). The inability of CO1 to work as a barcode in plants set off a race among botanists to find a more appropriate marker. A number of candidate gene regions have been suggested as possible barcodes for plants, but none have been widely accepted by the taxonomic community. This lack of consensus is in part due to the limitations inherent in a plastid marker relative to plant CO1, and also because a quantitative context for selecting a gene region as a barcode for plants has not been offered. Several factors must be considered and weighted in selecting a plant DNA barcode: (*i*) universal PCR amplification, (*ii*) range of taxonomic diversity, (*iii*) power of species differentiation, and (*iv*) bioinformatics analysis and application.

4.2. Molecular genetics reveals evolutionary relationships

Evolution results from the accumulation of inherited changes in populations. Because DNA is the molecule of heredity, evolutionary changes must be reflected in changes in DNA. Systematics have long known that comparing DNA within a group of species would be a powerful method for inferring evolutionary relationships, but for most of the history of systematics, direct access to genetic information was nothing more than a dream. Today, however, **DNA sequencing**—determining the sequence of nucleotides in segment of DNA – is comparatively cheap, easy, and widely available. *The polymerase chain reaction* (PCR) allows systematics to easily accumulate large samples of DNA from organisms, and automated machinery makes sequence determination a comparatively simple task.

4.3. Direct benefits of DNA barcoding undoubtedly include

i.   make the outputs of systematics available to the largest possible community of end-users by providing standardized and high-tech identification tools, e.g. for biomedicine (parasites and vectors), agriculture (pests), environmental assays and customs (trade in endangered species);

ii. relieve the enormous burden of identifications from taxonomists, so they can focus on more pertinent duties such as delimiting taxa, resolving their relationships and discovering and describing new species;

iii. pair up various life stages of the same species (e.g. seedlings, larvae);

iv. provide a bio-literacy tool for the general public.

Perhaps another advantage of DNA barcoding is that it will also facilitate basic biodiversity inventories. Indeed, from the premises of molecular phylogenetics to assembling the tree of life, DNA sequences in environmental sampling and reconstruction of phylogenetic trees to place sequences into an evolutionary context have been used in several inventories of cryptic biodiversity (e.g. soil bacteria or marine/freshwater micro-organisms).

New 'Genetic Bar Code' Technique Establishes Ability to Derive DNA Information from RNA

Science Daily (Apr. 8, 2012) — Researchers from Mount Sinai School of Medicine have developed a method to derive enough DNA information from non-DNA sources -- such as RNA -- to clearly identify individuals whose biological data are stored in massive research repositories. The approach may raise questions regarding the ability to protect individual identity when high-dimensional data are collected for research purposes.

A paper introducing the technique appears in the April 8 online edition of *Nature Genetics*.

DNA contains the genetic instructions used in the development and functioning of every living cell. RNA acts as a messenger that relays genetic information in the cell so that the great majority of processes needed for tissue to function properly can be carried out.

To date, access to databases with DNA information has been restricted and protected as it has long been considered the sole genetic fingerprint for every individual. However, vast amounts of RNA data have been made publicly available via a number of databases in the United States and Europe. These databases contain thousands of genomic studies from around the world.

In this study, authors developed a technique whereby a person's DNA could be inferred from RNA data using gene-expression levels monitored in any of a number of tissues. In contrast, most studies involving DNA and RNA begin with DNA sequences and then seek to associate expression patterns with changes in DNA between individuals in a population. This is the first time going from RNA levels to DNA sequence has been described.

"By observing RNA levels in a given tissue, we can infer a genotypic barcode that uniquely tags an individual in ways that enables matching the individual to an independently derived DNA sample,". Not only can genotypic barcodes be deduced from RNA, but RNA levels in some tissue can inform not only individual characteristics like age and sex, but on diseases such as Alzheimer's and cancer, as well as the risks of developing those diseases."

Author details

Hassan A. I. Ramadan* and Nabih A. Baeshen
Department of Biological Sciences, Faculty of Science,
King Abdulaziz University, Jeddah, Saudi Arabia

Hassan A. I. Ramadan
Department of Cell Biology, National Research Centre, Dokki, Cairo, Egypt

Acknowledgement

The authors gratefully acknowledge the financial support from the Deanship of Scientific Research (DSR) at King Abdulaziz University (KAU) represented by the Unit of Strategic Technologies Research through the Project number **(10-BIO1257-03)**.

* Corresponding Author

5. References

[1] Pascal G, Mahe S (2001) Identity, traceability, acceptability and substantial equivalence of food. Cell Mol Biol 47:1329–1342

[2] Skarpeid HJ, Kvaal K, Hildrum KI (1998) Identification of animal species in ground meat mixtures by multivariate analysis of isoelectric focusing protein profiles. Electrophoresis 19:3103–3109

[3] 3. Hsieh YH, Sheu SC, Bridgman RC (1998) Development of a monoclonal antibody specific to cooked mammalian meats. J Food Prot 61(4):476–487

[4] Ashmoor SH, Monte WC, Stiles PG (1998) Liquid chromatographic identification of meats. J Assoc Off Anal Chem 71:397–403

[5] Parson W, Pegoraro K, Niederstatter H, Fo"ger M, Steinlechner M (2000) Species identification by means of the cytochrome b gene. Int J Legal Med 114:23–28

[6] Hsieh HM, Chiang HL, Tsai LC, Lai SY, Huang NE, Linacre A, Lee JC (2001) Cytochrome b gene for species identification of the conservation animals. Forensic Sci Int 122:7–18

[7] Murray BW, McClymont RA, Strobeck C (1995) Forensic identification of ungulate species using restriction digests of PCRamplified mitochondrial DNA. J Forensic Sci 40(6):943–951

[8] Balitzki-Korte B, Anslinger K, Bartsch C, Rolf B (2005) Species identification by means of pyrosequencing the mitochondrial 12S rRNA gene. Int J Legal Med 119:291–294

[9] Rodrı´guez MA, Garcı´a T, Gonza´lez I, Asensio L, Herna´ndez PE, Martı´n R (2004) PCR identification of beef, sheep, goat, and pork in raw and heat-treated meat mixtures. J Food Prot 67(1):172–177

[10] Rodrı´guez MA, Garcı´a T, Gonza´lez I et al (2003) Identification of goose, mule, duck, chicken, turkey, and swine in foie gras by species-specific polymerase chain reaction. J Agric Food Chem 51:1524–1529

[11] Montiel-Sosa JF, Ruiz-Pesini E, Montoya J, Roncale´s P, Lo´pez- Pe´rez MJ, Pe´rez-Martos A (2000) Direct and highly speciesspecific detection of pork meat and fat in meat products by PCR amplification of mitochondrial DNA. J Agric Food Chem 48:2829–2832

[12] Ramadan HAI, El Hefnawi M (2008) Phylogenetic analysis and comparison between cow and buffalo (including Egyptian buffaloes) mitochondrial displacement-loop regions. Mitochondrial DNA 19(4):401–410

[13] Waugh J (2007) DNA barcoding in animal species: progress, potential and pitfalls. BioEssays 29(2):188–197

[14] Hebert PDN, Cywinska A, Ball SL, de Waard JR (2003) Biological identifications through DNA barcodes. Proc R Soc B 270:313–321

[15] Hebert PDN, Stoeckle MY, Zemlak TS, Francis CM (2004) Identification of birds through DNA barcodes. PLoS Biol 2:E312

[16] Ward RD, Zemlak TS, Innes BH, Last PR, Hebert PDN (2005) DNA barcoding Australia's fish species. Philos Trans R Soc Lond B 360:1847–1857

[17] Hajibabaei M, Janzen DH, Burns JM, Hallwachs W, Hebert PDN (2006) DNA barcodes distinguish species of tropical Lepidoptera. Proc Natl Acad Sci USA 103:968–971

[18] Vences M, Thomas M, Meijden AVD, Chiari Y, Vieites DR (2005) Comparative performance of the 16S rRNA gene in DNA barcoding of amphibians. Front Zool 2:5. doi:10.1186/1742- 9994-2-5

[19] Bradley RD, Baker RJ (2001) A test of the genetic species concept: cytochrome b sequences and mammals. J Mammal 82: 960–973

[20] Lemer S, Aurelle D, Vigliola L, Durand JD, Borsa P (2007) Cytochrome b barcoding, molecular systematics and geographic differentiation in rabbitfishes (Siganidae). C R Biol 330:86–94

[21] De Salle R, Egan MG, Siddall M (2005) The unholy trinity: taxonomy, species delimitation and DNA barcoding. Philos Trans R Soc Lond B 360(1462):1905–1916

[22] Hajibabaei M, Singer GAC, Clare EL, Hebert PDN (2007) Design and applicability of DNA arrays and DNA barcodes in biodiversity monitoring. BMC Biol 5:24. doi:10.1186/1741-7007-5-24)

[23] Ramadan HAI, Mahfouz ER (2009) Sequence of specific mitochondrial 12S rRNA fragment of Egyptian buffalo as a reference for discrimination between buffalo, cattle, sheep and goat. J Appl Biosci 21:1258–1264

[24] Ramadan, H.A.I (2011). Sequence of specific mitochondrial 16S rRNA gene fragment from Egyptian buffalo is used as a pattern for discrimination between river buffaloes, cattle, sheep and goats. Mol. Biol. Rep. 38 (6) 3929-3934.

[25] Ramadan, H.A.I., El-Banhawy, E.M., Hassan, A.A., and Afia,S.I. (2004): Genetic variation in the predacious phytoseiid mite, Amblyseius Swirskii (Acari: Phytoseiidae): Analysis of specific mitochondrial and nuclear sequences. Arab J. Biotech., 7, No.(2): 189-196.

[26] Ramadan, H. A. I, El-Banhawy, E. M and Afia, S. I. (2009). On the identification of taxa collected from Egypt in the species sub-group andersoni Chant and McMurtry: Morphological relationships with related species and molecular analysis of inter and intra-specific variations (Acari: Phytoseiidae). Acarologia (France), Vol 49, fasc 3-4: 115- 120

[27] Ramadan, H.A.I, Galal, A., Fathi, M.M., El Fiky, S.A., Yakoub, H.A. (2011): Characterization of Two Egyptian Native Chicken Breeds Using Genetic and Immunological Parameters. Biotechnology in Animal Husbandry 27 (1), p 1-16.

Physical Sciences, Engineering and Technology

Chapter 6

Image Processing for Spider Classification

Jaime R. Ticay-Rivas, Marcos del Pozo-Baños, Miguel A. Gutiérrez-Ramos,
William G. Eberhard, Carlos M. Travieso and Alonso B. Jesús

Additional information is available at the end of the chapter

1. Introduction

As is defined by UNESCO [22]: "Biological diversity or biodiversity is defined as the diversity of all living forms at different levels of complexity: genes, species, ecosystems and even landscapes and seascapes. Biodiversity is shaped by climatic conditions, the properties of soils and sediments, evolutionary processes and human action. Biodiversity can be greatly enhanced by human activities; however, it can also be adversely impacted by such activities due to unsustainable use or by more profound causes linked to our development models."

It is clear that climate change and biodiversity are interconnected. Biodiversity is impacted by climate change but it also makes an important contribution to both climate-change mitigation and adaptation through the ecosystem services that it supports. Therefore, conserving and sustainable managing biodiversity is crucial to meet the clime change.

Biodiversity conservation is an urgent environmental issue that must be specially attended. It is as critical to humans as it is to the other lifeforms on Earth. Countries of the world acknowledge that species research is crucial in order to obtain and develop the right methods and tools to understand and protect biodiversity. Thus, biodiversity conservation has became a top priority for researchers [1]

In this sense, a big effort is being carried out by the scientific community in order to study the huge biodiversity present on the planet. Sadly, spiders have been one of most unattended groups in conservation biology [2]. These arachnids are plentiful and ecologically crucial in almost every terrestrial and semi-terrestrial habitats [3] [4] [5]. Moreover, they present a series of extraordinary qualities, such as the ability to react to environmental changes and anthropogenic impacts [5] [6].

Several works have studied the spiders' behavior. Some of them focuses on the use of the way spiders build their webs as a source of information for species identification[7] [8]. Artificial intelligent systems have been proven to be of incalculable value for these systems. [9] proposed a model for spider behavior modeling, which provides simulations of how a specific spider specie builds its web. [10] recorded how spiders build their webs in a controlled scenario for further spatiotemporal analysis.

Because spider webs carry an incredibly amount of information, this chapter presents an study about its usage for automatic classification of spider species. In particular, computer vision and artificial intelligence techniques will be used for this aim. Moreover, the amount of information will be such, that it would be enough to perform the spider specie classification. This is, to authors extend, a novel approach on this problem.

The remainder of this paper is organized as follow. First, the database is briefly presented. Section 3 explains how images were preprocessed in order to extract the spider webs from the background. The feature extraction and classification techniques are introduced in sections 4 and 5. Next, experiments and results are shown in detail. Finally, the conclusions derived from the results are presented.

2. Database

The database contains spider web images of four different species named Allocyclosa, Anapisona Simoni, Micrathena Duodecimspinosa and Zosis Geniculata. Each class has respectively 28, 41, 39 and 49 images, which makes a total of 150 images. Some examples can be seen in Figure 1. Since the webs images were taken in both controlled and uncontrolled environments the lightness condition and background differ between classes.

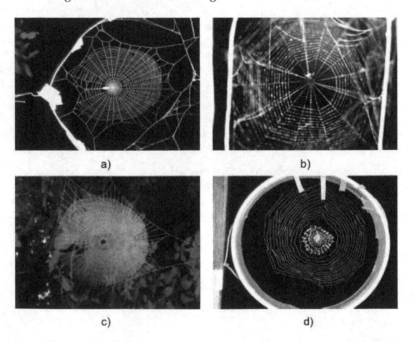

Figure 1. Spider web samples of the four species: *a) Allocyclosa. b) Anapisona Simoni. c) Micrathena Duodimspinosa. d) Zosis Geniculata.*

The images that correspond to *Allocyclosa* were assigned to Class 1 (C1). These were taken in a natural night time environment. The flash of the camera in conjunction with dark background enhanced the spider webs, resulting in a set of images with good quality for the processing stage.

Class	Number of samples	Size (pixels)	Bits number (color/gray)
1	28	1024x768	24 (True color)
2	41	2240x1488	8 (Gray scale)
3	39	2240x1488	24 (True color)
4	42	2216x2112	24 (True color)

Table 1. Tecnical features of the dabase.

Class 2 (C2) corresponds to the *Anapisona Simoni* images. Theses spider webs were built in a controlled environment, with the singularity that they were built as tents, causing overlapping thread and light reflections. This made the processing stage far more more complex than in C1.

Class 3 (C3) is composed by *Micrathena Duodimspinosa*. These images were also taken in a natural environment during the day. This scenario, with the presence of the sun light and natural elements such as leaves and tree branches, required a more complex treatment as well.

Finally, Class 4 (C4) corresponds to *Zosis Geniculata* images. Again, they were taken in a controlled environment, allowing the capture of images with black background and uniform light.

The technique features of images from each class are summarized in Table 1

3. Preprocessing

As can be seen in Figure 1, spider web images were taken in both controlled and uncontrolled environments. Thus, the preprocessing step was vital in order to isolate the spider webs and remove possible effects of background in the system.

Image processing techniques were employed in order to isolate the spider webs from light reflections and elements of the background of the image such as leaves or tree branches. Once it was applied, a new normalized database was obtained.

3.1. Spider web selection

Since the image collection was not taken for this research, there are information that does not provide valid data for the spiders' study. This information corresponds to any external element of the spiderwebs. This is why the region of interest of the image was manually selected. This is represented in figure 2.

Once the spider web has been select, an adjustment of the proportional ratio was necessary in order to obtain a proportional square image. This will be explained in detail in the following section.

3.2. Image contrast

To enhance the contour of cobweb's threads an increase of the color contrast was first applied. A spacial filtering was applied to enhance or attenuate details in order to obtain a better visual interpretation and prepare the data for the next preprocessing step. By using this filtering, the

a)   b)

Figure 2. Spider web Selection

value of each pixel is modified according to neighbors' values, transforming the original gray levels so that they become more similar or different to the corresponding neighboring pixels.

In general, the convolution of a image f with MxN dimensions with a h mxn mask is given by 1:

$$g(x,y) = \sum_{s=-a}^{a} \sum_{t=-b}^{b} f(x+s,y+t)h(s,t) \tag{1}$$

Where $f(x+s,y+t)$ are the pixel value's of the selected block, $h(s,t)$ are the mask coefficients and $g(x,y)$ is the filtered image. The block dimension is defined by $m = 2a+1$ and $n = 2b+1$. The effect of applying contrast enhancement filtering over a gray image can be observed in figure 3.

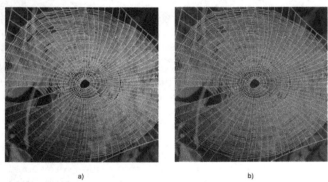

a)   b)

Figure 3. a) Original Image b) Contrast enhanced

3.3. Image binarization

The binarization process transforms the image to a black and white format in a way that it does not change the esential properties of the image. Equation 2 defines the binarization process, where $f(x,y)$ is the original image and $g(x,y)$ the obtained image:

$$g(x,y) = \begin{cases} 0 & f(x,y) < threshold \\ 1 & f(x,y) \geq threshold \end{cases} \tag{2}$$

This threshold is computed by using the very know Otsu's method [25], which assigns the membership of each pixel to a determined group by computing the optimal value from which carrying out that assignment.

3.4. Image denoising

Once the image has been binarized, a denoising process was used aiming to eliminate any irrelevant information. To achieve this goal, two specific techniques were applied: *Wiener Filtering* and *Morphological Operations*.

The Wiener Filtering applied a spatial filtering using statistical methods in order to reduce noise and smooth shapes. It gradually smooths the image by changing the areas where the noise is very apparent, but keeping the areas where the details are present and the noise is less apparent. The Wiener filter is adapted to the local image variance.

In this work, the algorithm *wiener2* [21] was used in order to compute the local mean and the variance around each pixel in the image a

$$\mu = \frac{1}{NM} \sum_{\eta_1, \eta_2 \in \eta} a(\eta_1, \eta_2) \tag{3}$$

$$\sigma^2 = \frac{1}{NM} \sum_{\eta_1, \eta_2 \in \eta} a^2(\eta_1, \eta_2) - \mu^2 \tag{4}$$

Where η is defined as the local neighborhood for each pixel NxM in the image a. An $2x2$ block has been chosen by euristics, i.e. this was the configuration that provided the best visual effect. *Wiener2* then filters the image using these estimates, where b is the resulting image.

$$b(n_1, b_2) = \mu + \frac{\sigma^2 - v^2}{\sigma^2} (a^2(\eta_1, \eta_2) - \mu) \tag{5}$$

If the noise variance is not given, *wiener2* uses the average of all the local estimated variances.

On the other hand, morphological operations are those transformations that modify the structure or shape of the objects in the image based on the their geometry and shape, simplifying the images. These techniques can be used to denoise an image, for feature extraction or processing specific regions.

An illustrative example of these operations is shown in figures 4 and 5, where noise and projections (in the inner circle) are removed, obtaining more uniform boundaries.

The resulting image after image denoise can be observed in figure 6.

3.5. Center of the spiderwebs

Finally, the center of the spiderwebs was used as the source of discriminative information in order to classify spiders. Thus, once the images were preprocessed this center area was selected to conform the experimentation database. Figures 7 and 8 show the center of the spiderwebs for each specie.

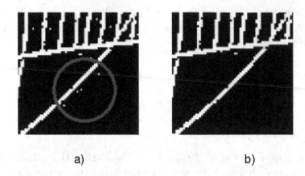

a)   b)

Figure 4. Example of applying morphological operations: elimination of isolate pixels

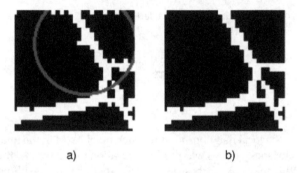

a)   b)

Figure 5. Example of applying morphological operations: smoothing of contour

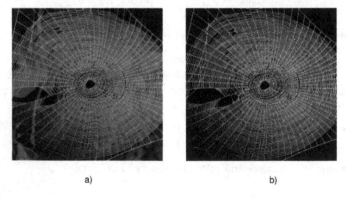

a)   b)

Figure 6. Image after denoising. a) Original image b) Image binarized

4. Features extractors

In general, feature extraction refers to the process of obtaining some numerical measures of images such as area, radio, perimeter, etc. Also it concerns to the process of transforming a set of original features; with dimension m, in another set of characteristics; usually with

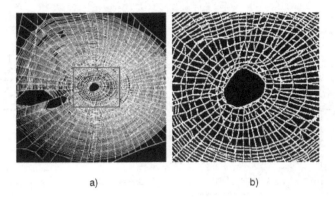

a) b)

Figure 7. Selecting the center of the spiderwebs.

Figure 8. Resulting center of each specie.

dimension $n < m$, which is termed as transformed domain techniques. This is the concept used in the present work.

Two well known techniques were used: Discrete Wavelet Transform (DWT) and Discrete Cosine Transform (DCT). These were selected as they have been successfully used in other biometric studies. Besides being able to reduce the dimensionality of data and reduce the computational requirements of the classifier stage, these techniques can improve the generalization of the information and the system's success rate.

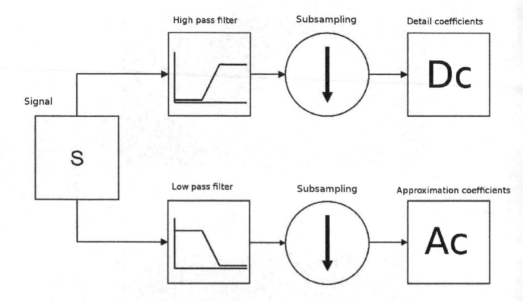

Figure 9. Algorithm for the Discrete Wavelet Transform

4.1. Wavelet transform

The discrete wavelet transform (DWT) is based on the idea of decomposing a signal in terms of displaced and dilated versions of a finite wave called mother wavelet. The Wavelet transform is a preprocessing and feature extraction technique which can be directly applied to the image of spiderwebs. The DWT is defined in [24] as follows:

$$C[j,k] = \sum_{n \in \mathbb{Z}} f[n]\psi_{j,k}[n] \tag{6}$$

where $\psi_{j,k}$ is the transform function:

$$\psi_{j,k}[n] = 2^{-j/2} \cdot \psi[2^{-j}n - 1] \tag{7}$$

In the wavelet analysis is common evaluate the results as approximations and details. The approximations are the low frequency components of the signal and the details are the high frequency components. For many signals the most important information is content in the low frequencies . This content is what gives identity to the signal. The following is the diagram that has been used in this work, which would be one-dimensional DWT. The filtering process to obtain the approximations and detail of the discrete wavelet transform shown in the following figure 9:

The application of different mother families on pre-processing (artifacts elimination) and on the feature extraction has a set of good and discriminate parameters. Unlike the Fourier transform, the wavelet transform can be implemented on many bases. The different categories of wavelets (continuous, discrete, orthogonal, etc..) and various types of wavelet functions within each category provide a wide number of options for analyzing a signal. This allows selection of the base functions whose shape better approximates the characteristics of the signal to be represented or analyzed. On this work, the families Daubechies1 (db1), Biorthogonal 3.7 (bior3.7) and Discrete Meyer (dmey) were used.

4.2. Discrete Cosine Transform

Discrete Cosine Transform (DCT) was applied for noise and details of high frequency elimination [23]. Besides, this transform has a good energy compaction property that produces uncorrelated coefficients, where the base vectors of the DCT depend only on the order of the transformation selected, and not of the statistical properties of the input data.

Another important aspect of the DCT is its capacity to quantify the coefficients utilizing quantification values, which are chosen of visual way. This transformation has had a great acceptance inside the image digital processing, as there is a high correlation among elements for the data of a conventional image.

5. Classification: Support Vector Machine

Once the images were transformed to a set of features, the classification stage tried to produce an answer to the spider identification problem. In this work, the well known Support Vector Machine (SVM) technique has been used.

The SVM is a method of structural risk minimization (SRM) derived from the statistical learning theory developed by Vapnik and Chervonenkis [17]. It is enclosed in the group of supervised learning methods of pattern recognition, and it is used for classification and regression analysis.

Based on characteristic points called Support Vectors (SVs), the SVM uses an hyperplane or a set of hyperplanes to divide the space in zones enclosing a common class. Labeling these zones the system is able to identify the membership of a testing sample. The interesting aspect of the SVM is that it is able to do so even when the problem is not linearly separable. This is achieved by projecting the problem into a higher dimensional space where the classes are linearly separable. The projection is performed by an operator known as kernel, and this technique is called the kernel trick [18] [19]. The use of hyperplanes to divide the space gives rise to margins as shown in figure 10.

In this work, the Suykens' et. al. LS-SVM [20] was used along with the Radial Basis Function kernel (RBF-kernel). The regularization parameter and the bandwidth of the RBF function were automatically optimized by the validation results obtained from 10 iterations of a Hold-Out cross-validation process. Two samples from each class (from the training set) were used for testing and the remaining for training as we saw that the number of training samples has a big impact in the LS-SVM optimal parameters. Once the optimal parameters were found, they were used to retrain the LS-SVM using all available training samples.

6. Experiments and results

To sum up, the proposed system normalized all the images to 10x10 pixels. This system used the first M features obtained from the DCT projection of the spider webs images and the outcome of the DWT transformation of the spider webs images as inputs for a RBF-kernel LS-SVM with regularization and kernel parameters. The former parameter (the number of features) was varied during experimentation, while the later two parameters (the

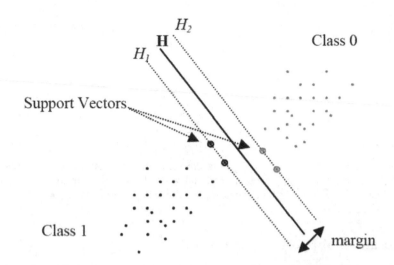

Figure 10. Example of a separate hyperplane and its Support Vectors and margin for a linear problem.

regularization and the kernel parameters) were automatically optimized by iteration using validation results. To obtain more reliable results, the available samples were divided into training and test sets, so that the system was trained and tested with totally different samples.

The well known K-Folds cross-validation and Hold-Out cross-validation techniques were used to obtain the final results. In particular, experiments with K equal 3, 5, 7, and 10 were run. The percent of training samples in Hold-Out cross-validation was 50, 40, 30, 20, 10 respectively. It is worth it to mention that the training and testing sets were computed for each class individually, having into account that each class has different number of samples. These experiments were performed for both datasets, i.e. using the whole spiderwebs and only the center area.

6.1. DCT results

In order to obtain the optimal number of coefficients, 30 experiments were performed using the Hold-Out cross validation technique. As the size of image was normalized to 10x10 pixels the total number of characteristics corresponded to 100, therefore, in this phase, the number of coefficients was swept from 1 to 100 coefficients. Figure 11 represents the mean of those 30 experiments. It can be observed that 60 is the optimal number of coefficients.

Table 2 shows the results reached for K-Fold cross-validation and Hold-Out cross-validation using the optimal number of coefficients.

6.2. DWT results

In this case, the length of the feature vector depends on the wavelet family. Windows *db1* and *bior3.7* return an image with half the size of the original image while *dmey* returns an image with the same dimension. In all cases, the reconstructed image was used for classification, ignoring the horizontal, vertical and perpendicular components.

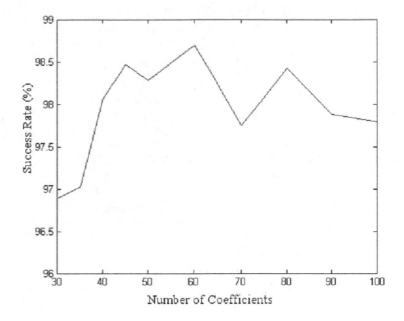

Figure 11. Evolution of success rate when number of coefficients is incremented for the system DCT-based.

K-Fold cross-validation (K)	Success rate (%)	Hold-Out cross-validation (% of training samples)	Success rate (%)
10	98.75% ± 0.18	50	98.69% ± 1.20
7	98.56% ± 0.58	40	98.45% ± 1.38
5	98.44% ± 0.76	30	96.89% ± 2.06
3	98.07% ± 0.52	20	94.29% ± 4.07
-	-	10	79.75% ± 6.54

Table 2. Results obtained for K-Fold cross-validation and Hold-Out cross-validation using DCT

The Table 3, 4, 5 shows the results reached for K-Fold cross-validation and Hold-Out cross-validation for each DWT family.

7. Discussion and conclusions

This work has faced the problem of spider web recognition improving the results obtained by the previous work [11]. It is important to note that, to the authors extend, these are the only published works using the proposed technique.

Images were preprocessed to isolate the center of the spider web and remove the effects of the background in the system. The resulting images were then transformed by using DCT and DWT. For the former, the optimal number of DCT coefficients M was found by

K-Fold cross-validation (K)	Success rate (%)	Hold-Out cross-validation (% of training samples)	Success rate (%)
10	99.40% ± 0.53	50	98.42% ± 1.74
7	99.31% ± 0.27	40	96.89% ± 1.93
5	99.18% ± 0.31	30	96.26% ± 2.40
3	98.47% ± 0.12	20	94.38% ± 2.89
-	-	10	91.08% ± 4.28

Table 3. Results obtained for different Ks of a K-Fold cross-validation and procedure using wavelet db1

K-Fold cross-validation (K)	Success rate (%)	Hold-out cross-validation (% of training samples)	Success rate (%)
10	99.40% ± 0.53	50	98.42% ± 1.74
7	99.31% ± 0.27	40	96.89% ± 1.93
5	99.18% ± 0.31	30	96.26% ± 2.40
3	98.47% ± 0.12	20	94.38% ± 2.89
-	-	10	91.08% ± 4.28

Table 4. Results obtained for different Ks of a K-Fold cross-validation procedure using wavelet bior3.7

K-Fold cross-validation (K)	Success rate (%)	Hold-Out cross-validation (% of training samples)	Success rate (%)
10	98.62% ± 0.36	50	97.70% ± 2.10
7	98.30% ± 0.50	40	97.12% ± 1.54
5	98.55% ± 0.25	30	95.66% ± 2.29
3	97.89% ± 0.35	20	94.77% ± 2.28
-	-	10	89.14% ± 6.28

Table 5. Results obtained for different Ks of a K-Fold cross-validation procedure using wavelet dmey

euristics, while families *db1*, *bior3.7* and *dmey* were tested for the DWT. Finally, the resulting characteristics were classified using an LS-SVM. In this case, regularization and kernel parameters were automatically optimized by the system dividing the training samples in training and validation sets and retraining the system with the optimal configuration using all available training data.

The results confirmed the improvement compared to [11], where only three species (versus the four species used in thius work) were classified with a maximum success rate of 95%. Thus, tables 2, 3, 4 and 5 show that the new system reached performance of around 99% on K-Fold cross-validation and 98% on Hold-Out cross- validation. Moreover, the obtained standard deviation was significantly low, although, as expected, slightly higher on Hold-Out as the number of samples for training lowered. All in all, the standard deviation achieved in both K-Fold and Hold-Out procedures are smaller than those obtained on [11].

When comparing DCT and DWT, the DWT provided a better behavior for this problem. It is worth it to emphasize that the images have been normalized to a size of 10x10, this is, quite compressed, considering the spatial distribution of the threads in the spider webs.

The results achieved by this work support the conclusions derived from [11] stating that the center of the spiderwebs provide enough discriminative information to recognize different species of spiders. However, it is still necessary to run more experiments with a larger database and execute a more detailed study on which parts of the spiderweb provide the most discriminative information before make stronger conclusions. On the other hand, this will allow to test the system's performance with larger training sets, which will be interesting having into account that the system clearly improved when the number of training samples increased.

Acknowledgements

This work has been supported by Spanish Government, in particular by "Agencia Española de Cooperación Internacional para el Desarrollo" under funds from D/027406/09 for 2010, and D/033858/10 for 2011

Author details

Ticay-Rivas Jaime R., del Pozo-Baños Marcos, Gutiérrez-Ramos Miguel A., Travieso Carlos M. and Jesús B. Alonso
Signals and Communications Department, Institute for Technological Development and Innovation in Communications, University of Las Palmas de Gran Canaria, Campus University of Tafira, 35017, Las Palmas de Gran Canaria, Las Palmas, Spain

Eberhard William G.
Smithsonian Tropical Research Institute and Escuela de Biologia Universidad de Costa Rica, Ciudad Universitaria, Costa Rica

8. References

[1] Sytnik, K.M., Preservation of biological diversity: Top-priority tasks of society and state (2010) Ukrainian Journal of Physical Optics, 11 (SUPPL. 1), pp. S2-S10.
[2] Carvalho, J. C., Cardoso, P., Crespo, L.C., Henriques, S., Carvalho,R., Gomes, P., "Biogeographic patterns of spiders in coastal dunes along a gradient of mediterraneity." Biodiversity and conservation (2011):1-22.
[3] Johnston, J. M. 2000. The contribution of microarthropods to aboveground food webs: A review and model of belowground transfer in a coniferous forest. American Midland Naturalist 143: 226-238
[4] Peterson, A. T., D. R. Osborne, and D. H. Taylor. 1989. Tree trunk arthropod faunas as food resources for birds. Ohio Journal of Science 89(1): 23-25.
[5] Cardoso P, Arnedo MA, Triantis KA, Borges PAV (2010) Drivers of diversity in Macaronesian spiders and the role of species extinctions. J Biogeogr 37:1034"1046
[6] Finch O-D, Blick T, Schuldt A (2008) Macroecological patterns of spider species richness across Europe.Biodivers Conserv 17:2849"2868

[7] Eberhard,W.G., Behavioral Characters for the Higher Classification of Orb-Weaving Spiders, Evolution, Vol. 36, No. 5 (Sep., 1982), pp. 1067-1095, Society for the Study of Evolution

[8] Eberhard,W.G., Early Stages of Orb Construction by Philoponella Vicina, Leucauge Mariana, and Nephila Clavipes (Araneae, Uloboridae and Tetragnathidae), and Their Phylogenetic Implications, Journal of Arachnology, Vol. 18, No. 2 (Summer, 1990), pp. 205-234, American Arachnological Society

[9] Eberhard,W.G., Computer Simulation of Orb-Web Construction , J American Zoologist , pp. 229-238, February 1, 1969

[10] Suresh, P. B., Zschokke, S., A computerised method to observe spider web building behaviour in a semi-natural light environment. 19th European colloquium of arachnology, Aarhus, Denmark, 2000.

[11] Ticay-Rivas, Jaime R.; del Pozo-Baños, Marcos; Eberhard, William G.; Alonso, Jesús B.; Travieso, Carlos; Spider Recognition by Biometric Web Analysis. IWINAC 2011, Part II, LNCS 6687, pp. 409-417, 2011.

[12] Jing Hu; Si, J.; Olson, B.P.; Jiping He; , "Feature detection in motor cortical spikes by principal component analysis," Neural Systems and Rehabilitation Engineering, IEEE Transactions on , vol.13, no.3, pp.256-262, Sept. 2005.

[13] Qingfu Zhang; Yiu Wing Leung; , "A class of learning algorithms for principal component analysis and minor component analysis," Neural Networks, IEEE Transactions on , vol.11, no.1, pp.200-204, Jan 2000.

[14] Langley, P.; Bowers, E.J.; Murray, A.; , "Principal Component Analysis as a Tool for Analyzing Beat-to-Beat Changes in ECG Features: Application to ECG-Derived Respiration," Biomedical Engineering, IEEE Transactions on , vol.57, no.4, pp.821-829, April 2010.

[15] Haibo Yao; Lei Tian; , "A genetic-algorithm-based selective principal component analysis (GA-SPCA) method for high-dimensional data feature extraction," Geoscience and Remote Sensing, IEEE Transactions on , vol.41, no.6, pp. 1469- 1478, June 2003.

[16] Nan Liu; Han Wang; , "Feature Extraction with Genetic Algorithms Based Nonlinear Principal Component Analysis for Face Recognition," Pattern Recognition, 2006. ICPR 2006. 18th International Conference on , vol.3, no., pp.461-464, 0-0 0.

[17] V. Vapnik, "The Nature of Statistical learning Theory." Springer Verlag, New York, 1995.

[18] Vojislav Kevman. "Learning and Soft Computing: Support Vector Machines, Neural Networks, and Fuzzy Logic models", Puiblished by The MIT Press, 2001.

[19] B. Schölkopf y A.J. Smola. "Learning with Kernels. Support Vector Machines, Regularization, Optimization, and Beyond", Published by The MIT Press, 2002 .

[20] J.A.K. Suykens, T. Van Gestel, J. De Brabanter, B. De Moor, J. Vandewalle, "Least Squares Support Vector Machines", World Scientific, Singapore, 2002 (ISBN 981-238-151-1)

[21] Lim, Jae S., Two-Dimensional Signal and Image Processing, Englewood Cliffs, NJ, Prentice Hall, 1990, p. 548

[22] http://www.unesco.org/new/en/natural-sciences/special-themes/biodiversity-initiativ Las visit in March 2011.

[23] Ahmed, N. Natarajan, T. Rao, K.R "Discrete Cosine Transform", IEEE transactions on Computes, pp. 90-93;1974

[24] Mallat, S., "A theory for multiresolution signal decomposition: the wavelet representation", IEEE Pattern Analysis and Machine Intelligence, Vol. 11, no. 7, pp. 674-693. (1989).

[25] Otsu , N.; "A threshold selection method from gray-level histograms". IEEE Trans. Sys., Man., Cyber. 9 (1): 62-66; 1979

Chapter 7

Integrated Measurements for Biodiversity Conservation in Lower Prut Basin

Florin Vartolomei

Additional information is available at the end of the chapter

1. Introduction

To establish protected areas together with an efficient management is a necessity as: protected areas are representative of natural and semi-natural ecological ecosystems that can be valued and monitored to a certain degree in relation to their state as well. Such ecosystems are the main components of the natural capital providing resources and services for the socioeconomic development; protected areas are zones where it is possible to develop the knowledge necessary for passing from the transition period to a sustainable development model; protected areas are *"out-door lessons"* of education on nature's role and the necessity of nature's conservation and sustainable development. Differences of terminology can be eliminated simply, by using IUCN system of classification whose main aim is to manage the protected area. In the system there are 6 categories of protected areas, which also involve a varying degree of human intervention – from nonexistent (category I-a and I-b) to a higher degree (category V). All categories are the same importance and relevance for biodiversity conservation. According to the Urgency Ordinance No. 236/2000 on the system of protected natural areas, conservation of natural habitats and wild flora and fauna, in our country the accepted *categories of natural protected areas* defined depending on the assigned management goals are the following: *scientific reserve, natural reserve, national park, natural park, natural's monument, biosphere reserve, wet area importance (RAMSAR site), site of the world natural inheritance.*

2. About study area

The Prut River is the last largest tributary of the Danube, before it discharges into the Black Sea (through the Danube Delta). The Prut river springs in Ukraine, flows through the border area between Romania and Republic of Moldova, down to the point where it enters the Danube (Figure 1).

The Prut river is an allochthon river that originates in the Woody Carpathians in Ukraine. Between Orofteana and the confluence with the Danube on the lengh of 946 km, it drains a basin of 28,463 sqkm. At Czernowitz (in Ukraine) the Prut river has a multi-annual average flow of 73.62 mc/s, it grows up to Fălciu at 103.48 mc/s, and then it comes down to 85.3 mc/s, because of the lateral losses [1].

3. Review of current river management practices and assessment of anthropogenic activities

From hydrological point of view the following data have been gathered for the Prut River Basin, part on the Romanian territory (figure 1). Main attitude at the entrance in Romania is about 140 m (above Black Sea. level). Main altitude at the discharge in Danube about 15m (Black Sea). Average multiannual runoff is 88 m³/s, maximum multiannual runoff (1952-2010) is 4240 m³/s in 2008 at Radauti-Prut hydrometric station. Minimum multiannual runoff (1952-2010) is 7.6 m³/s. On the Romanian territory (at the entrance in the country) for 1952-2010 period the data referring to the flow characteristics are included in the table 1 below:

Hydrometric station	Rădăuți-Prut	Ştefăneşti	Ungheni	Dorohoi	Todireni	Bădeni-Hârlău	Cărpinați-Victoria
river	Prut	Başeu	Prut	Jijia	Sitna	Bahlui	Jijia
m³/s	78,28	1,94	85,97	0,66	2,03	0,44	6,65
Hydrometric station	Todireni	Podu Iloaie	Podu Iloaie	Iaşi	Iaşi	Murgeni	Fârțăneşti
river	Jijia	Bahlui	Bahluieț	Bahlui	Nicolina	Elan	Covurlui
m³/s	2,21	1,18	1,08	3,01	0,43	0,49	0,56

Table 1. The flow characteristics during 1952-2010 in the Romanian section of the basin

As the lower Prut area is situated in the Eastern part of the country, the climate is influenced by the vicinity with the Carpathian Mountains in the West. Moldavian plain in the East and the river Danube meadow the South. This creates some small changes in comparison with the average level at the normal level for temperature, precipitation and winds. The average precipitation is between 690 mm (1989) - 515 mm (1993)-632,27 mm (1996) with an average net flow of 174 mm (1988) -130 mm (1992) and 159,45 mm (1996).

The main human activities for which water use demands normally identified are: hydrotechnical works and flood protection schemes, source for water supply, for urban and rural use, electrical energy, agriculture, fisheries, industry and others [2].

4. General data

Prut basin, through the variety of geomorphic elements, of flora and fauna, is permanently subjected to the researches of various experts from Romania, Republic of Moldova or other countries.

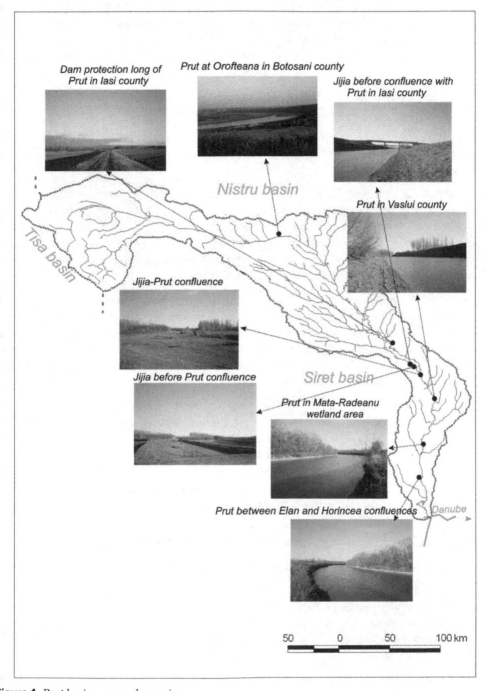

Figure 1. Prut basin – general overview

With regard to forests diversity, different kinds of wood trees can be found which are mostly specific for *silvostepa,* with alternation of plains, hills and forests. The structure of species composition is the following: *resinous trees* 30% (spruce fir, fir, pine tree and others) and *deciduous trees*-70% (fagus, representatives of Querqus family., different soft and hardwoods).

Owing to the Law no 18/1991 (also called *Land Restitution Law*) within the Galati county a surface of 5,283 ha was given back to the former land forest owners.

The percent of forestation within the county is very low about 8% of the county total surface, being necessary to increase the forest surface through the forestry plans works or through ecological reconstruction [3].

Along the lower Prut river there is a forested corridor very well preserved and managed with typical tree species, playing a multifunctional role. Forest functions have actually diversified over time and as a result forest management practices are increasingly having to meet multipurpose requirements including such environmental functions as sustainability nature conservation and water buffer zone for flood management, as well as community (recreation, culture and aesthetics) and production functions.

With regard to the fauna, the most specific species are determined by the *steppe* and *silvostepa* zones. The most comune are rabbits, around squirrel, hamster and many others. From the bird's family can be mentioned bustard, quail, partridge and large types of flying birds. As representatives from other faunistic families can be mentioned fox, wolf, wild boar, as well as dear, squirrels, etc. There is a number of species (flora, fauna and forests) and locations which have been protected by law starting from 1930, which are organized as protected areas and monuments of nature.

The economic, activities in the region during the last several decades deteriorated the environmental balance in the basin area. Urbanization, industrial development, intensive agricultural practices, high density of railway and road networks, new reservoirs and other hydrotechnical works in the flood plain of the river, brought about a considerable loss of floodplain habitat and wetlands. Prut River management and rehabilitation of the cross-border wetlands can only be approached internationally.

5. Key species of terrestrial and aquatic area of the Prut river

The key species are represented by plants, butterflies, fish and birds [4].

Plants: the *Tamarix ramoissima* occurs on virgin soil habitats along the lower Prut. This area also offers narrow riparian pioneer arcas that are favorable to white willow (*Salix alba*) succession. The water chestnut (*Trapa natans*) occurs on the lower Prut, due to water eutrophication it has dramatically decreased and counts among the Red List species of this area.

Butterflies: the distribution map of the *Large Copper* shows many occurrences along the central Prut valley, a continuous occurrence along the whole river is very probable. The occurrence of the *Danubian Purple Emperor* is very probable for the lower Prut, because it has been found in the both adjacent lower Siret floodplain and the Danube Delta.

Fish: European *mud-minnow* occurs on the lower Prut where the floodplains provide sufficient adequate habitats. *Blue bream* occurs regularly in the river downstream of the dam Stanca Costesti (a well-operating communication between the river and floodplain waters is guaranteed in this section). *Crucian carp* and *mud load* occur in the lower river, where adequate aquatic habitats for the Crucian carp have been preserved.

Birds: about the *White-tailed eagle* there is no breeding occurrence on the Prut.

6. General frame and measurements for biodiversity protection

The Lower Meadow of Inferior Prut River Natural Park includes all the flood meadow of Prut river on the administrative territory of Galați county. The Lower Meadow of Inferior Prut River Natural Park has the endorsement of CMN 19/Cj/18.02.2003.

The planning maps of UP V Prut Meadow were drawn up by SILVAPROIECT, in 1995. The area of the Natural Park fits the type of habitat formed of: Natural eutrophic lakes with a *Magnopotamion* or *Hydrocharition*-type vegetation (Natura 2000 code of habitats: 3150) and lakes or dirty grey to blue – green water ponds, more or less turbid, especially rich in basic substances (pH usually higher than 7), many freely floating *Hydrocharition* communities being present at the surface or, in case of the deep systems and open water surfaces, the *Hydrocharition* communities are associated to the submersed vegetation formed of large cormophytes (*Magnopotamion*).

For all the types of existing habitats housing a large variety of fauna (especially avifauna), sedentary as well as migrating or passing fauna, the Mața – Rădeanu humid area, with a surface of 386 ha, is similar to the special preservation areas from the Danube Delta. Other areas on Prut river may be considered similar the this one (Pochina lake, the area where dams are being built, between Vlădești and Giurgiulești customs point, Prut Isle, Brateș lake) [5].

Romania is part of the the Natura 2000 European Network (SPAs – Special Protection Areas and SCIs - Sites of Community Importance) aiming to protect wildlife and its habitats,2 whose surface is not definitively established.

Also, the national authority responsible for the protected areas in Romania is hardly starting the process (January, 2010) of handing out the Natura 2000 sites to different legal entities (NGOs, economic agencies, research institutes, local authorities etc.) in order to manage them.

GIS techniques and GPS means were used for the inventory and Land Register records of these types of surfaces and also for the integration in digital formats of the protected area limits at a European level [6, 7].

7. Inventory of wetlands and floodplain habitats

In 1998 "Romanian Waters" National Company the main manager of the water resources from Romania has started the preparation at the request of the Ministry of Water Forest and Environment Protection the inventory of the wetland and floodplains at national level including the potential for restoration according with the particular case from Romania where the process of land restitution to the previous owners is in the second step of application.

In order to determine the wetland conservation potential in the Danube River Basin, an evaluation study of wetlands and floodplains areas was done by an international consortium under the UNDP/GEF Assistance.

Also, at the national level an inventory of the wetlands and floodplains was done including all existing natural wetlands or wetlands for which the initial situation was changed.

In both reports the Prut catchment area was presented with a large number of existing wetlands and also with a large restoration potential.

Out of about 200 wetlands recorded for whole Prut basin (many of them are less than 1 sqkm surface) a number of 19 wetlands were selected and discussed in the inception phase. This is included in the table 2. Some of these wetlands are still under the natural conditions (10) and the rest were modified to be used by agriculture [8].

It has to be mentioned that several wetlands which in present are in natural stage are included or will be included in the List of Protected Areas under the legislation preservation. In this regard the planning of wetlands and floodplains rehabilitation is underdevelopment and will depend by the finalization of the land restitution action. In the lower Prut basin within the Vaslui and Galati counties the following protected areas are to be mentioned (Table 3) [9].

Among the protected areas within Galati county, according to the criteria of habitat identification, three of them (**Prut Ostrov, Lower Prut river meadow and Vlascuta swamp**) have been indicated to include some wetlands as well (Figure 2) [10].

To this point bellow is presented a more detailed situation and characterization of them.

a. For the **Prut Ostrov** which is an eyot and has been included in the 4th category of protected areas by the County Council from 1994. The Prut Ostrov is located in the lower Danube river, near to the Prut river. As types of vegetation, the forest and specific wetland vegetation were identified.

Regarding the main habitats of the protected area, habitats of freshwater/wetlands and forest are present. The surface of main types and surface of habitats is described in table 4.

The fauna is represented by mammals, birds, reptiles, frogs, fishes and representatives of *Nevertebrata Fillum.*

No.	County	Location	River	Surface (sqkm)	Wetlands conditon/usage
1	Iasi	Dranceni	Drinceni-	2.70	agriculture
2	Iasi	Albita-Falciu	Poganesti	24.3	agriculture
3	Vaslui	Albita-Falciu	Stanilesti	36.8	agriculture
4	Vaslui	Albita-Falciu	Banului lake	56.55	agriculture
5	Vaslui	Albita-Falciu	Berezeni	40.20	agriculture
6	Vaslui	Albita-Falciu	Falciu	28.70	agriculture
7	Vaslui	Bata- Rinzesti	Ranzesti	3.00	agriculture
8	Vaslui	Urlati	Elan	0.75	natural
9	Vaslui	Gusitei	Elan	2.30	natural
10	Vaslui	Poste Elan	Elan	0.75	natural
11	Vaslui	Paicani	Elan	1.25	natural
12	Vaslui	Giurcani	Elan	0.30	natural
13	Vaslui	Murgani	Elan	0.50	natural
14	Galati	Galati - Vadoni	Prut	75	natural
15	Galati	Rogojani	Horincea	1.50	natural
16	Galati	Vladesti	Prut	269	agriculture
17	Galati	Bratesul de Sus	Prut	58.01	agriculture
18	Galati	Bratesul de Jos	Prut	97.47	agriculture
19	Galati	Badalani	Prut and Danube	17.86	agriculture
20	Galati	Prut Ostrov (Prut island)	Prut and Danube	56.6	natural
21	Galati	LowerPrut floodplain	Prut	5,480.41	natural
22	Galati	Vlascuta swamp	Prut	41.8	natural

Table 2. The wetlands within the Prut catchment area

Regarding the activities which occur within this protected area fishing, sporadic deforestation, hunting, different types of poaching. The impact of fishing, hunting and poaching activities according to action time the protected area is sporadic.

b. The Lower Prut river meadow. The vegetation of forest, pasture, floodplain and peat swamp, are characteristic for the vegetation type. Regarding the main habitats of the protected area, habitats of freshwater/wetlands/floodplain, herbal associations, lawns and bushes, forest, the Prut river on the Romanian territory are present (Figure 3).

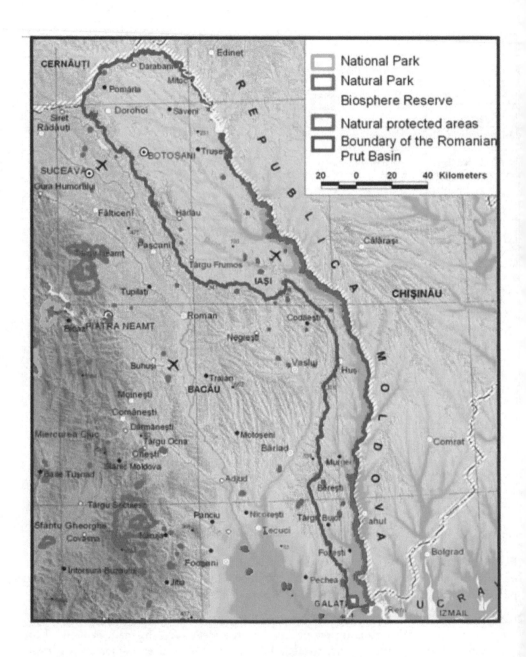

Figure 2. Natural Protected Areas in the Romanian Prut Basin

Vaslui county			
No.	Natural reservation	Location (commune)	Surface (ha)
1	Fossil area Malusteni	Malusteni	1,000
2	Fossile place Nisiparia Hulubat	Vaslui	250
3	Hill of Burcel	Miclesti	1,200
4	Tanacu – Coasta Rupturile	Tanacu	600
5	Badeana forest	Tutova	12,670
6	Harboanca forest	Stefan cel Mare	4,310
7	Balteni forest	Balteni	2,200
8	Hay meadow Glodeni	Glodeni- Negrilesti	600
Galati county			
No.	Natural reservation	Location (commune)	Surface (ha)
1	Sand dunes – Hanu Conachi	Hanul Conachi	199.3
2	Garboavele forest	Galati	220.4
3	Breana – Roscani forest	Baneasa	78.3
4	Fossil place Tirighina/ Barbosi	Galati	1.0
5	Fossil area rates	Tecuci	1.5
6	Fundeanu forest	Draguseni	110.7
7	Talasmani forest	Beresti	20.0
8	Buciumeni forest	Buciumeni, Brahesesti	71.2
9	Prut Ostrov	Ghimia Prut	56.6
10	Potcoava swamp	Branistea	49.0
11	Talabasca swamp	Tudor Vladimirescu	130.0
12	Lacul fusilier Beresti	Beresti	49.0
13	Lower Floodplain Prut	Cavadinesti	5,480.41
14	Pochina swamp/lake	Suceveni	74.8
15	Vlascuta swamp/lake	Mascatani	41.8
16	Pogonesti forest	Suceveni	33.5

Table 3. The protected areas within Lower Prut river countries

Main types	Surface (ha)
Habitats of freshwater/Wetlands	20.5
Forests	35.5
Total	56

Table 4. The surface of main types and surface of habitats in Ostrovul Prut

The surface of main types and surface of habitats are described in the table 5 bellow:

Main types	Surface (ha)
Habitats of freshwater/Wetlands	25.5
Herbal associations/lawns and bushes	31.5
Forests	2,573.43
The Prut river on the Romanian territory	225

Table 5. The surface of main types and surface of habitats in The Lower Prut river meadow

The flora is specific floodplains and swamps being represented by *Salix alba, Populus alba, Rosa canina, Satix fragila, Eqtasetum limosum, Typha angustifolia, Nymphae alba, Sagitaria sagitifolia*, etc. The fauna is represented by species of birds, fish, reptiles, frogs, insects, mammals which are characteristic for wetlands biotopes.

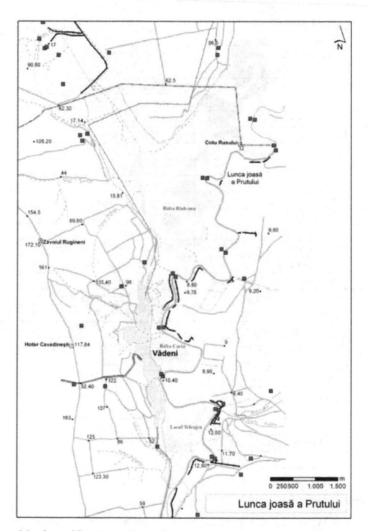

Figure 3. Lower Meadow of Prut River Natural Reserve

The Lower Prut river meadow represents an excellent habitat for more than 230 birds species/nesting, feeding and halt places, many of birds being protected by the international conventions. Fish species are represented by freshwater species, such as *Misgurnus fosilis. Titca tica, Esox lucius, Cyprinu carpio, Silurus glanis*,etc. The mammals are presented in a small number of species *Ondrata yibethica, Vulpes vulpes*.

Regarding the activities which occur within this protected area fishing, deforestation, hunting, different types of poaching. The impact of fishing, deforestation according to their action time on the protected area is periodical hunting and poaching activities have a sporadically time action.

There have been recorded a number of activities with accidental character and which take place outside of the protected area and which have a negative impact on it (table 6 below):

Impact Source	Pollutant	Types of affected environmental factors
Fish ponds	Wastes	Water, soils
Mata/Radeanau, Sovarca	Wastes	Water
Vladesti, Brates	nutrients	Water

Table 6. Types of affected environmental factors in The Lower Prut river meadow area

This protected area is affected by periodical floods especially during the springs. The natural reserve Lower Prut river meadow has been proposed by the Environmental Protection Agency in 1999, in the frame "Green Corridor Of the Lower Danube" which is coordinated by the Research & Design Institute "Danube Delta" and the Direction of Nature Conservation and Biological Diversity from the Romanian Environmental Ministry to be presented to the Romanian Academy and County Council in order to be official declared. In present is protected according to Annexe 1 of Law no 5/2000.

c. **Vlăşcuţa swamp**

The vegetation of this protected area is typically for wetlands (100%).The flora is represented by *Typha angustifolia, Nymphae alba, Sagitaria sagitifolia, etc.* The fauna, is characteristically for shallow swamp being made up of invertebrates, frogs, mollusks, reptiles, fish, birds and mammals communities. Regarding the activities which occur within this protected area fishing, hunting, different types of poaching have a sporadic character. There have been recorded a number of activities with accidental character and which take place outside of the protected area and have a negative impact on it (table 7 below).

Impact Source	Pollutant	Types of affected environmental factors
Crops culture	Fertilizers	Soils, groundwater
Mesteacanis village	Wastes	Water, soils

Table 7. Types of affected environmental factors in Vlăşcuţa swamp area

8. The special avifaunistic protected area

8.1. Stânca-Costeşti lake

The Stânca-Costeşti Dam and hydropower station was built between 1974–1978 years, as a Romanian-Russian common project and is located between Costeşti (Moldova) and Stânca (Romania) (Figure 4). The main goal of building this power station was to protect villages down the Prut River from annual floods.

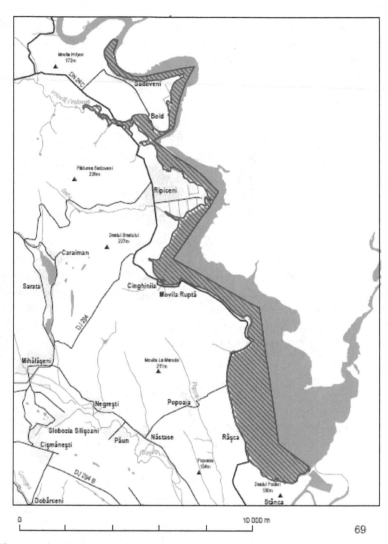

69

Figure 4. The special avifaunistic protected area - Stânca-Costeşti lake

The formation of the Stânca-Costeşti reservoir determined important changes of the ichthyofauna of the area. The main lake has lost the rheophil species (barbell, chub or wheatear), thus developing species with a mixed profile.

In Romanian section, a lot of hydrotechnical projects have been executed (damming, irrigation and draining canals etc.) on the Lower Prut, mainly on the territory of Galati county, downstream of Targu Bujor city. All these hydraulic works have affected the biodiversity of the aquatic and riverbank ecosystems, have restrained the natural habitats, have influenced the flow patterns of the river, have modified the local climate and have influenced the life of the local population.

The development of industrial and agricultural activities in the whole catchment area of Prut have led to increased pollution by nutrients, organic substances, detergents, pesticides, heavy metals etc. All these factors adversely affected the water quality of the Prut river and its tributaries, and of the Lower Danube river and therefore have a direct impact upon the Black Sea.

The formation of the Stânca-Costeşti accumulation lake determined important changes of the ichthyofauna of the area.

The main lake has lost the rheophil species (barbell, chub or wheatear), thus developing species with a mixed profile.

The special avifaunistic protection area of Stânca-Costeşti lake stretches on 2,950 ha and comprises partially the localities of Mitoc, Liveni, Manoleasa, Manoleasa-Prut, Sadoveni, Ripiceni, and Stânca from Botoşani county.

The special avifaunistic protection area of Stânca-Costeşti lake has the endorsement of C.M.N. no. B 939/07.06.2004.

The pressure of economic development from the last 50 years in the area of Prut river, the protection measures against floods by building dams in the major river bed and the building of the hydrotechnical knot Stânca-Costeşti have been the causes of the transformation of the typical habitats in the humid areas at the border of Prut river, thus of the flood area favouring the breeding of fish and birds, endangering the ecological integrity of the area eco-system complex (Figure 5).

Figure 5. Stanca-Costesti lake (general view from Romanian bank)

The reduction in quantity and in quality of the aquatic fauna has been noticed, and this situation cannot be justified by pollution, because Prut river has a good status from a physical and chemical point of view, the concentrations not exceeding the standards set forth by Directive 98/83/EC.

At the moment, Prut river is facing a reduction of the fish quantities, according to the records made in 1947 when there were 37 species of fish compared to the 26 found nowadays [11].

The birds, more than any group of vertebrates, have a large living surface in the area presented in terms of the number of species as well as the number of individuals. The largest part of the birds found in the area are migrating species (44 species) representing international natural resources; among these, we mention the white stork, the bee-eater, the little egret, great crested grebe etc.

The implementation of some efficient measures for the protection of these populations and their habitats is necessary in order to comply with conventions and international agreements for the protection of wild flora and fauna, ratified by Romania. 31 sedentary species and 18 passing species are still living in this area.

From the total number of the species, 22 represent community interest, 27 species whose preservation needs the designation of avifaunistic protection areas and 39 species found under strict protection, according to Directive 79/409/EEC on the preservation of wild birds.

The project implementation for the declaration of the Stânca-Costeşti lake area as special avifaunistic protected area shall aim to preserve and restore the viable populations of birds specific to the humid climates, according to the stipulations of the Convention on the humid areas of international importance, especially as a habitat of aquatic birds, concluded at Ramsar in 1971 and ratified by Romania through law no. 5/1991.

In this context, the preservation of humid areas includes an important cross-border component, in the sense of the collaboration and adoption of a common and unitary strategy regarding the integrated administration and the management of the humid areas, the great economic, natural, scientific and recreational value [12].

Conferring in the future the official status of "cross-border protected site" to this area shall allow the establishment of restrictive measures for the protection and the preservation of biodiversity. Moreover, it shall encourage the preservation of the traditional practices of sustainable revaluation of the natural resources and the preservation of the social and cultural values of the river communities.

The European Union promotes a clear policy of cross-border collaboration for saving and perpetuating the natural and cultural heritage, supporting the interstate collaboration initiatives, in order to preserve the natural eco-systems, whose borders do not depend on the political or administrative ones.

8.2. The Maţa – Rădeanu complex of lakes

The Maţa – Rădeanu complex of lakes and ponds from the Lower Meadow of Prut River reservation (position 2.414 from appendix I of Law 5/2000) is an area that defines the establishment in the southern part of Prut basin of a special protection unit such as a Natural Park, as part of the *Green Corridor* of the Danube.

The complex of lakes and ponds in the north part of the Natural Park called Lower Meadow of the Inferior Prut River is developing through the fittings set on a surface exceeding 640 hectares. The initial usage category: pond (360 ha), meadow (71 ha), swamp (50 ha), unproductive (approx. 160 ha). This aquatic complex is located on the right bank of Prut river, between km 113/landmark 1255 and 121 + 400 m/landmark 1252, in the junction area with Elan river, landmark 1253, on the territory of the commune of Cavadineşti – the village of Vădeni. In a natural status, at the maximum level of free floodwaters, the surface of the ponds exceeded 568 ha. The fishing fittings (568 ha) and the agricultural ones (78 ha) were built in the 1980. The land was divided in two separate groups by Elan river [13].

9. The natural park from the southern sector of Prut basin

Declaring the LOWER MEADOW OF THE INFERIOR PRUT RIVER NATURAL PARK in the southern part of Prut hydrographical basin is the result of the interaction between human activities and nature over time.

This protected area was created as a distinct area with a significant landscape value and with a great biological diversity where, through the maintenance of a harmonious interaction of man with nature and through the protection of the diversity of habitats and landscape, the traditional use of lands and some activities by the local population are encouraged (Figure 6-7).

In addition, the public is offered recreational activities and tourism and may unfold scientific, educational and cultural activities in the area.

9.1. Ecological considerations

The north group (Maţa pond, at the border of Vaslui county) made of two ponds (135 ha representing the low area of the pond and 57 ha – the high area of the pond) and the south group (Rădeanu pond: 342 ha of water + 78 ha of agricultural surface).

Regarding Rădeanu pond, from the south-western part of Elan stream (inferior course), this presents features of integral natural area within which the water alternates with the reed, backwater, swampy surface, where the colonies of birds are present all year round.

The proposed surface as a special avifaunistic protection area is of 194 hectares.

Practically, because of the inadequate exploitation of numerous hydrotechnical works, especially because of the lack of financial resources, at the moment there are 148 ha of fishing ponds still functional (Figure 8).

The area of the Natural Park fits the type of habitat formed of: Natural eutrophic lakes with a *Magnopotamion* or *Hydrocharition*-type vegetation (Natura 2000 code of habitats: 3150) and lakes or dirty grey to blue – green water ponds, more or less turbid, especially rich in basic substances (pH usually higher than 7), many freely floating *Hydrocharition* communities being present at the surface or, in case of the deep systems and open water surfaces, the *Hydrocharition* communities are associated to the submersed vegetation formed of large cormophytes (*Magnopotamion*) (Figure 9).

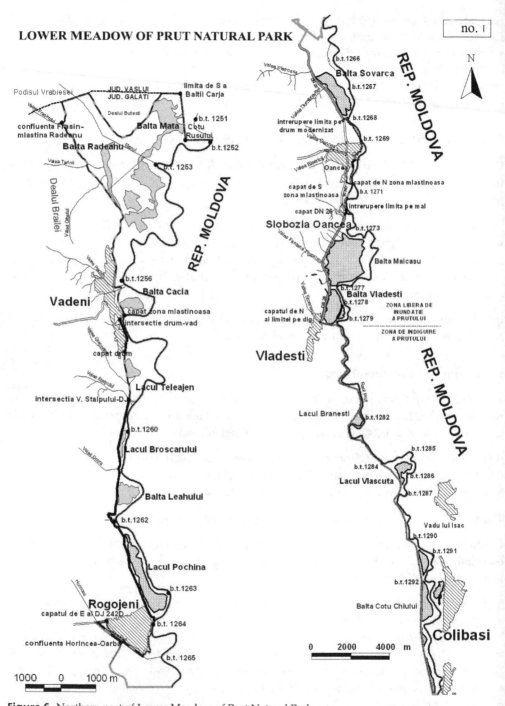

Figure 6. Northern part of Lower Meadow of Prut Natural Park

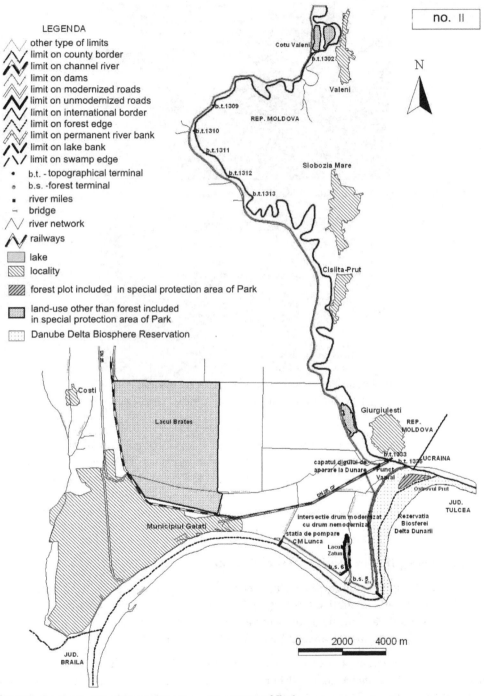

Figure 7. Southern part of Lower Meadow of Prut Natural Park

Figure 8. Biodiversity aspect in Mata-Radeanu complex of lakes

Figure 9. Natural potential for touristic activities in Lower Meadow of the Inferior Prut River Natural Park

The specific plants of the area are the duck weed (*Lemna, Spirodella, Wolffia*), *Hydrocharis morsus – ranae* (frog bit), *Stratiotes aloides* (water soldier), species of bladderwort (*Utricularia australis, Utricularia vulgaris, etc.*), *Aldrovanda vesiculosa* (waterwheel plant), ferns of *Azolla* type, aquatic moss of *Riccia* and *Ricciocarpus* types, different submersed cormophytes like the species of shining pond weed, such as *Potamogeton lucens, Potamogeton praelongus, Potamogeton zizii, Potamogeton perfoliatus,* etc.

9.2. The description of special preservation areas

The Lower Meadow of Inferior Prut River Natural Park includes all the flood meadow of Prut river on the administrative territory of Galați county. The Lower Meadow of Inferior Prut River Natural Park has the endorsement of CMN 19/Cj/18.02.2003. The planning maps of UP V Prut Meadow were drawn up by SILVAPROIECT, in 1995.

The special preservation area of the Lower Meadow of Inferior Prut River Natural Park includes: Brateş lake, including the fishery and the reed area and paludicolous vegetation, etc.; Prut Isle, with u.a. 82 from UP V Prut Meadow of OS Galați, between forest landmarks

166 and 167; the area with dams of Prut river between Giurgiuleşti customs point, (topographical landmark 1333 on Prut river and the forest landmark 23, OS Galaţi) and Vlădeşti (topographical landmark 1297 on Prut river) which includes the forest plots and subplots 11 - 81 A, including lakes, ponds and reed area.

The flood area of Prut river that includes Pochina – Rogojeni lake, including the associated reed area and paludicolous vegetation, and Vădeni area located between the junction of Stâlpului stream with Prut river (Figure 10).

Figure 10. Biodiversity aspect in Pochina lake area

This lake is on the north border of Galaţi county with Vaslui county, between topographical landmarks 1260 and 1252, on Prut river, which includes Teleajen lake and Cacia, Maţa and Rădeanu Ponds and the reed areas, agricultural surfaces, meadows and border forests; the area of Prut river between the river bank and the thalweg of the water flow on a distance of 122.4 km, between the junction of the Danube right next to the forest landmark 21, OS Galaţi, respectively the topographical landmark 1335, on Prut river, and the topographical landmark 1252, from Cotu Rusului, on Prut river [14].

10. Proposal for wetland management and solutions

We propose the following potential demonstration sites for wetland management and restoration projects in the Lower Prut basin:

- Prut flood plain, downstream of Sovarca swamp, up to the mouth point in the Danube;
- Brates Lake located NE from Galaţi city, connected with the Prut.river by the valley of Ghimia brook;
- Horincea hydrographical basin (Figure 11).

Based on the analysis of the premises and conclusions that have emerged there can be extracted a series of proposals to solve the problem. The most important aspects are the technical (technical works for effective exploitation of water resources and the role improvement works combined with fish farming biotechnology) and organizational ones (decision-makers involved and the specific tasks) [15].

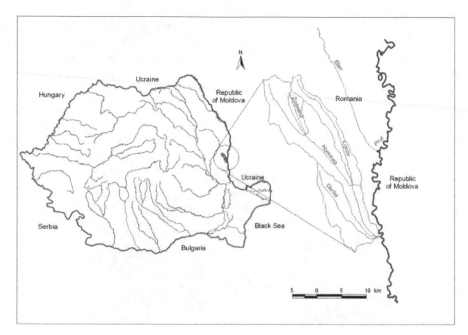

Figure 11. Location of Horincea sub-basin

10.1. Technical aspects

Rivers crossing the plain area as is it is the case of Prut river are not adequate for the partitioning of the river bed of their basins, because of the hydrologic regime with large flow variations. They may however be used as power sources for the system units created as a result of improvement works on the former marshes or for natural marshes as well as for economic and social utilities. Water use in these two cases requires the installation of pumping stations in the Giurgiulesti Oancea-area location, water and wastewater treatment. Their location will be dictated by the population exodus from town to village and the development of small rural industries.

The hydro potential of the sub-basins of this stream opens the door to elaborating an unified scheme in line with landscape features, hydrology and geology of the area. In this respect a complex arrangement of Horincea subbasin, Oancea, Bisericii and Stoenesei valleys which are located in an area of moisture deficit becomes essential. To complete the sub-basin planning, they should be viewed as indivisible natural units.

Developing works on these basins should start from the watershed line and include all works required for combating and preserving soil erosion and the total elimination of the harmful effects of the flood water. The accumulation of water thus created can store the flood waves and can also have a complex use: agro-fishery, water supply for livestock farms, for recreation.

Regardless of the type of use, they must perform the following functions: to not allow the water flooding downstream, to ensure a guaranteed minimum flow during periods of low fluid potential and to ensure efficient use of water resources. Possible locations of accumulation for Horincea sub-basin can be completed in a subsequent step, with accumulations in its lower sector thereby ensuring effective control of the flow of the whole basin. The investment costs will be higher because of the fact that in this area Horincea stream has a riverbed requiring a dam of approximately 6 km. Another future possibility would be that of transferring water from Prut although it would involve higher costs. This option would be justified in case the population in this area will grow up and small industry would develop.

The entire range of hydrotehnical works in the sub-basin Horincea of Oancea, Bisericii and Stoenesei valleys aimed at regulating the water stream in order to to avoid negative effects of flooding must nevertheless respect the principles of ecological planning in order to avoid failures occurring after the completion of this type of works respectively: the disappearance of flooded area which increases the speed of the water drainage because of the fact that the river beds, after the improvement works are performed, they become channels, thus the riparians can only use the water for a short time; increasing speed also leads to a gradual deepening of the river causing a general lowering of groundwater in the area leading to depletion of water from wells and land dryness.

When considering the environmental planning, one should start from the principle that the streams in the Prut basin represent simultaneously ways of circulation, tanks and complex ecological zones which are in strict interaction with the surrounding areas. Based on the data presented so far, respectively the abiotic and biotic components of the climatic, hydrologic regime, soil structure, vegetation, the intensity of erosion processes, profiles, the first steps that are recommended are: cutting the steep banks, which immediately reduces erosion, creating low gradient banks, stabilization of river' s bottom current by adding of rocks and boulders and planting both grass and shrub vegetation on the banks in order to stabilize the soil.

Regarding fisheries ponds built on the old marshes that belonged to the lower basin of the Prut River - they need to be restructured to ensure the optimal application of new technologies that are characteristic to the competitive business environment specific and market economy. The restructuring of fisheries facilities take the following general technical aspects:

- Sensible dimensioning of fishery farms to focus on technological activity on small areas of, easier to control and more effective;
- Use of hydraulic pressurized systems for the water supply of the units;
- Mechanization of the main technological phases for fishing, nutrition, maintenance - using recirculation systems for intensive fish farming.

The restructuring of fisheries facilities open the prospect of achieving some strategic objective of the sector, namely: the application intensive fish farming of the valuable species in demand on the internal and external markets, the application of biotechnology in acclimatised spaces, mechanization and automation of piscicultural technologies.

Regarding the natural marshes that can still be found in the Prut meadow the best solution would be that they preserve their current form and they should become natural reservations. Preserving these areas will lead to the conservation of the biological balance and biodiversity of the area. Moreover, under the present circumstances, opens the perspective towards a new approach: the Prut meadow would enter the international circuit of protection and development.

10.2. Organisational issues

The central authorities that have specific responsibilities in environmental protection are the Ministry of Agriculture and Rural Development, Ministry of Environment and Forests and the two national companies - Romanian Waters National Company and National Company Romsilva).

Ministry of Environment and Forests has major responsibility for environmental protection in Romania, its main tasks being related to water management of river basin planning for the reclaim of new water sources, coordinates the preparation of plans and frameworks for developing the hydrographical basins, approves the water-related works, establishes forecast and information activities in the field of water management and hydrology, etc.

Ministry of Agriculture and Rural Development has specific responsibilities the field of protection of soil, terrestrial and aquatic ecosystems. Also, it elaborates and sets up priority programs for improvement of works and financing, preventing and combating animal diseases, plant protection and phyto-sanitary quarantine, quality control of seeds and seedlings.

This ministry approves land improvement, conservation and environmental protection programs and it elaborates regulations regarding agricultural systems, technologies of plant cultivation and animal husbandry, forest regeneration, harvesting, collection and transport, and soil quality standards in order to maintain and improve it, to remove the negative consequences on aquatic and terrestrial ecosystems to ensure conservation of specific functions, biodiversity and natural habitats, and communicates with the central environmental authorities.

Ministry of Agriculture and Rural Development keeps track of land rendered unfit for agricultural production and provides upon the request of their owners specialized technical assistance for land improvement works.

Romanian Waters National Company manages water resources (surface and groundwater) and prepares and monitors the implementation of programs for meeting the water demands of the population and economy, exploitation of new water sources, rational use and protection against depletion and pollution, complex planning of water in accordance with current and future requirements. It is also the Romanian Waters National Company that correlates the water works with land reclamation works.

National Forest Company Romsilva is required to perform all the works of ecological restoration, regeneration, plantation and maintenance.

10.3. Law aspects

In Legislation of many states definition „Natural Environment" was transfonned through last decade into „Naturally-Anthropogenic" one. This change reflects attempts to find more precise equilibrium between the present-economical development and future generation's survival [16].

River basins became the main „indicators" to attain such equilibrium. And their problems are focal for further Sustainable Development (and for success of concrete modern approaches, such as Spatial Planning, Environmental Management, Technology Foresight, Pollution Prevention, Cleaner Production, Eco-efficiency, Life Cycle Assessment etc.

For Water Management Systems on the cross-border flows this reality brings up a huge knot of multilevel problems. Their step-by-step resolution will be possible when, instead to struggle with consequences, authorities of all levels will do away with causes.

Further absence if causal-investigatory connections between the economical and environmental aspects in this area as well as in activity of authorities will aggravate the situation.

Therefore the „survival" of existent and new enterprises under the new conditions as well as their attraction for necessary investments and international support, will directly depend of systems accounting, appraisal, risks assessment and Audit implementation.

Today in Lower Prut basin there are some examples of the enterprises reporting completely harmonised with the EU regulations. But parallel to the Environmental Inspectors the separately collected data is referred to other control bodies (Sanitary Service, Workmen's Protection, Emergency Planning, Statistics Office, Municipal Structures, Water Management etc.). But at the source of information is absent the interior self-organisation accordingly to the „process approach", foreseen by International Standards of Quality ISO 9000:2000 and Environmental Management ISO 14000.

Other experience such as Ukrainian-Austrian-Romanian former project already demonstrate, that such approach is profitable for the enterprises, regional executive authorities and local self-Governments, whereas it concentrates limited resources for the key (weak) points and sectors. And simultaneously it generates a good opportunity for joint revealing and agreement of win-win solutions [17].

On the other hand the same principles becomes now a basis for safe development of business and investments in the Eastem Europe, today and in the future as well.

11. Irrigations and works for this purpose in lower Prut basin related to wetlands

The total surface included in the irrigation system in the Prut catchment area is about 100,000 ha and the drainage systems is about 130,000 ha.

In the irrigation system about 40,000 mc water are used, which are contributing to the groundwater sources increase. The dykes created near the river have reduced the floodplain wetlands, has disconnected the land from the river and creates problem with salinity and stepisation.

A part of the reservoirs mentioned above are used for water supply for irrigations. Other sources are the rivers and groundwater from the region and for distribution a large network of channels and ditches were constructed in the years 70s and 80s.

This has influenced the water flow regime as well as the supply of wetlands and floodplains of the river.

As a result of the drainage and irrigation systems created 2/3 of the Lake Brates (located in soth part of the basin) has disappeared its well as a lot of swamps, meadows, reed beds, which represented habitats for a high biodiversity, especially because it was closed located to another big wetlands Danube Delta and the Danub-Macin complex [18].

This was mainly happening inside of the protected area, in the agricultural land.

In the management plan for wetlands as well as in the proposal for future actions these channels are proposed for creations of micro-area for nature conservation. Fortunately a floodplain area was saved having about 5,000 ha in the lower part of Prut and located between embankments and river which in some places has few hundred meters width.

This is characterized by a flood plain with good water dynamics, flood plain gallery forest with different species, floodplain small lakes with macrophytes, meadows and reed beds which still preserves a number of birds and animal species, but which is characterized by a mainly economically-oriented management and less in the favor of biodiversity.

A recent study produced by WWF Danube - Carpathian Programme within the Danube Pollution Reduction Programme has indicated that about 51,000 ha. are possible to be restored in the locations of the previous wetlands and flood plains from the lower Prut area.

This will allow the restoration of floodplain by reconnecting areas to the water river dynamics having as effect improvement of water quality, restoration of wetland vegetation and biodiversity conditions enhancement.

12. Effects of anthropogenic activities on wetlands and floodplains

The human pressure has been led to the changes of structure and functionality of the wetlands floodplains and of the Prut river, characterized through the apparition of a strongly anthropizated environment [19].

The key issues of anthropogenic activities which have had effects on wetlands and floodplains of the river Prut are represented in table 8.

Regarding the industrial and diffuse pollution which have an impact on Prut ecology and water quality, this being closely linked with the ecology of the floodplain forested corridor and the existing wetlands, the situation is much severe in the upper part of the basin, based on a large concentration of industry, agriculture and human settlements in comparison with the lower Prut where the impact is much smaller.

Key issues	Impact
Hydrotechical works and reservoirs 8 non-permanent or semi-permanent reservoirs which keep 40 million m² 329 km of embankments along of 480 km banks on the Prut river and some tributaries to protect of near 100.000 ha of agricultural land, 25 localities and 120 industrial units.	Reduction of wetlands and floodplains. Reduction of efficiency of nutrients sink. Reduction of the water running area of the river which increases the flow velocity and affects the flood forest and embankments. Reduction of water quality self-dynamics. Reduction of suspended solids transport efficiency. Increasing of erosion un downstream areas. Decreasing of groundwater recharge capacity. Water shortages. Disappearance of diverse habitats which offered life conditions for a high vegetal and animal. Birds biodiversity. Decrease of economical (fisheries, wood, reeds, hunting) and touristic potential. Disappearance of micro – climate effects.
Agriculture and rural land use	Increase of nutrients concentration in river water. Increase of toxic substance loads. Salinisation. Stepisation. Disappearance of swamps, meadows, reed beds.
Fisheries	Disappearance of spawning conditions. Reduction of fish quantity. Reduction of economically high quality species.

Table 8. The key issues of anthropogenic activities which have had effects on wetlands and floodplains

13. Conclusions and recommendations about biodiversity protection and water resources management in lower Prut basin

In present we recommend that surveys are carried out to evaluate the current state of biological and landscape diversity of the lakes, especially those with ornithological importance and so that a long-term integrated monitoring system is devised for the lakes and wetlands in the Lower Danube Region.

Also it is recommend that scientific evidence is compiled for assigning nature conservation status to Lakes Kagul, Kartal, and Kugurlui and prepare proposals for designating Lake Kagul as a wetland of international significance, so that the point and non-point sources of pollution in the lake basins are identified and assessed.

At least we recommend that dynamic modeling is used as an aid for lake management taking full account of any inherent limitations in such models.

The spirit of Environment protection Law and Water Law seen as interior documents for each enterprise aimed to order their own knowledge about the flows of waste (losses, sewage, discharges and package materials). The general outside task of this study becomes the universal primary source of information for further accounts to Environmental and Water Management Structures, Emergency Planning, Labour Safety, Sanitary and Municipal Services, Statistics Office etc.

In general, the implementation of this study is the first step, which brings together the interests of the enterprises, Local, Regional and National authorities for the Natural Resource Conservation, safe Water Management and Waste Minimisation and Competitiveness of Productions and Services as core elements of Sustainable Spatial Development in Lower Prut basin.

Author details

Florin Vartolomei
Faculty of Geography, SPIRU HARET University, Bucharest, Romania

14. References

[1] Bălteanu D., Badea L., Buza M., Niculescu Gh., Popescu C., Dumitraşcu Monica (eds.), Romania. Space, Society, Environment, Romanian Academy Publishing House, Bucharest, ISBN 973-27-1275-9, 2006.

[2] Pop Iuliana, Madalina-Teodora Andrei, Charlotte-Valentine Ene, Radita Alexe, Florin Vartolomei, Petronela-Sonia Nedea, The strategy of European Union regarding climate change in *International Journal Of Energy And Environment,* Issue 4, Volume 5, 2011, ISSN 1109-9577, pp. 558-565.

[3] Pop Iuliana, Madalina-Teodora Andrei, Charlotte Valentine Ene, Florin Vartolomei, Petronela-Sonia Nedea, Radita Alexe, E.U. environmental policies: a document-based

qualitative research, in *Recent researches in Energy & Environment*, Cambridge, U.K., February, 2011, ISBN 978-960-474-274-5, ISSN 1792-8230, pp 372-377.

[4] Florin Vartolomei, Măsuri superioare de protecţie a biodiversităţii naturale în bazinul hidrografic Prut, in *Revista Mediul Înconjurător*, no. 1 din 2006, pp. 59-65, ISSN 1453-3944, Bucureşti, 2006.

[5] Florin Vartolomei, Parcurile Naturale şi Naţionale ale României, in *Geograful*, no. 2, pp. 25-35, www.apgr.eu, ISSN 2067 - 4090, e-ISSN 2068-9977, Bucureşti, 2010.

[6] Florin Vartolomei, Stabilirea limitelor ariilor protejate prin utilizarea tehnicilor G.P.S. şi G.I.S. Studiu de caz: Cheile Lucavei-jud. Suceava, in *Volumul de Rezumate al Sesiunii anuale de comunicări ştiinţifice a Institutului de Speologie Emil Racoviţă din cadrul Academiei Române*, intitulată "85 de ani de la înfiinţarea Institutului de Speologie Emil Racoviţă", pp. 24-25, ISBN 973-0-04349-3, Bucureşti, 2006.

[7] Florin Vartolomei, Despre utilizarea tehnicilor GPS la stabilirea limitelor ariilor protejate din România, in *Revista Mediul Înconjurător*, Nr. 1 din 2003, pp. 40-44, ISSN 1453-3944, Bucureşti, 2003.

[8] Florin Vartolomei, Rezervaţiile naturale din bazinul hidrografic al Prutului, in *Annals of SPIRU HARET University Geography Series*, No. 5, pp. 135-140, ISSN 1453-8792, Bucureşti, 2002.

[9] Florin Vartolomei, Corneliu Râclea, Arii naturale protejate şi monumente ale naturii din judeţul Vaslui, ISBN (10) 973-87523-5-3; ISBN (13) 978-973-87523-5-1, Edit. Mondoro, Bucureşti, 2006.

[10] Vartolomei Florin, Madalina-Teodora Andrei, Iuliana Pop, Petronela-Sonia Nedea, Radita Alexe, Biodiversity protection actions in lower Prut basin, in *International Journal Of Energy And Environment*, Issue 5, Volume 5, 2011, ISSN 1109-9577, pp. 678-685.

[11] Florin Vartolomei, Aspecte asupra calităţii apei în acumularea Stânca-Costeşti (bazinul hidrografic Prut), in *Annals of SPIRU HARET University Geography Series*, No. 6, pp. 59-64, ISSN 1453-8792, Bucureşti, 2003.

[12] Florin Vartolomei, Environment integration of ecological and biological influences in the lakes from lower section of Prut river, in *Romanian Journal of Limnology "Lakes, reservoirs and ponds"*, Editura Transversal, nr. 1-2, pp. 157-165, ISSN 1844-6477, Târgovişte, 2008.

[13] Florin Vartolomei, Tehnici G.I.S. utilizate la valorificarea potenţialului turistic natural în bazinul hidrografic Prut, in *Revista Simpozionului cu participare internaţională de la Universitatea Româno-Americană „Turismul durabil românesc în contextul integrării economice europene"*, ediţie pe CD, ISBN (10) 973-8994-67-5, ISBN (13) 978-973-8994-67-6, pp. 225-237, Bucureşti, 2006.

[14] Mădălina Andrei, Florin Vartolomei, Landscape's evolution. GIS application approach, in *Applied Geography in Theory and Practice*, Book of Abstracts, International Scientific Conference, 5-6 Noiembrie 2010, Universitatea din Zagreb, Facultatea de Ştiinţe – Departamentul de Geografie, p. 27, ISBN 978-953-6076-22-2, Zagreb, 2010. http://atlas.geog.pmf.hr/~nbuzjak/konf/Zagreb2010_Abstracts.pdf

[15] Vartolomei Florin, Radita Alexe, Madalina-Teodora Andrei, Petronela-Sonia Nedea, Iuliana Pop, Development of a pilot wetland area to protect biodiversity in the southern

basin of Prut river, in *Recent researches in energy, environment, devices, systems, comunications and computers*, Venice, Italy, March, 2011, ISBN 978-960-474-284-4, ISSN 1792-863X, pp 112-117.

[16] Florin Vartolomei, Mădălina-Teodora Andrei, Legislația pentru protecția factorilor de mediu din România, cu privire specială asupra biodiversității, in *Volumul Simpozionului cu participare internațională al Universității Româno-Americane*, 7-8 noiembrie 2008, București, intitulat "Performanță, riscuri și tendințe în turismul mondial", Editura Pro-Universitaria, pp. 159-169, ISBN 978-973-129-326-4, București, 2008.

[17] Florin Vartolomei, Natural protected areas from Vaslui county, in *Annals of SPIRU HARET University Geography Series*, No. 10, pp. 163-168, ISSN 1453-8792, București, 2007.

[18] Bălteanu D., Dumitrașcu Monica, Ciupitu D., Maxim I., România, Ariile naturale protejate, Map 1:750 000 scale , Editura CD Press, București, 2009.

[19] Surd Vasile, Veronica Constantin, Camelia-Maria Kantor, Strategic vision and concept of regional planning and sustainable development in Romania based on the use of geospatial solutions, in *International Journal Of Energy And Environment*, Issue 1, Volume 5, 2011, ISSN 1109-9577, pp. 91-101.

Ecosystem, Social and Humanity Sciences

Chapter 8

Dynamic Informatics of Avian Biodiversity on an Urban and Regional Scale

Wei-Ta Fang

Additional information is available at the end of the chapter

1. Introduction

Birds often constitute the most diverse and abundant species in a large-scale range (Rutschke, 1987; Virkkala, 2004). As such, and because their specific richness is relatively high in winter seasons, they may provide a useful raw material to evaluate predictive methods in wintering groups migrating through over an enormous range of environments (Bradley & Bradley, 1983). In the avian ecology, because they are highly mobile, and often yet ordinarily forage, breed, and stop in very specific habitats, birds indicate their specific behaviors of habitat selection (Jokimäki & Suhonen, 1998; Paillisson et al., 2004; Silva et al., 2004). Thus, they are also an ideal subject for habitat studies.

Indeed, the lives of migrants and residents are replete with habitat biological choices: where to stop to replenish depleted fat stores, or where to build a rest site to molt feathers during migration, etc. (Erni et al., 2002; Glimcher, 2002; Wiltschko & Wiltschko, 2003). Since birds' mobility may lead them to exploit different habitats depending on whether they are anchoring or shortly stopping, the relative importance of habitat characteristics may therefore be selected by birds spatially and temporally. Therefore, birds are likely to be candidates for habitat condition assessments for their specific bio-choices where to select for approximate stopover sites.

In order to understand their choices, avian community studies both in microhabitat-scale and landscape-scale are required (Buckley & Forbes, 1979; Palmer & White, 1994). Specifically, assessing which habitat elements are associated with bird communities during the non-breeding season (i.e., winter) may require a larger spatial scale than in the breeding season. During non-breeding and migrating seasons, individual birds spread and forage over larger spatial scales (Williams et al., 2003). At the larger scale, landscape configuration becomes crucial factors accounting for the variation in wintering bird species richness and diversity. However, there are many controversial issues to form avian refuges on larger scales. In the following section, concept of avian refuges for area and habitat issues is described.

2. Concept of diversity and habitat

Species respond to the size of patches when one considers designing avian refuges. Given the variable situation in a fragmented area perplexed by urbanization, it is not surprisingly that there is no best size to fit carrying capacities for avian habitats. Similarly, the increase in bird species individuals with area of habitat islands is attributed to minimum area requirements as interacting effects of competition or food demanding. Therefore, "how big is big" became issues to bring debates for the requirement to build such refuges. Many debates relate to use the island biogeographic concept to generate optimal refuge designs (Diamond, 1975; Gilpin & Diamond, 1980; Higgs & Usher, 1980). Several "principles" were provided by island biogeographic hypothesis (MacArthur & Wilson, 1967). First, refuges should be designed as large as possible, or a single large reserve is better than several small ones. Second, refuges should be close together as possible. Third, refuges should be as circular as possible to avoid "peninsula effect", in which species individuals reduces in an elongate areas compared with the circular areas in the same size (Forman, 1995, pp.108). However, there are many debates of this concept to design refuges related to the "species-area relationship" and "species-habitat relationship" (Simberloff & Abele, 1976; Forman et al., 1976; Forman, 1995; Pelletier, 1999; Oertli et al., 2002). In this chapter, I have reviewed several issues to document as below:

2.1. Species-area relationships

The concept of species-area relationship idea dates back to Arrhenius who studied data from a number of quadrat samples in plant associations (Arrhenius, 1921). Gleason (1922; 1925) came to the conclusion that a straight-line relationship was obtained. However, his theory was developed empirically to find a graph to fit certain observed results, and this rule was not based on mathematical reasoning. Later, Preston (1948; 1962) studied large amounts of empirical data to fit this model. He created an equation named Arrhenius equation as follows:

$$S = cA^z \tag{1}$$

Species area curves were then calculated for each plot using the equation, such as:

$$\log S = z \log A + \log c \tag{2}$$

Where S is species richness, z is the slope, A is the area, and c is the constant. Such a general pattern is important not only for fundamental aspects of ecological concept but also for ecological designs for refuges. Preston concluded that if the number of species (S) are recorded in different areas (A), there are almost an increase in S with increasing A. However, there were so many debates to regard this model as merely a computationally convenient method to fit observed data, despite some undesirable properties.

Forest pattern is the first patch to study the relationships between avian communities and areas (Forman et al., 1976). Martin (1988) declared that species numbers are related to forest foliage cover. He confirmed that foliage provides the substrate to protect the nests from predators. In addition, it also can influence the thermal environment to warm bird bodies in cold winter. Other studies have found birds responded to woody cover, shrub cover, grass cover, and litter cover.

There were many critiques for this hypothesis (Simberloff & Abele, 1976; Sismondo, 2000). In nature, area-per-se hypothesis was expected to be observed only within a certain intermediate range of areas, not at all spatial scales. At small spatial scales, the species-area relationship is not governed by equation (1) but is curvilinear on a log-log plot; and at landscape scales, the species-area relationship bends upward toward a limiting slope of unity (Durrett and Levin 1996). Second, species differ ecologically, thus not all units of species (S) are equal. Since some habitat generalists are widespread, most species in small patches associated with surrounding matrix are generalists which choose between major habitats and edge habitats, whereas in large patches are specialists only which finitely choose at interior habitats. These studies indicated that spatially and taxonomically different species differ from one another in their responses to area. Different avian communities are likely to yield different land-use patches.

2.2. Species-habitat relationships

Debates between field domains of the area-per-se hypothesis and species-habitat hypothesis have lasted for almost forty years. However, there were still no conclusion to generalize principles in ecological designs and no final upshot on which hypothesis was better (MacArthur & Wilson, 1967; Forman & Godron, 1986). Birds respond to both food and rest sites in habitat selection as above-mentioned. Species individuals are correlated with the need of lawn, mudflat, open shore, and canopy or water surface for horizontal heterogeneity. Bird-habitat relationships, thus, are the results of responses that bird use habitats for different activities, such as foraging, molting, and resting in winters.

There are many habitats for bird to select in pondscape configuration, majoring as water regimes. Recently on pond-core studies, Lane & Fujioka (1998) found species-habitat hypothesis works. They declared that watercourses, connecting by ditches around rice fields, affect shorebirds in shallow ponds. Elphick & Oring (2003) suggested that water depth significantly affected bird communities in flooding mudflats. The experiment explained this phenomenon and confirmed that if pond's water level was too deep, often causing respiration to slow down in bottoms due to a lack of oxygen exchange. They found that the species in constructed wetland was worse than those of natural wetland in comparison of the ecological integrities. Therefore, constructed wetland required to regulate water level from an ecological view, according to the demands for the principles of ecological designs. Taking into consideration of design criteria, reducing water level to promote shorebird's habitat quality could also increase in other avian diversity (Johnsgard 1956; Tamisier & Grillas, 1994; Bird et al., 2000; Fujioka et al., 2001; Quan et al., 2002; Ravenscroft & Beardall, 2003). Hattori & Mai (2001) declared that high water levels (equal to deep at 1 m or at more), reducing ecological diversity, only attracted water-edge's species (i.e., families Ardeidae, etc; such as egrets) in often. As Green et al. (2002) said, constructed wetland could not replace the value of natural wetland because the water level in constructed wetland was too deep, causing avian community worse than that of natural wetland. They suggested that water level in constructed wetland had been regulated so well about 10- 15 cm as to attract shorebirds (families Charadrii and Scolopaci) more. If water level reduction caused an increase of shorebirds, then, how many influences are running counter to interior waterbird individuals, like ducks (i.e., family Anatidae)? Taft et al. (2002) recommended that, if drained continuously, reducing water level would drop waterfowl individuals (i.e., family Anatidae). So, how to control water level, adjust mudflat area in order to observe changes of avian diversity, became the major subject of farm-pond management in habitat-scale studies.

2.3. Anthropogenic disturbances

As mentioned in the previous section, there were many studies focused on avian community with microhabitats as well as anthropogenic disturbance, such as drawdown, etc. Anthropogenic disturbances may be of beneficial or harmful to avian communities (Mustachio and Cousin, 2001). Focused on disturbed and undisturbed habitat, authors claimed the species that located in undisturbed habitats were much higher than in the sites from the highly disturbed habitat (Bolder et al., 1997; Chamberlain & Fuller, 2000). Most cases insisted that intensive anthropogenic influences caused avian decline due to negative edge effects (e.g., habitats adjacent to road paving, traffic flows, and urban development), and habitat fragmentation effects (e.g., habitat loss or segmentation). Edge effect, defined as the "juxtaposition of natural habitat and human modified habitat", may cause habitat less favorable and species likely to become locally extinct. For example, farm-pond roadside hedges were distinguished from non-roadside hedges in several analyses. The rationale for this distinction is that proximity of traffic may be a factor reducing habitat quality for some landbirds in roadside habitats. Roadside hedges may be poorer in species and less preferred by several bird species than non-roadside hedges. Other direct and indirect influences from anthropogenic disturbances are indicated as following tables, such as: (1) habitat loss or fragmentation; (2) introduced exotic species; (3) pollution (air, water, or soils); (4) population loss of specialists; (5) over population of generalists. Regarding to the impact of anthropogenic disturbance on habitats, the characteristics of birds categorized as roughly "specialist" or "generalist" as well as grouped as detailed "guilds" to illustrate habitat relationships are described in the following section.

3. Concept of diversity in a regional scale

Regional ecosystem is the number of avian species it contains. Therefore, avian community turns to indices of a habitat examination in a given area. Different levels of edge disturbance have different effects on avian communities. If the goals were to preserve biodiversity in microhabitats as well as in a landscape scale, to understand how diversity was impacted by different management strategies is required. Because diversity indices provide more information than simply the number of species present (i.e., they accounted for some species being rare and others being common), they serve as valuable tools that enable to quantify diversity in avian communities and describe their numerical structure. However, many debates between taxonomic diversity were around the entire groups and taxonomic diversity in specific guilds. Since Howell (1971) started to use five functional groups to examine avian residency in forests, many avian ecologists used "guilds" to avoid errors from large amount of species counts involved. They critiqued that the taxonomic approach of avian studies could not be commensurate with landscape scales. Alternatively, studies using aggregate species richness or diversity indices were over-simplified, too (Karr 1971; Emlen, 1972). In the following section, some approaches to calculate species diversity in all species and in specific functional groups are described and compared, therefore, a suitable approach to fit for avian community in farm-ponds would be carefully selected.

3.1. Species diversity

Population ecology was generally defined as "the scientific study of the abundance and distribution of species" (Fisher et al., 1943; Brown, 1984). With the two topics of relative abundance of species (diversity) and distribution along gradient zonation (guilds), one should start to find with effects (avian community), and then move on to causes (landform changes in gradient zonation) (Terborgh, 1977). Species diversity in the entire groups focuses attention upon the first topic. Then, the next guild topic is to dissect the environmental factors that affect that avian distribution in microhabitats and in a region.

Diversity provides information about rarity and commonness of species in an avian community (May, 1975; Karr, 1976). The ability to quantify diversity was an analytical tool for biologists trying to understand environmental quality, such as anthropogenic disturbance and environmental change (Rosenzweig, 1995). After the term "biodiversity" defined at the Rio Convention in 1992, there was a sudden shift in the literature towards the search for indicators of biodiversity itself (Duelli & Obrist, 2003). Since then, however, the term biodiversity has sometimes been used to indicate some aspect of environmental quality by diversity indices.

A diversity index is a mathematical measure of species in a community (Buckley & Forbes 1979; Magurran, 1988). It provides more information about community composition than simply species richness (i.e., the number of species present); and more, it also provides mixed counts of the relative abundances as well as species richness. There are several equations to calculate the indices of diversity. For example, Shannon-Wiener diversity index (also named for Shannon index or Shannon-Weaver index) is one of many diversity indices used by biologists (Shannon & Weaver, 1949). Others include the Simpson diversity, and so on. Each of these indices has strengths and weaknesses. An ideal index would discriminate clearly and accurately between samples, not be greatly affected by differences in sample size, and be relatively simple to calculate. In the avian survey project, well-designed indices should be considered to take advantages of the strengths of each and developed a more complete understanding of avian community structure. In this section, the above-mentioned indices of species diversity were discussed.

3.1.1 Shannon-Wiener diversity index (H'): is an index that is commonly used to characterize species diversity in an avian community. This index accounts for both abundance and evenness of the species present. The proportion of species (i) relative to the total number of species (P_i) is calculated, and then multiplied by the logarithm of this proportion ($\log_2 P_i$). The resulting product is summed across species, and multiplied by -1, such as:

$$H' = -\sum_{i=1}^{S} P_i \log_2 P_i \qquad (3)$$

S: avian species richness
P_i: The percentage of the i species in avian community

3.1.2 Shannon-Wiener evenness index (J): is a measure of the relative abundance of the different species making up the richness of an area (Hill, 1973). The Shannon-Wiener evenness index for a given number of species can be calculated as:

$$\text{Shannon} - \text{Wiener}\ (J) = H' / H_{max} = H' / \ln S \qquad (4)$$

3.1.3 Simpson's dominance index (C): if the greater the C value, the more dominant species among avian community.

$$C = \sum_{i=1}^{n} \left(\frac{Ni}{N} \right)^2 \qquad (5)$$

Ni: individual numbers of the i species
N: individual numbers of avian community

3.1.4 Simpson' diversity index (D):

$$D = 1 - C \qquad (6)$$

C: Simpson's dominance index

3.2. Species guilds

All avian species have their specific bio-choices to select a suitable habitat for diet for food or water, shelter from weather and predators, and a place to raise offsprings. In addition, each species has its own special requirement. Muller's Barbet (*Megalaima oorti*), for example, nests in tree cavities; while Chinese Bamboo-partridge (*Bambusicola thoracica*) finds bush covers to be underneath forest layers for their chicks. The species are jointed in such a manner so that every group should consist of similar characteristics such as "generalists" and "specialists". Named for generalist species, some avian species select what they need in a variety of habitats in farm ponds. They can cope with a large range of water and vegetation types, and sooner adapt to different diets and environmental conditions. Examples of such species are Tree Sparrow (*Passer montanus*), Chinese Bulbul (*Pycnonotus sinensis*), and Japanese White-eye (*Zosterops japonica*), etc. Duelli and Obrist (2003) suggested that generalists may not good indicators to illustrate biodiversity for the full visions of entire species. It is fundamentally an environmental indicator for broad range including habitat edges rather than a biodiversity indicator as the interior species in pond cores.

However, "real" biodiversity indicators may be needed to measure the impact of anthropogenic influences. Such an assessment is different from measuring the impact of lead on a selected taxonomic group from habitat specialists, which had been chosen because it is especially sensitive to lead rare and threatened. Specialists are less common than generalist and can thrive only in a narrow range of habitat quality. These animals have limited microhabitat ranges and hardly adapt well to new diets or environmental conditions. These species include many types of birds, such as waterfowl (families Anatidae and Podicipedidae) and shorebirds (families Charadrii and Scolopaci), etc. In conclusion, specialists rather than generalists are most likely to be those that are poisoning.

The concept of "*guild*" provides a beneficial approach of divided avian habitat selection into groups with environmental quality according to landscape configuration. Root (1967), the first avian scientist to form guild concept, defined a guild was as "a group of species that exploit the same class of environmental resources in a similar way". He focused the Blue-gray Gnatcatcher (*Polioptila caerulea*) associated with other species in California oak woodlands. Recognizing that the traditional taxonomic approaches was failed to categorize avian communities, he described a "foliage-gleaning guild" that feeds from foliage and occasionally from branches. This group included five species having similar diet, foraging location, and feeding behavior.

Since Root defined functional groups based on the traditional guilds: diets and foraging strategies, some authors followed his approaches (Emlen, 1972; Terborgh, 1977; Karr, 1980; May, 1982; Blake, 1983) to study avian behavior and foraging strategies; other authors studied nesting, resting, singing, or residential locations (Howell, 1971; Karr, 1971; Karr, 1976; Emlen, 1977; Riffell et al., 1996; Canterbury et al., 2000; Skowno & Bond, 2003); or they studied both, such as foraging strategies and singing location (Recher et al., 1983). However, most studies using functional groups have tended to: (1) group species by subjective criteria, or by a single behavior; (2) focus on just one or some groups; and (3) apply only at a single, or at a small spatial scale. Indeed, bio-choices based on entire species studies produced an objective result in a regional scale. Rather, microhabitat selection due to bio-choices reflects partitioning of spatial variation in a heterogeneous landscape. Clearly explained landscape configuration patterns, "*guilds*" based on bio-choices would be likely formed as indicators to monitor microhabitat quality. The "*guilds*", used to judge environmental conditions, were examined within heterogeneous landscape. Not assumed as the same definition as the first "*guild*" defined by Root (1967), habitat preference was to use to define functional groups later (Recher & Holmes, 1985; French et al., 2002). French et al. (2002) declared that wintering birds were related to land uses by grouping to generalists and specialists. In order to avoid the problems of landscape complexity, avian grouping was a useful approach to decide avian diversity in the microhabitat perspectives. Due to a lack of prior information about the necessary environmental factors that affected avian guilds, cluster analysis was applied in avian studies. It was used to study for grouping avian community of similar kind into respective functional groups. As a set of methods for building groups (clusters) from multivariate data, their aim was to identify groups with habitat preferences for microhabitats. Then, groups were made as homogenous as possible to reduce the differences between them as large as possible. This obtained a result for existing data correlation hierarchy and expected numbers of functional groups

4. Materials and methods

4.1. Dynamic informatics

I selected ecologically significant Taoyuan Tableland associated irrigation ponds as my study area because one fifth of all the bird species find home on these ponds in Taiwan (Chen, 2000; Fang, 2004a, b; Fang & Chang, 2004; Fang et al., 2009; Fang & Huang, 2011; Fang et al., 2011). This tableland, at an area of 757 km² in size, comprises an area of 2,898 ha of irrigation ponds on the northwestern portion of Taiwan. Located approximately 30 km from the capital city of Taipei, this rural area was easily converted to urban lands due to the aggregated effects of urbanization and commercialization. Socioeconomic benefits are driving public opinion which is urging the government to approve land-use conversion from farmlands into urban uses. The Taoyuan Tableland lies between the northern border of the Linkou Tableland (23°05'N, 121°17'E) and the southern border of the Hukou Tableland (22°55'N, 121°05'E); it borders the town of Yinge in the east (22°56'N, 121°20'E) and the Taiwan Strait in the west (22°75'N, 120°99'E) (Department of Land Administration, Ministry of the Interior, 2002)(see Fig. 1.). It sits at elevations from sea level to 400 m and is composed of tableland up to 303 m and hills with sloping gradients from 303 to 400 m. It runs in a southeast-to-northwest trend, abutting mountains in the southeastern corner and the shore of the Taiwan Strait at the far end. With a high average humidity of 89%, the tableland is located in a subtropical monsoon region with humid winters and warm summers. January temperatures average 13 °C, and July temperatures average 28 °C. Annual average precipitation ranges from 1,500 to 2,000 mm.

The tableland gradually rose approximately 180,000 years ago. At that time, the Tanshui River had not yet captured the flow from the ancient Shihmen Creek, which directly poured out of the northwestern coast forming alluvial fans. Eventually, foothill faults caused by earthquakes during the same era, resulted in the northern region of Taiwan abruptly dropping by 200 m, and thus, the Taipei basin was born. Since the Taipei area had subsided, the ancient Shihmen Creek which meandered across the Taoyuan Tableland was captured by northward-flowing rivers some 30,000 years ago. The middle streams changed their courses because of the subsidence in the Taipei basin. The resulting Tahan Creek, became the upstream portion of the Tanshui River in the Taipei Basin. Due to blockage of water sources, downstream areas on the Taoyuan Tableland were deficient in water. This caused high flushing and drops in water yields. Historically, it was difficult to withdraw and supply irrigated surface water from rivers due to the tableland's unique topography, thus, forming an obstacle for the development of agriculture (Agricultural and Forestry Aerial Survey Institute, 2010) .

This area has a population density of 2,331 persons/km² and its population is increasing at a rate of 2,000~3,000/month. Population pressures have contributed to reductions in historical areas of farmlands and irrigation ponds (Fang, 2001). Losses of farm-pond and farmland habitats have had series effects on a range of avian communities as well as other fauna and flora. On the Taoyuan Tableland, agricultural practices are intensifying, which is reducing the heterogeneity of the existing landform, and adding pollutants, also resulting from industrial practices (Fang et al., 2011).

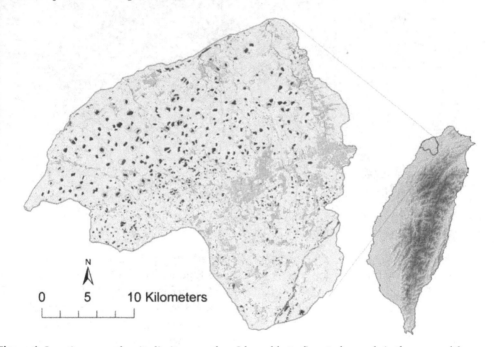

Figure 1. Location away the city limits more than 2 km of forty-five study ponds in the range of the tableland (see also as Fang et al., 2011).

4.2. Waterbirds sampled

Avian observers recorded all bird species seen within a 100-ha radius at 564.19-m basal radius of the bird census point at pond edge associated with line transects along pond-edge trails during 30-minute periods (one case of irrigation ponds see Fig. 2.). Sites were visited four times in the winter seasons between November and February. To reduce the effects of bird-observer bias, three to four observers were grouped and rotated between ponds. The observers counted birds that were in any habitats. All counts were conducted between 7:00 a.m. and 10:00 a.m. on days without rainy days when visibility was good (Bookhout, 1996).

Figure 2. Avian observers recorded all bird species seen within a 100-ha radius at 564.19-m basal radius of the bird census point at pond edge (photo by Wei-Ta Fang).

Foliage-loving species was also recorded followed the point-count method. Avian presence/absence on foliage strata was recorded in each pond at each of the following height intervals: edge ground, wetland grasses (< 0.5 m in height), bushes (> 0.5- 2.5 m in height), trees (> 2.5 m in height). Points were sampled at 10-m internals along edge trails established down each side of each pond. Waterbirds were grouped into microhabitat guilds based on actual observations on the sites. Foliage-loving species were initially classified into four height categories: pond-edge ground, low foliage (< 0.5 m in height), middle foliage (> 0.5- 2.5 m in height), and high foliage (> 2.5 m in height). Species were subsequently classified into two groups: understory (ground and low foliage groups) and canopy (middle and high foliage groups).

I calculated the number of individuals detected of each species at each pond for each month. Then, we calculated mean values of these variables for each study microhabitat across all study ponds in a wintering season (Fang et al., 2011).

4.3. Waterbird diversity

There are two traditional bird analyses for entire avian communities and specific avian groups, richness, and diversity. Differences in the characteristics of avian groups and pondscape configuration may vary according to species-area relationships among regions. Therefore, to find differences in the response of species to habitat area and isolation, studies must include multiple analytical approaches to detect which analysis was better based on an entire community, or on a specific group.

Descriptive statistics for entire communities were used as the first stage of statistical avian data processing. The main aim was initial analysis of the distribution of avian communities sooner, such as an average individual value and; or a guild value was described for specific groups later. Afterwards, avian diversity was described in the result of diversity indices for all communities or a single group. To detect species evenness and abundance, we used Shannon-Wiener diversity index (H') (also named for Shannon index or Shannon-Weaver index), which is given a measure of the richness and relative density of a species to calculate diversity (Shannon and Weaver, 1949). This diversity measure conducted by Shannon and Weaver which originally came from information theory and measures the order observed within a particular system. Regarding to my studies, this order was characterized by the number of avian individuals observed for each species in the sampling ponds. The first step was to calculate P_i for each category (i.e., avian species), and then we multiplied this number by the log of the number. The index was computed from the negative sum of these numbers. In short, the Shannon-Wiener index (H') is defined as (7):

$$H' = -\sum_{i=1}^{S} P_i \log^2 P_i \qquad (7)$$

S: avian species richness

P_i: The percentage of the i species in avian community

This index reflected bird richness in species and evenness amongst the avian community. The benefits of H' was sensitive by the change in threatened birds by avian study than that of Simpson's diversity index (D)(Dean et al., 2002). If the value of H' is higher, it means that species is abundant, or species distribution is even. However, species diversity is sometimes difficult to see relationships with spatial heterogeneity by limited survey data. Grouping and classification are required as well as for spatial heterogeneity reduction from the analyzed variables. It is the main procedure in this methodology for invoking avian groups with similar attributes of spatial behavior. The main approach in cluster analysis application is based on the idea to represent the grouping structure by avian data classification, based on the similarity in guilds between the species.

4.4. Simulation for dynamic informatics

Studies of variation in species individuals with relative abundances have been conducted by using species diversity. Although diversity may be measured most directly as the individual numbers, but it has been expressed the interplay of species richness and abundance into a single value (Shannon & Weaver, 1949; MacArthur & MacArthur, 1961; Dean et al., 2002). In this study, diversity was considered over a wide spectrum of spatial scales, from variation across a single pond scale to a regional scale, where temporal patterns were consequences of individual habitat selection. The diversity was measured all species. Four regional diversity variations were mapped from experimental semivariogram for avian communities in contour maps. On these maps a successional gradient was indicated to document concentric rings in bird diversity for spatial-temporal analysis (see Equation 8):

$$\gamma(h) = \frac{1}{2N(h)} \left\{ \sum_{i=1}^{N(h)} [z(x_i + h) - z(x_i)]^2 \right\}$$ (8)

The best linear unbiased estimator (BLUE) will be achieved to (9) :

$$\hat{z}(x_0) = \sum_{i=1}^{n} \lambda_i z(x_j)$$ (9)

while λ_i : weighting of detections; and $\sum_{i=1}^{n} \lambda_i = 1$

The estimation value is equal to the true value, such as:

$$E[\hat{z}(x_0)] = E[z(x_0)]$$ (10)

I introduced μ (Lagrange multiplier), then

$$L = Var[\hat{z}(x_0) - z(x_0)] - 2\mu(\sum_{i=1}^{n} \lambda_i - 1) \Rightarrow min[L]$$ (11)

$$\begin{cases} \dfrac{\partial L}{\partial \lambda_i} = 0, & (i = 1,2,\ldots,n) \\[2mm] \dfrac{\partial L}{\partial \mu} = 0 \end{cases}$$ (12)

$$\begin{cases} \sum_{j=1}^{n} \lambda_j \gamma(|x_i - x_j|) + \mu = \gamma(|x_0 - x_i|) \\[2mm] \sum_{j=1}^{n} \lambda_j = 1 \end{cases} \quad (i = 1,2,\ldots,n)$$ (13)

5. Results and discussion

5.1. Dynamic informatics for individual frequencies

The avian survey detected ninety-four species in 45 point-count locations associated with line transect of this investigation as a 2003-2004 example (see also Fang et al., 2011). In Taoyuan, forty-five species (48%) species were wintering migrants; forty species (43%) were permanent residents. Five short-transit species (5%) were encountered on the farm-pond sites, one species (1%) was not present at the site previously, defined "missing"; and three species (3%) were escaped from captivity. The total number of species in the winter seasons in the study area varied. I found greater species richness in wintering migrants (48%) compared with permanent residents (45%). In the microhabitat scale, the species in water regime (vertical structure from water table to aerial space) and waterfront edge were encountered most frequently.

Avian individual frequencies of occurrence were surveyed (see Table 1). I found significantly higher abundances of ten species, accounted for 74% of the entire species abundance, such as: Black-crowned Night-Heron (*Nycticorax nycticorax*) (occurrence frequency 2,363, occurrence rate of 15.7%, resident species), Little Egret (*Egretta garzetta*)(occurrence frequency 1,883, occurrence rate of 12.5%, resident species), Grey Heron (*Ardea cinerea*) (occurrence frequency 1,829, occurrence rate of 12.2%, wintering migrant species), Light-vented Bulbul (*Pycnonotus sinensis*) (occurrence frequency 1,575, occurrence rate of 10.5%, resident species), Eurasian Tree Sparrow (*Passer montanus*)(occurrence frequency 1,125, occurrence rate of 7.7%, resident species), Great Egret (*Casmerodius alba*)(occurrence frequency 726, occurrence rate of 4.8%, wintering migrant species), Red Collared-dove (*Streptopelia tranquebarica*)(occurrence frequency 509, occurrence rate of 3.4%, resident species), Japanese White-eye (*Zosterops japonica*)(occurrence frequency 504, occurrence rate of 3.3%, resident species), Little Ringed Plover (*Charadrius dubius*)(occurrence frequency 316, occurrence rate of 2.1%,wintering migrant species), and Little Grebe (*Tachybaptus ruficollis*)(occurrence frequency 304, occurrence rate of 2%, resident species), respectively. Other kinds of avian abundance, 84 species, were accounted for the total abundance of 36%. There were 23 species of which above 100 individuals were detected in the entire survey records, fewer than 10 individuals of 40 species were detected throughout the survey (see also the detection of 2003-2004 in Fang et al., 2009).

Place	Common Name	Scientific Name	Individual Frequency	Ratio of Frequency
1	Black-crowned Night-heron	*Nycticorax nycticorax*	2,363	15.7%
2	Little Egret	*Egretta garzetta*	1,883	12.5%
3	Grey Heron	*Ardea cinerea*	1,829	12.2%
4	Light-vented Bulbul	*Pycnonotus sinensis*	1,575	10.5%
5	Eurasian Tree Sparrow	*Passer montanus*	1,125	7.7%
6	Great Egret	*Casmerodius alba*	726	4.8%
7	Red Collared-dove	*Streptopelia tranquebarica*	509	3.4%
8	Japanese White-eye	*Zosterops japonica*	504	3.4%
9	Little Ringed Plover	*Charadrius dubius*	316	2.1%
10	Little Grebe	*Tachybaptus ruficollis*	304	2.0%
Totals			11,134	74.1%

Table 1. The individual frequency and their frequency of ten abundant species.

5.2. Dynamic informatics for biodiversity

Based on the point-count locations used random samplings in Taoyuan Tableland, the Shannon-Wiener index (H') by the data of ornithology have been caculated from December 2008, January 2009, and Febuary 2009 in migrating winters. This list with only 7 point-count locations within the entire points of the value of $H' > 2$ can be detected during December 2008, such as: No. 2 (2.522), No. 5 (2.152), No. 15 (2.128), No. 24 (2.127), No. 44 (2.062), No. 33 (2.057), No. 39 (2.022), respectively. This is also 7 point-count locations within the entire points of the value of $H' > 2$ can be detected during January 2009, such as: No. 32 (2.351), No. 27 (2.267), No. 7 (2.259), No. 40 (2.205), No. 19 (2.134) No. 2 (2.123), No. 5 (2.038), respectively. During the February 2009, a total of 14 point-count locations that the value of $H' > 2$ can be detected at the list, such as the numbers of No. 23 (2.575), No. 44 (2.528) No. 40 (2.516), No. 15 (2.360) No. 1 (2.357), No. 20 (2.320), No. 24 (2.312) No. 2 (2.282), No. 36 (2.281), No. 5 (2.219), No. 37 (2.145), No. 30 (2.046), No. 23 (2.042), No. 34 (2.007), respectively. The average value from three months was calculated at a lower value of 1.603 ± 0.494. This represents some seasonal dynamic informatics currently at a relative peak of H' in the month of Febuary on an urban and regional scale from anthropogenic influences during migratory seasons.

5.3. Modelling by biodiversity

Studies of variation in species individuals with relative abundances have been conducted under the calculation of species diversity. Although diversity may be measured most directly as the individual numbers, but it has been expressed the interplay of species richness and abundance into a single value (Shannon & Weaver, 1949; MacArthur & MacArthur, 1961; Dean et al., 2002). In this study, diversity was considered over a wide spectrum of spatial scales, from variation across a single pond scale to a regional scale, where temporal patterns were consequences of individual habitat selection. The spatial scale on diversity was measured, depended on the pondscape mosaic upon the moment of all species considered from December 2008, January 2009, and Feburary 2009. Three regional diversity variations were mapped for avian communities in contour maps (Figs. 3, 4, & 5). These maps were indicated a successional gradient to document concentric rings in bird diversity for spatial-temporal analysis.

Based on the experimental semivariogram for avian communities in contour maps. Diversities (H'), markly indicate from this anthropogenic influenced trends decreased with increasing dysfunctional pondscapes (monthly variations, see Figs. 2, 3, & 4). This is to say that pondscape configuration is so important in this situation. Indeed, three-month surveys demonstrated monthly diversity oscillations that horizontal heterogeneity might still occur at microhabitats. Species were able to select their proper habitats and then either over-wintered or undertook long migrations by different groups as well as by local generalists in huge assembleges. I, thus, hypothesized that diversities, at meso-scale, varied among different guilds of species from habitat selection. The occurrence rates detected by observers on avian communities were intriguing this hypothesis in different microhabitats, and they were largely examined and classified into groups in the section that follows.

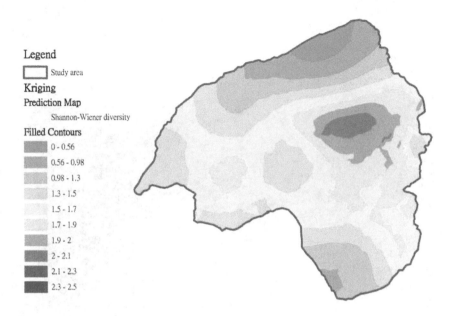

Figure 3. Scenario Model in Shannon-Wiener Diversity by Kriging Approach (December 2008).

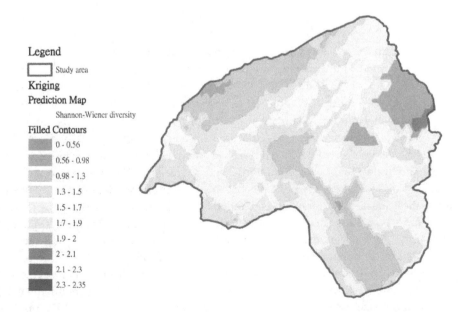

Figure 4. Scenario Model in Shannon-Wiener Diversity by Kriging Approach (January 2009).

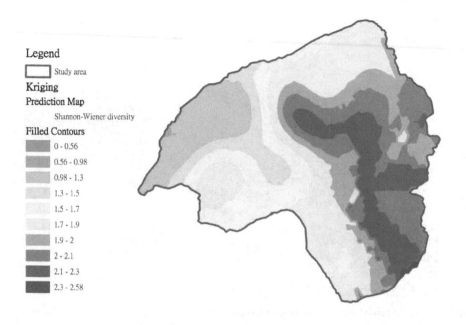

Figure 5. Scenario Model in Shannon-Wiener Diversity by Kriging Approach (Febuary 2009).

5.4. Finding

Despite their agribusiness value, farm ponds appear to have great influences on the make-up of avian communities in urbanized areas, especially for water-edge avian community (See Figs. 6, 7, & 8). I compared the following community characteristics against the corresponding ratio of constructed area value associated with pond configuration of each site for all functional groups: cumulative waterfowl, cumulative shorebirds, cumulative landbirds, cumulative air feeders, and cumulative water-edge species. Pondscape was a strong and/or moderate correlate in any birds of the ordinations (i.e., water-edge birds, shorebirds, and waterfowl) beyond landbirds and air feeders. The presence of adjoining natural and/or urbanized habitats was probably the most important determinant of wetland avifauna in these areas. Regarding to this detailed study, there may be a number of reasons why some farm ponds do not become a refuge for the more sensitive species. First, the ornamental vegetation covers used for surrounding areas are often too few, and they may support a small insect population. Second, anthropogenic structure is subjected to concrete construction without native trees, and this may make it unattractive to water-edge species that require an intact shrub layer, dead wood, or generally undisturbed microhabitats. Third, small pond size associated with curvilinear shape is not optimum to support for preserving and attracting water-edge birds and other avifauna.

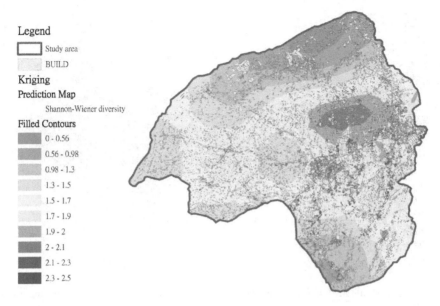

Figure 6. Scenario Model in Shannon-Wiener Diversity by Kriging Approach within building areas (December 2008). Based on the experimental semivariogram for avian communities in contour maps, the same overlapped map layers as Fig. 7 & Fig. 8.

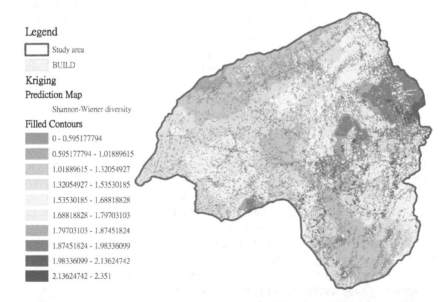

Figure 7. Scenario Model in Shannon-Wiener Diversity by Kriging Approach withing building areas (January 2009).

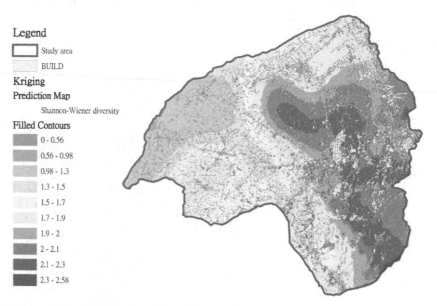

Figure 8. Scenario Model in Shannon-Wiener Diversity by Kriging Approach within building areas (February 2009).

6. Conclusion

Farm ponds generally represent a habitat condition between natural and urban environments, and have great potential for supporting quite varied bird communities (Fang et al.,2009; Fang, 2011; Fang & Huang, 2011; Fang et al., 2011). This chapter characterizes species richness and community structure over a habitat size gradient to a landscape gradient of a farm-pond complex. In my study, forty-five ponds were surveyed ranging in area from 0.2 to 20.47 ha within a landscape complex in the Taoyuan Tableland. An avian survey, detected ninety-four species and individuals, was conducted on three occasions. Contrasting responses to pond configurations at the functional group level, the relationship between the landscape and birds were calculated the effects of pond size and shape within the complex on species richness and community composition. Seven types of avian functional groups, representing locally major species, were identified over urbanized characteristics. Some generalists, like Chinese Light-vented Bulbul (*Pycnonotus sinensis*) and Japanese White-eye (*Zosterops japonica*), have been detected in the urbanized areas. The presence of adjoining natural and/or constructed habitats from anthropogenic influences was probably the most important determinant of avifauna in study areas. This study was used a simulation of diversity of birds with kringing methods beyond a lower mean *H'* value of 1.603 ± 0.494 during migratory seasons for three occasions from 2008 to 2009 in Taoyuan, Taiwan. Studies of variation in species individuals with relative abundances have been conducted by using species diversity for further results and findings in several months. This study will be conducted for decade for more advanced findings in urbanized rural areas.

Author details

Wei-Ta Fang

Graduate Institute of Environmental Education, National Taiwan Normal University, Taiwan, Republic of China

7. References

Agricultural and Forestry Aerial Survey Institute. (2010). *Aerial Photographs*, 1:5,000 of scale in digital debase forms, Taipei, Taiwan, ROC.

Arrhenius, O. (1921). Species and area. *Journal of Ecology* Vol. 9, 95-99, ISSN: 1365-2745.

Begon, M; Harper, J.L. & Townsend, C.R. (1996) *Ecology: Individuals, Populations and Community*, Third Edition, Blackwell Science, ISBN: 0632038012, Oxford, UK.

Bird, J.A.; Pettygrove, G.S. & Eadie, J.M. (2000). The impact of waterfowl foraging on the decomposition of rice straw: mutual benefits for rice growers and waterfowl. *The Journal of Applied Ecology* Vol. 37, 728-741, ISSN: 1365-2664.

Blake, J.G. (1983). Trophic structure of bird communities in forest patches in east-central Illinois. *The Wilson Bulletin* Vol. 83, 416-430, ISSN:1938-5447.

Bolger, D.T.; Scott, T.A. & Rotenberry, J.T. (1997). Breeding bird abuandance in an urbanizing landscape in coastal southern Califorina. *Conservation Biology* Vol. 11, 406-421, ISSN:0888-8892.

Bookhout, T.A. (1996). *Research and Management Techniques for Wildlife and Habitats*, The Wildlife Society, ISBN:093586881, Bethesda, Massachusetts, USA.

Bradley, R.A. & Bradley, D.W. (1983). Co-occurring groups of wintering birds in the lowlands of southern California. *The Auk* Vol. 100, 491-493, ISSN: 0004-8038.

Brown, J.H. (1984). On the relationship of distribution and abundance of species. *American Naturalist* Vol. 124, 255-279, ISSN:0003-0147.

Buckley, G.P. & Forbes, J.E. (1979). Ecological evaluation using biological habitats an appraisal. *Landscape and Planning* Vol. 5, 263-280, ISSN:0169-2046.

Bookhout, T.A. (1996) *Research and Management Techniques for Wildlife and Habitats*. The Wildlife Society, , Bethesda, Massachusetts, USA.

Canterbury, G.E.; Martin, T.E.; Petit, D.R.; Petit, L.J. & Bradford, D.F. (2000). Bird communities and habitat as ecological indicators of forest condition in regional monitoring. *Conservation Biology* Vol. 14, 544-558, ISSN:0888-8892.

Chamberlain, D.E. & Fuller, R.J. (2000). Local extinctions and changes in species richness of lowland farmland birds in England and Wales in relation to recent changes in agricultural land-use. *Agriculture, Ecosystems and Environment* Vol. 78, 1-17, ISSN: 0167-8809.

Chen, C.J. (2000) *The Change about Culture Landscape of Irrigation Reservoir and Pond in the Taoyuan Terrace, and the Establishment of Sustainable Environment Waterways*, National Science Council, Taipei, Taiwan, ROC. (in Chinese)

Cheng, Q. & Agterberg, F.P. (1995) Multifractal modeling and spatial point processes. *Mathematical Geology* Vol. 27, 831-845, ISSN (electronic): 1573-8868.

Dean, W.R.J.; Anderson, M.D.; Milton, S.J. & Anderson, T.A. (2002). Avian assemblages in native Acacia and alien Prosopis drainage line woodland in the Kalahari, South Africa. *Journal of Arid Environments* Vol. 51, 1-19, *ISSN:0140-1963*.

Department of Land Administration, Ministry of the Interior. (2002) *Taiwan's Geographic Aerial Map*, 1:5,000 of scale in digital database forms, Taipei, Taiwan, ROC. (in Chinese)

Diamond, J. (1975). The island dilemma: lessons of modern biogeographic studies for the design of natural reserves. *Biological Conservation* Vol.7, 129-146, *ISSN: 0006-3207*.

Duelli, P. & Obrist, M.K. (2003). Biodiversity indicators: The choice of values and measures. *Agriculture, Ecosystems and Environment* Vol. 98, 87-98, *ISSN:0167-8809*.

Durrett, R. & Levin, S. (1996). Spatial models for species area curves. *Journal of Theoretical Biology* Vol. 179, 119-127, *ISSN:0022-5193*.

Elphick, C.S. & Oring, L.W. (2003). Conservation implications of flooding rice fields on winter waterbird communities. *Agriculture, Ecosystems and Environment* 94, 17-29, *ISSN: 0167-8809*.

Emlen, J.T. (1972). Size and structure of a wintering avian community in southern Texas. *Ecology* Vol. 53, 317-329, *ISSN:0012-9658*.

Erni, B.; Liechti, F. & Bruderer, B. (2002). Stopover strategies in passerine bird migration: A simulation study. *Journal of Theoretical Biology* Vol. 219, 479-493, *ISSN:0022-5193*.

Fang, T.-Y. (2001) *The Study of the Spatial Structure Change of Water Land in Taoyuan Terrace*, Master Thesis, Department of Bioenvironmental Systems Engineering, National Taiwan University, Taiwan, ROC. (in Chinese)

Fang, W.-H. (2004a) *Threaten Birds of Taiwan*, Second Edition. Wild Bird Federation Taiwan, *ISBN:9867415817*, Taipei, Taiwan, ROC. (in Chinese)

Fang, W.-T. (2004b) The ecological drawdown assessment for avian communities in Taoyuan's constructed wetlands. In: Proceeding of 2004 Annual Symposium of Environmental Education, Yeh. S.-C. (ed.), pp. 861-869, Chinese Society for Environmental Education, Kaohsiung, Taiwan, ROC. (in Chinese)

Fang, W.-T. (2011) Creating pondscapes for avian Communities: an artificial neural network experience beyond urban regions. In Hong, S.-K.; Wu, J.; Kim, J.-E.; Nakagoshi, N. (Eds.) *Landscape Ecology in Asian Cultures*. Springer, New York, pp. 187-200.

Fang, W.-T. & Huang, Y.-W. (2011) Modelling geographic information system with logistic regression in irrigation ponds, Taoyuan Tableland, In: *Proceeding of 2011 International Conference on Fuzzy Systems and Neural Computing* Vol.2, 26-30.

Fang, W.-T. & Chang, T.-K. (2004) The scientific exploring of ecoscape-design on constructed wetlands in Taoyuan, In: *The Proceeding of Water Source Management of Taoyuan Main Canal*, Lee, C.-J. (ed.), pp. 345-369, Taoyuan Irrigation Association, Taoyuan, Taiwan, ROC. (in Chinese)

Fang, W.-T.; Chu, H. & Cheng, B.-Y. (2009). Modelling waterbird diversity in irrigation ponds of Taoyuan, Taiwan using an artificial neural network approach. *Paddy and Water Environment* Vol. 7, 209-216, *ISSN: 1611-2490*.

Fang, W.-T.; Loh, K.D., Chu, H.-J. & Cheng, B.-Y. (2011). Applying artificial neural network on modelling waterbird diversity in irrigation ponds of Taoyuan, Taiwan. In: *Artificial*

Neural Networks – Application, Hui, C.L.P. (ed.), pp. 423-442, InTech Press, *ISBN:978-953-307-188-6*, Rijeka, Croatia.

Fisher, R.A.; Corbet, A.S. & Williams, C.B. (1943). The relation between the number of species and the number of individuals in a random sample of an animal population. *Journal of Animal Ecology* Vol. 12, 42-58, *ISSN:1365-2656*.

Forman, R.T.T. (1995). *Land Mosaic: The Ecology of Landscape and Regions*, The University of Cambridge, *ISBN:9780521474627*, Cambridge, UK.

Forman, R.T.T. & Godron, M. (1986). *Landscape Ecology*. John Wiley & Sons, *ISBN:0471870374*, New York, USA.

Forman, R.T.T.; Galli, A.E. & Leck, C.F. (1976). Forest size and avian diversity in New Jersey woodlots with some land use implications. *Oecologia* Vol.26, 1-8, *ISSN:0029-8549*.

French, D.D. & Picozzi, N. (2002). 'Functional groups' of bird species, biodiversity and landscapes in Scotland. *Journal of Biogeography* Vol. 29, 231-259, *ISSN:1365-2699*.

Fujioka, M., Armacost Jr, J.W., Yoshiko, H. & Maeda, T. (2001). Value of fallow farmlands as summer habitats for waterbirds in a Japanese rural area. *Ecological Research* Vol. 16, 555-567, *ISSN:1440-1703*.

Gilpin, M.E. & Diamond, J.M. (1980). Subdivision of nature reserves and the maintenance of species diversity. *Nature* 285:567-568, *ISSN:0028-0836*.

Gleason, H.A. (1922). On the relation between species and area. *Ecology* Vol. 3, 158-162, *ISSN: 0012-9658*.

Gleason, H.A. (1925). Species and area. *Ecology* Vol. 6, 66-74, *ISSN: 0012-9658*.

Glimcher, P.W. (2002). Decisions, decisions, decisions: choosing a biological science of choice. *Neuron* Vol. 36, 323-332, *ISSN:0896-6273*.

Green, A.J.; Hamzaoui, M.E.; Agbani, M.A.E & Franchimont, J. (2002). The conservation status of Moroccan wetlands with particular reference to waterbirds and to changes since 1978. *Biological Conservation* Vol. 104, 71-82, *ISSN:0006-3207*.

Hattori, A. & Mai, S. (2001). Habitat use and diversity of waterbirds in a costal lagoon around Lake Biwa, Japan. *Ecological Research* Vol. 16, 543-553, *ISSN:1440-1703*.

Higgs, A. J. & Usher, M.B. (1980). Should nature reserves be large or small? *Nature Vol* 285:568-569, *ISSN:0028-0836*.

Hill, M.O. (1973). Diversity and evenness: a unifying notation and its consequences. *Ecology* Vol. 54, 427-432, *ISSN:0012-9658*.

Howell, C.A.; Latta, S.C.; Donovan, T.M; Poreluzi, P.A.; Parks, G.R. & Faaborg, J. (2000). Landscape effects mediate breeding bird abundance in midwestern forests. *Landscape Ecology* Vol.15,547-562, *ISSN:0921-2973*.

Howell, T. R. (1971). An ecological study of the birds of the lowland pine savanna and the adjacent rain forest in northeastern Nicaragua. *The Living Bird* Vol. 10, 185-242, *ISSN:0459-6137*.

Johnsgard, P.A. (1956). Effects of water fluctuation and vegetation change on bird populations, particularly waterfowl. *Ecology* Vol. 37, 689-701, *ISSN:0012-9658*.

Jokimäki, J. (1999). Occurrence of breeding bird species in urban parks: Effects of park structure and broad-scale variables. *Urban Ecosystems* Vol. 3, 21-34, *ISSN:1083-8155*.

Karr, J.R. (1971). Structure of avian communities in selected Panama and Illinois habitats. *Ecological Monographs* Vol. 41,207-233, *ISSN:0012-9615*.

Karr, J.R. (1976). Within- and between- habitat avian diversity in African and Neotropical lowland habitats. *Ecological Monographs* Vol. 46,457-481, *ISSN:0012-9615*.

Karr, J.R. (1980). Geographical variation in the avifaunas of tropical forest undergrowth. *The Auk 97,283-298, ISSN:0004-8038*.

Lane, S.J. & Fujioka, M. (1998). The impact of changes in irrigation practices on the distribution of foraging egrets and herons (Ardeidae) in rice fields of central Japan. *Biological Conservation* Vol. 83, 221-230, *ISSN:0006-3207*.

MacArthur, R.H. & MacArthur, J.W. (1961). On bird species diversity. *Ecology* Vol. 42,594-598, *ISSN:0012-9658*.

MacArthur, R.H. & Wilson, E.O. 1967. *The Theory of Island Biogeography*. Princeton University Press, *ISBN:0691088365*, Princeton, New Jersey, USA.

Magurran, A. E. (1988). *Ecological Diversity and its Measurement*, ISBN:9780691084916 Princeton University Press, Princeton, NJ, USA.

Martin, T.E. (1988). Habitat and area effects on forest bird assemblages: Is nest predation an influence? *Ecology* Vol. 69, 74-84, *ISSN:0012-9658*.

May, P.G. 1982. Secondary secession and breeding bird community structure: patterns of resource utilization. *Oecologia* Vol. 55,208-216, *ISSN:1432-1939*.

May, R.M. (1975). Patterns of species abundance and diversity. In: Cody, M. L. & Diamond, J.M (eds.), pp. 81-120, *Ecology and Evolution of Communities*, ISBN:9780674224469. Harvard University Press, Cambridge, Massachusetts, USA.

Musacchio, L.R. & Coulson, R.N. (2001). Landscape ecological planning process for wetland, waterfowl, and farmland conservation. *Landscape and Urban Planning* Vol. 56, 125-147, *ISSN:0169-2046*.

Oertli, B.; Joye, D.A.; Castella, E.; Juge, R.; Cambin. D. & Lachavanne, J. (2002). Does size matter? The relationship between pond area and biodiversity, *Biological Conservation* Vol. 104, 59-70, *ISSN:0006-3207*.

Ozesmi, U.; Tan, C.O.; Ozesmi. S.L. & Robertson, R. J. (2006). Generalizability of artificial neural network models in ecological applications: Predicting nest occurrence and breeding success of the red-winged blackbird Agelaius phoeniceus, *Ecological Modelling*, Vol. 195, No. 1-2, 94-104, *ISSN:0304-3800*.

Paillisson, J.-M., A. Carpentier, J. L. Gentil, and L. Marion. (2004). Space utilization by a cormorant (Phalacrocorax carbo L.) colony in a multi-wetland complex in relation to feeding strategies. *Comptes Rendus Biologies* Vol. 327, 493-500, *ISSN 1631-0691*.

Palmer, M.W., and P.S. White. (1994). Scale dependence and the species-area relationship. *American Naturalist* Vol. 144, 717-740, *ISSN:0003-0147*.

Palmer, M.W. (1990). The estimation of species richness by extrapolation. *Ecology* Vol. 71, 1195-1198, *ISSN:0012-9658*.

Pelletier, J.D. (1999). Species-area relation and self-similarity in a biogeographical model of speciation. *Physical Review Letters* Vol. 82, 1983-1986, *ISSN:0031-9007*.

Preston, F.W. (1948). The commonness, and rarity, of species. *Ecology* Vol. 29, 254-283, *ISSN: 0012-9658*.

Preston, F.W. (1962). The canonical distribution of commonness and rarity: part I. *Ecology* Vol. 43, 182-215, *ISSN: 0012-9658*.

Quan, R.-C.; Wen, X. & Yang, X. (2002). Effect of human activities on migratory waterbirds at Lashihai Lake, China. *Biological Conservation* Vol. 108, 273-279, *ISSN:0006-3207*.

Ravenscroft, N.O.M. & Beardall, C.H. (2003). The importance of freshwater flows over estuarine mudflats for wintering waders and wildfowl. *Biological Conservation* Vol. 113, 89-97, *ISSN:0006-3207*.

Recher, H.F. & Holmes, R.T. (1985). Foraging ecology and seasonal patterns of abundance in a forest avifauna. In: *Birds of Eucalypt Forests and Woodlands: Ecology, Conservation, Management*, Keast, A.; Recher, H.F.; Ford, H. & Saunders, D. (eds.), pp. 79-96, Surrey Beatty and Sons, *ISBN:9-780-949324061*, Chipping Norton, N.S.W., Austeria.

Recher, H.F.; Gowing, G. & Kavanagh, R.; Shields, J. & Rohan-Jones, W. (1983). Birds, resources and time in a tablelands forest. *Proceedings of the Ecological Society of Australia* Vol. 12, 101-123, *ISSN:0070-8348*.

Riffell, S.K.; Gutzwiller, K.J. & Anderson, S.H. (1996). Does repeated human instruction cause cumulative declines in avian richness and abundance? *Ecological Applications* Vol. 6, 492-505, *ISSN:1051-0761*.

Root, R.B. (1967). The niche exploitation pattern of the Blue-gray Gnatcatcher. *Ecological Monographs* Vol. 37, 317-350, *ISSN:0012-9615*.

Rosenzweig, M.L. (1995). *Species Diversity in Space and Time, ISBN:0-521-49618-7*, Cambridge University Press, Cambridge, UK.

Rutschke, E. (1987). Waterfowl as bio-indicators. pp.167-172. In: *The Value of Birds*, Diamond, A. W. & Filion, F. (eds.), International Council for Bird Preservation, *ISBN:0946888108*, Norfolk, UK.

Shannon, C.E. & Weaver, W. (1949). *The Mathematical Theory of Communication*. University of Illinois Press, *ISBN:0252725484*, Urbana, Illinois, USA.

Silva, J.P., M. Pinto & Palmeirim, J.M. (2004). Managing landscapes for the little bustard Tetrax tetrax: lessons from the study of winter habitat selection. *Biological Conservation* Vol. 117, 521-528, *ISSN:0006-3207*.

Simberloff, D.S. & Abele, L.G. (1976). Island biogeographic theory and conservation practice. *Science* Vol. 191:285-286, *ISSN:0036-8075*.

Sismondo, S. (2000). Island biogeography and the multiple domains of models. *Biological and Philosophy* Vol. 15,239-258, *ISSN:0169-3867*.

Skowno, A.L. & Bond, W.J. (2003). Bird community composition in an actively managed savanna reserve, importance of vegetation structure and vegetation composition. *Biodiversity and Conservation* Vol. 12, 2279-2294, *ISSN:1572-9710*.

Taft, O.W.; Colwel, M.A; Isola, C.R. & Safran, R.J. (2002). Waterbird responses to experimental drawdown: implications for multispecies management of wetland mosaics. *Journal of Applied Ecology* Vol. 39, 987-1001, *ISSN:1365-2664*.

Tamisier, A. & Grillas, P. (1994). A review of habitat changes in the Camargue: An assessment of the effects of the loss of biological diversity on the wintering waterfowl community. *Biological Conservation* Vol. 70, 39-47, *ISSN:0006-3207*.

Terborgh, J. (1977). Bird species diversity on an Andean elevational gradient. *Ecology* Vol. 58, 1007-1019, *ISSN: 0012-9658.*

Virkkala, R. (2004). Bird species dynamics in a managed southern boreal forest in Finland. *Forest Ecology and Management* Vol. 195, 151-163, *ISSN:0378-1127.*

Wiltschko, R. and W. Wiltschko. (2003). Avian navigation: from historical to modern concepts. *Animal Behaviour* Vol. 65, 257-272, *ISSN:0003-3472.*

Williams, J.C.; ReVelle, C.S & Bain, D.J. (2003). A decision model for selecting protected habitat areas within migratory flyways. *Socio-Economic Planning Sciences* Vol. 37, 239-268, *ISSN:0038-0121.*

Chapter 9

Genus Lists of Oribatid Mites – A Unique Perspective of Climate Change Indication in Research

V. Gergócs, R. Homoródi and L. Hufnagel

Additional information is available at the end of the chapter

1. Introduction

In most habitats oribatid mites account for the biggest part of microarthropods (e.g. Schenker, 1986, Johnston and Crossley, 2002). They can be found in most terrestrial microhabitats: in soil, leaf litter, moss, underwood, foliage and in aquatic habitats as well (Behan-Pelletier, 1999). They can be found mostly in great species richness and abundance in their habitats (Behan-Pelletier, 1999). They play a significant role in decomposition processes because they fragment the organic matter and influence the biomass and species composition of fungi and bacteria (Wallwork, 1983; Seastedt, 1984; Yoshida and Hijii, 2005). As this group plays a significant role in soil processes, it is necessary to get to know its spatial pattern and the causes of pattern generation, which can be used later for indication (Behan-Pelletier, 1999).

Applicability of Oribatid mites as an indicator group has been emphasized by researchers for several decades. These organisms possess such kind of extraordinary characteristics by which (considered even separately or as a whole) they are able to indicate different changes in their environment. These characteristics have been summarized in several reviews, most thoroughly in the works of Lebrun and van Straalen (1995), Behan-Pelletier (1999) and Gulvik (2007).

Oribatid mites can be found in almost every kind of habitats worldwide: on land, water and most importantly in the layers of soil containing organic materials, but they also conquered several other kind of microhabitats (e. g. lichen, moss, treebark etc.). Apart from the diversity of habitats, their excessive adaptational ability is also shown by great abundance and species richness. In most habitats, they constitute the largest proportion of microarthropods. These characteristics mentioned above can be primarily used in the application of coenological methods.

Oribatid mites consume mainly living or dead parts of plants or fungi, however there are some predators and scavengers to be mentioned as exceptions (Behan-Pelletier, 1999). As a consequence, they consume variuos kinds of food, and as such, they participate in numerous ways in the structure of the food web (Lebrun and van Straalen, 1995). Thus they are in strong interaction with their microenvironment (e. g. Ca-storage, heavy metal accumulation (Norton and Behan-Pelletier, 1991, Behan-Pelletier 1999), play an important role in the forming of soil structure and decomposition processes (Behan-Pelletier, 1999). These features can be applied for indicating the effects of chemical or heavy metal pollutions, and disturbances in the succession of decomposition processes (Lebrun and van Straalen, 1995).

The reproduction biology and life cycle of Oribatid mites can be considered extraordinary among arthropods from several aspects. There are some species/populations with sexual and asexual reproduction, and the proportion of species with obligate thelytokous parthenogenesis is very high – around 10% (Lebrun and van Straalen, 1995). Iteroparity and multiannual life cycle are also quite prevalent among the species, especially in moderate and cold climate zones (Norton, 1994, Luxton, 1981, Behan-Pelletier, 1999). The slow development, low fecundity and long larval stage of Oribatid mites can help indicating long-term disturbances. Their low dispersion ability (Lebrun and van Straalen, 1995) is also quite important, since these mites can hardly flee from sites affected by some kind of stress. Oribatid mites are classified as a „K-selected" group; this can be lead back to their slow metabolism according to Norton (1994). Based on the characteristics listed above, many researchers think that this group is quite promising since it can be used for various indication purposes.

Nowadays there are several methods to describe the natural state of a habitat; the focus is mainly on the measuring of biodiversity. However, uncertainty can arise when measuring biodiversity, as several questions can be raised already as to the explanation of the term, starting from which level it should be considered on (genetic, taxon, ecological diversity), to – if the taxon level has been chosen – the decision on which taxon the focus should be.

The main goal of this study is to set up a comparison scale based on genus-level presence-absence lists of oribatid mite communities (Acari: Oribatida) of habitats examined on different spatial and temporal scales. The secondary goal – and this time the precondition as well – is to get a reliable picture of the indication strength of the distances to be used, i.e. the information content included.

The indication suitability of the order of oribatid mites for describing the state of their habitat is justified by the special characteristics of the group. Oribatid mites can be found in almost all kinds of habitats: on land and in water; first of all in soil layers containing organic matter as well, however, they have penetrated into different other microhabitats, too (e.g. lichens, moss, bark etc.), which is mainly due to their indeed various food sources (e.g.

organic debris, fungi, other mites etc.). Besides the diversity of habitats, their high adaptation ability is shown by their enormous abundance and species richness as well. The above characteristics can be mainly used in the case of coenological methods (Lebrun and van Straalen, 1995; Behan-Pelletier, 1999; Gulvik, 2007; Gergócs and Hufnagel, 2009).

The choice of the genus level can be explained by different aspects. In the analysis by Caruso and Migliorini (2006) it was shown that there were not any significant changes in data examining anthropogenic disturbance on oribatid mites when switching from species level to genus level. Our study has a similar goal as we would like to show potential habitat changes with our method. Podani (1989) had a similar observation in case of plants, according to which switching to genus level does not mean a significant change when comparing the examined habitats. Osler and Beattie (1999) carried out a meta-analysis similar to ours, which confirmed their expectation that taxonomic levels above species are more suitable for comparing habitats. This research showed further that habitats can be chosen on family level in case of oribatid mites, therefore our study covers besides the genus level the family level as well. There were also some other arguments for our decision, namely that the number of databases used could be considerably extended in this way, in addition, taxonomical processing became faster and more reliable in our field studies as well. Genus-level identification of oribatid mites is solved on the basis of the work by Balogh and Balogh (1992) on a global scale, too. However, species-level identification is only possible in case of some zoogeographical regions and only some taxa on a global scale as the related literature is not properly synthesized yet (e.g. Balogh and Mahunka, 1983; Olsanowski, 1996).

2. Review of literature – Suitability of oribatid mites as indicators

Research into oribatid mites goes back to the 1880s, the work of A. Berlese, who invented the Berlese funnel and made it possible to extract and examine soil mesofauna more precisely. His lifework was carried on by several renowned taxonomists, such as Grandjean, Hammer, Beck, Aoki, Wallwork, Engelbrecht, Corpus-Raros, Lee, Pérez-Inigo, Baggio, Bhattacharia and Haq (Balogh et al., 2008) with taxonomical descriptions of oribatid mites covering the bigger part of terrestrial habitats. Due to the above researches, nowadays it has become possible to examine oribatid mites from different indication aspects on community level.

One part of the studies on indication possibilities compares natural habitats. In these studies the goal is to reveal spatial and temporal pattern generation characteristics of habitats. Temporal change is examined in few studies (Irmler, 2006) and in case of spatial examinations different approaches are used: on substrate level (e.g. Fagan et al., 2006; Lindo and Winchster, 2006), examining altitudinal zonation of mountains (Migliorini and Bernini, 1999; Reynolds et al., 2003; Jing et al., 2005) and only seldom on the level of habitat types (Balogh et al., 2008). These studies do not always yield consistent results, however, the examinations prove that patterns exist.

2.1. Comparison of natural habitats

These examinations try to explore what properties of habitats play a role in pattern generation, among which spatial and temporal changes can be distinguished. Observations on seasonality have not yielded considerable results (Reynolds *et al.*, 2003, Noti *et al.*, 1996, Badejo *et al.*, 2002, Moldenke and Thies, 1996). Habitats and sampling frequencies are quite different and hardly comparable. Currently we do not possess any satisfying results on seasonal dynamics. A number of studies (Reynolds et al., 2003) surveying temporal changes measured the total abundance of the community. Measuring the changes in the number of individuals of larger groups does not mean thorough examination. It is worth to survey the temporal structures of the entire community on such places where seasons are well discernible. One of the most important studies has been made by Irmler (2006), who studied the seasonal changes of an Oribatid community living in the OL and OF layer of a beech forest. It was found that only the annual mean temperature had significant effect on the structure of the community. The study yielded more results when Irmler surveyed the seasonal dynamics of individual species. Mainly the amount of precipitation affected the abundance of certain species, but some species had been affected more significantly by temperature (primarily by the mean temperature in January). The significance of species-level examination was confirmed by the fact that certain species reacted differently on the surveyed parameters.

Spatial comparisons applied different scales; part of them compared soil and foliage of forests. These studies revealed that Oribatids of the soil showed greater α-diversity and species richness, but β-diversity proved to be greater in the foliage, which means difference among samples taken from individual trees has been greater than that of the samples collected from the soil (Lindo and Winchster, 2006, Fagan *et al.*, 2006).

Comparison of elevations above sea level attracted great attention: primarily the abundance and species richness of Oribatids have been studied in zones of different altitudes. However, obtained data are not concordant: according to Migliorini and Bernini, (1999) and Fagan et al., (2006) the abundance of Oribatids decreased with altitude, but Jing et al., (2005) and Reynolds *et al.*, (2003) observed an opposing tendency. Fagan *et al.*, (2006) pointed out a decrease in species richness, Migliorini and Bernini, (1999) observed a growth in diversity as a function of increasing altitude. It has to be mentioned by these contradictory results that altitudes of sampling and habitats are hardly comparable, and even if they were, this would not guarantee consistent results. This has been pointed out by Andrew *et al.*, (2003) in an extended series of studies conducted on different altitudes in Australia and New Zealand.

Beside altitude, vegetation also changes greatly when progressing upwards on a hill. Studies mentioned above did not lay an emphasis on vegetation. The work of Balogh et al.(2008) however demonstrates altitude as a difference in the type of vegetation: rainforest, moss forest and paramo. Samples were taken from the mountains of Brazil, Costa-Rica and New-Guinea. This work showed that the structure of Oribatid mite communities was primarily determined by the type of vegetation and not by the distance of several thousand kilometres, which means that climate and ecological conditions have stronger effects than zoogeographical connections (Balogh *et al.*, 2008).

2.2. Abundance, species richness and diversity

Studies examining Oribatid communities almost always measure which Oribatid species and in what quantity are present in samples taken from the given area. Species composition, abundance, total abundance, species richness, diversity and the uniformity of the community can be calculated from these data. In most cases, changes in the communities are examined using these variables.

When given the same climate, abundance, species richness and diversity of the Oribatids are greater in natural areas (forest or habitats not strongly affected by human activity) than in areas affected by agriculture (e.g. plant production or animal husbandry) or forestry (e.g. clear-felling, burning etc.) (Bedano *et al.*, 2006, Osler *et al.*, 2006, Cole(et al 2008, Olejniczak 2004, Arroyo and Iturrondobeitia, 2006, Migliorini *et al.*, 2003, Altesor *et al.*, 2006). The observation of Bedano et al. (2006) can be mentioned as an exception: it was found that the abundance of pastures was higher than that of natural forests.

Decrease in abundance can be caused by hard frost (Sulkava and Huhta, 2003) and serious heavy metal pollution (Seniczak *et al.*, 1995). According to Osler *et al.* (2006), mainly the number of individuals is lower in the initial state of succession. Decrease in abundance could be pointed out primarily as a result of water deficiency (O'Lear and Blair, 1999, Lindberg *et al.*, 2002), but contradictory results had been also obtained (O'Lear and Blair, 1999, Melamud *et al.*, 2007). Lindberg and Bengtsson, (2006) showed that community regeneration following drought can not be satisfactorily measured by the sole application of total abundance. Decrease in the abundance of Oribatids can also be caused by ash treatment of sour, acidic soils (Liiri *et al.*, 2002). In Japanese coniferous forests it has been shown that the abundance of Oribatids was greater in mixed litter (litter of several tree species) than in litters consisting of only one tree species (Kaneko and Salamanca, 1999). Kovács *et al.*, (2001) explored positive correlation between the nutrient content of the soil and abundance, but it was contradicted by several other studies (e.g. Osler and Murphy, 2005).

Removal of winter snow cover lead to a decrease in species richness, since the mesofauna of the soil has been exposed to greater fluctuation of temperature (Sulkava and Huhta, 2003). Response of species to heavy metal pollution varied greatly, sometimes even moderate pollution resulted in the highest species richness (Skubala and Kafel 2004). Drought generally decreased species richness (Tsiafouli et al., 2005), but there were several examples for growth as well (Melamud *et al.*, 2007). Ash treatment lowered abundance and also species richness (Liiri *et al.*, 2002). In mixed litter, both species richness and abundance were higher (Kaneko and Salamanca, 1999). Fagan *et al.*, (2006) found in Canadian coniferous forests that species richness of Oribatids in the soil had been greater when comparing Oribatid communities of the foliage and soil.

Diversity data can be found primarily in agricultural and forestry studies. It has been pointed out that irrigation (enhancing the moisture content of the soil) increased the diversity of Oribatid communities, because it raised the individual numbers of rare species (Tsiafouli *et al.*, 2005). Drought had a detrimental effect on diversity (Lindberg *et al.*, 2002). Studies of Taylor and Wolters (2005) pointed out that Oribatid diversity had been greater in a more decomposed beech litter than in fresh litter. Seniczak *et al.*, (2006) concluded that Oribatid diversity can be increased by increasing the number of ponds of forest habitats, since this means more ecotones and leads to the presence of such kind of species which prefer humid habitats and are normally absent from forest habitats. Age of temperate deciduous forests did not affect diversity (Erdman *et al.*, 2006). Growth in the diversity of tree species did not increase the diversity of Oribatids living in the soil of these forests (Kaneko et al, 2005. However, growing diversity of the litter not only increased abundance and species richness, but diversity as well (Coleman 2008). (*Table 1.*)

abundance	species richness	diversity
artificial disturbance↓	artificial disturbance↓	artificial disturbance↓
hard frost↓	snow cover removal↓	irrigation↑
drought↓	drought↓	drought↓
early stage of succession ↓	ash↓	number of ecotones↑
diverse litter mix↑	diverse litter mix↑	diverse litter mix↑
ash treatment↓		soil > foliage
heavy metal pollution↓		
organic mater content↑		

Table 1. Strongly abridged summary of information from studies on characteristics of Oribatid communities. (↑=increases or greater; ↓=decreases or lower)

With the overview of available studies, it can be clearly explored how various characteristics of Oribatid communities are modified due to changes in moisture, temperature, heavy metal concentration, organic matter content and level of disturbance. The most important question concerning the application of Oribatids as indicators is to clarify what kind of information content does natural Oribatid coenological patterns possess from the aspect of bioindication. Most of the variables listed above can be directly measured, since rapid methods are available to quantify temperature, heavy metal content etc. of the soil. Responses of Oribatids are worth to study in a more complex approach. Even now we have an expansive (but far from satisfactory) knowledge on how communities change due to modifications of different factors. These pieces of information necessitate the elaboration of such methods which render Oribatid communities suitable for the task to prognosticate what extent the given site can be considered near-natural or degraded, based on the Oribatid composition of a single sample taken from the given area. Raising further questions will be possible only after obtaining the answer for this problem. However, answering this problem needs extensive and coordinated work: approriate reference sites need to be appointed to clarify the concept of naturality, sampling and processing methods need to be standardized internationally – in conformity with the given environmental conditions – and the field of data processing methods also has to be developed. Definition and testing of Oribatid-based (or mesofauna-based in a broader sense) coenological indicators are also undoubtedly needed. The usefulness of Oribatid characteristics summarized in the introduction had been recognized long ago, now it is time to conduct research in a way that enables to explore and exploit the actual advantages Oribatid mites provide.

3. Similarities of genus lists on different scales

By setting up the spatial and temporal scales, we expected that the order based on the genus lists later should correspond to the real spatial and temporal scales, either the farther and qualitatively the more different habitats our lists originate from, the greater difference there should be among similarities inside the given categories. However, if data originate from the same site, the difference among the examined samples should be greater in case of the lists which are farther in time from each other.

The main goals of the present study are the following:

1. Developing a spatial and temporal scales reference based on the genus -level taxon lists with the help of similarity functions.
2. Examining the degree of distances in the similarity order used for indication.

3.1. Examination of the suitability of the genus level

Our analysis related to the notion that the genus level does not mean great data loss compared to the species level was carried out based on the databases by Marie Hammer. The work of Hammer was chosen due to the homogeneity and very extensive geographical cover of the databases. The series originate from two different sites of six different countries accordingly (Hammer, 1952, 1958, 1961, 1962, 1966, 1972). Besides the species and genus level the family level was analyzed as well, according to the taxonomical classification in the work by Balogh and Balogh (1992).

For our examination comparisons on genus level are sufficient as switching from species level to genus level did not cause a significant change regarding the distance and position of habitats according to the results of the ordinations. On family level inconsistency is caused by losing information. Using species-level data would be impractical due to taxonomical uncertainty on the one hand and lack of reliable databases on the other hand, and thirdly, due to unjustified increase in distance caused by genera with large number of species.

3.2. Categories of the genus lists

In order to be able to determine to which spatial and temporal distance the oribatid mite genus lists of two samples/sites examined by us correspond, different categories had to be defined. The categories were set up considering which combination of the given spatial and temporal scales the examined genus list pairs originate from. Regarding the time (Ti), differences between 0, 2, 12, 24 and 52 weeks and due to a study (Melamud *et al.*, 2006(2007) were able to consider six years i.e. 312 weeks, too. In space the smallest distinguishable unit was the different substrate (S), then the different types of habitats/sites (H), the different topographicums (T) follow, and finally the largest unit was the zoogeographic kingdom (K). Substrate is the lowest vegetation level such as soil, förna, leaf litter, moss, bark etc. Site means habitat types such as rainforest, mossy forest, páramo etc. Topographicum is practically a country such as Papua New Guinea or Chile. When differentiating between zoogeographic kingdoms, six kingdoms found in the work by Balogh and Balogh (1992) were considered: Holarktis, Neotropis, Aethiopis, Orientalis, Australis (there Notogea) and Archinotis (there Antarctis), which is the modified version of Müller's system (1980).

3.3. Sources of the genus lists

Genus lists of the different categories were collected from various sources. The first category means the similarity between genus lists of samples collected from the same zoogeographic kingdom (SaK), the same topographicum (SaT), the same type of site (SaH), the same substrate (SaS) and at the same time (Ti-0) (SaK/SaT/SaH/SaS/Ti-0). One part of these genus lists was obtained from our own research. From the soil of a dry oak forest in Törökbálint (Hungary), 9×300 cm³ förna sample was collected and the mesofauna was extracted from it, the oribatid mites were sorted out and identified to genus level according to the works by Balogh (1965); Balogh and Balogh (1972, 1992); Balogh and Mahunka (1980, 1983) and Olsanowski (1996). Further data for this category were collected by studying the scientific legacy of the late János Balogh, member of the Hungarian Academy of Sciences.

Data for the following four categories were also collected from our research. Samples were collected in 2005 and 2006 in a given quadrat of 100 m² in a dry oak forest in Törökbálint, Hungary (N 47°25'38" E 18°54'16") and they were surveyed every two weeks. Every time samples were obtained from three types of substrates: from 500 cm³ leaf litter, 300 cm³ förna (from under the leaf litter) and 0.5 dm² hypnum moss (*Hypnum cupressiforme*) living on tree trunks. Oribatid mites were extracted with the help of a Berlese-Tullgren funnel (Coleman et al., 2004) and identified on genus level. This examination made it possible to set up categories on pattern levels meaning a distance of two, 12, 24 and 52 weeks, in which substrate (S), site (H), topographicum (T) and zoogeographic kingdom (K) were the same (Sa). Abbreviations of these categories are: SaK/SaT/SaH/SaS/Ti-2, SaK/SaT/SaH/SaS/Ti-12, SaK/SaT/SaH/SaS/Ti-24 and SaK/SaT/SaH/SaS/Ti-52. A study by Melamud et al. (2007) was implied as well, in which samples were collected at different altitudes of Mount Carmel in Israel from the same sites with a difference of six years (312 weeks) (SaK/SaT/SaH/SaS/Ti-312).

Regarding spatial differences, the smallest change in scale is the difference in the substrate: SaK/SaT/SaH/DS/Ti-0, i.e. the substrate is different (D), however, there is no change in time (Ti-0). Genus lists belonging to this category originate from our own database and the above mentioned manuscripts by Balogh. Databases of three further studies were used as well (Behan-Pelletier and Winchester, 1998; Fagan et al., 2006; Lindo and Winchester, 2006).

In case of the following seven categories, only spatial scales "above" substrate change, substrate and time are not differentiated any more so they are marked "X". Abbreviation of the same type of sites which can be found in the same zoogeographic kingdom and in the same topography is SaK/SaT/SaH/XS/Ti-X. Genus lists belonging to this category were obtained from the manuscripts by János Balogh, the study by Migliorini *et al.*, (2005) and the studies by Hammer (1958, 1961, 1962, 1966). Abbreviation of the category of different sites is SaK/SaT/DH/XS/Ti-X. Sources of the series belonging to this category are: studies by Noti et al. (1996); Migliorini et al. (2002); Osler and Murphy (2005); Skubala and Gulvik (2005); Arroyo and Iturrondobeitia (2006); Osler et al. (2006), manuscripts by János Balogh, published series by János Balogh (Balogh et al., 2008) and studies by Hammer (1958, 1961, 1962, 1966). A series belonging here originates from samples collected by Levente Hufnagel in Australia (2006, Australia: QLD, Cairns)

In case of genus lists originating from different topographicums, we considered the point if they originate from the same (SaK/DT/SaH/XS/Ti-X) or different sites (SaK/DT/DH/XS/Ti-X) and if the two topographicums can be found in the same or different zoogeographic kingdoms (DK/DT/SaH/XS/Ti-X, DK/DT/DH/XS/Ti-X). These series come from studies by János Balogh and Marie Hammer.

In the last category only the zoogeographic kingdom can be interpreted as the complete genus lists of the six zoogeographic kingdoms were compared in it according to the work by Balogh and Balogh (1992) (DK/XT/XH/XS/Ti-X).

3.4. Data processing methods

The lists created from the Hammer-databases were analysed with Ochiai, Jaccard and Sørensen distance functions and non-metric ordination using the software Syn-tax 2000 (Podani, 2001).

From the other databases we did not consider all possible list combinations which fit the category, only the ones having at least nine genera. After our complete genus list database was set up, the number of genera of the two lists and the number of the common genera were determined considering the genus list pairs in each category. As we had only presence-absence data and the value "d" of the contingency table was not considered in case of the genus list pairs, the Ochiai and Jaccard functions were used as distance functions (Podani, 1997). The similarity data of each category was calculated from the means of the values of the distance functions for the genus list pairs.

As our data were not always independent within a category, it was determined with a complex method to what extent the means of the categories differ from each other. We had several distance function values within each category as we. We had 106 genus list pairs within one category on average. From among the distance function values of each category fifteen values were chosen randomly with the help of a random number generator in the Excel software. It was carried out ten times in case of each category. In this way we got 10 series containing 15 values for each category. Series of the data table containing 10×15 values in case of each of the 14 categories were now independent and since normal distribution could not be observed within each category, the data were analysed with the Kruskal-Wallis statistical test. Each of the 14 series were analysed with the Mann-Whitney post hoc test as well, so we got ten tables containing 14×14 post hoc test results. One table was made out of these ten, which shows 95% confidence interval of the appropriate values of the ten tables. Based on this we were able to decide which categories differ from each other significantly. These statistical tests were carried out using PAST software (Hammer *et al.*, 2001).

3.5. Order of the genus list categories

As we got nearly the same results using both distance functions (Ochiai and Jaccard), only the results calculated with the help of the Ochiai function are discussed further. *Fig.* 2 displays intervals with defined standard error around the Ochiai distance means in case of each category.

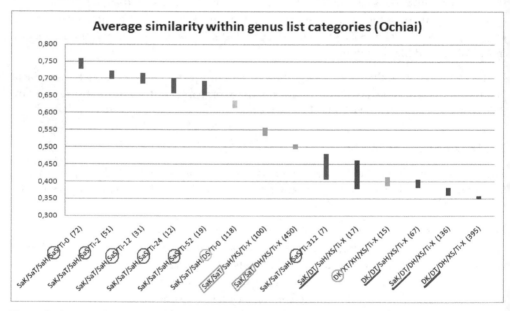

Figure 1. Average distance within genus list categories using Ochiai function. Next to the category codes, the number of genus list pairs used for calculating the average can be seen in brackets. Remarkable code parts are highlighted

In *Fig.* 2 it can be seen that the largest similarity between samples can be observed in the category where all criteria are the same (SaK/SaT/SaH/SaS/Ti-0) i.e. where the samples were collected at the same time and from the same substrates. Similarities of genus lists originating from different time but the same substrates are the next: first the two-week-difference, then the 12-, 24- and finally the 52-week-difference. Among comparisons on sample level the last one marked with yellow colour is the similarity of genus lists originating from different types of substrates.

As expected, within the same topographicum there is larger similarity between genus lists coming from the same type of sites than between those coming from different sites. This is where the sample series meaning six-year-difference (SaK/SaT/SaH/SaS/Ti-312) turn up. This considerable difference is amazing within one given habitat.

Among the last five there are the four categories in which the difference between lists originating from different topographicums (DT) is measured. Regarding the order of these four categories it is remarkable that the same type of site shows larger similarity than different sites, irrespective of the fact whether the different topographicums are in the same or different zoogeographic kingdoms. This corresponds with the results of a former study conducted with other methods (Balogh et al., 2008). The category DK/XT/XH/XS/Ti-X fits in the above mentioned categories in the order. Using the Jaccard distance function this is the only category position that changes places with the category DK/DT/SaH/XS/Ti-X.

The order set up with the help of genus lists based on the complete database met our expectations, so it can be definitely an appropriate reference in indication researches.

The significance of distances between genus list categories was tested by a Mann-Whitney tests. Our first remarkable result is that the average distances between genus lists originating from the same substrate but from different dates within a year does not differ from each other significantly. According to our former assumptions time difference could have been detected regarding a one-year-difference, however, it could not be detected from the substrates of the temperate dry oak forest studied. Consequently, if genus lists of the complete habitat type were examined with time lags less than a year, no change could be detected on site level, either. Irmler (2006) had a similar result on species level in a long-term European study.

The difference between oribatid mite communities originating from different substrates (DS/Ti-0) can be significantly larger in case of certain distance functions than the difference between communities coming from the same type of substrate (SaS/Ti-0). Besides, there is no large difference between samples collected within a year from a given type of substrate. It follows from these two statements that if genus lists originate from different types of substrates, there is larger difference between them than if samples are collected within a year from the same type of substrate. Consequently, the quality of substrate in a given habitat type is a more important factor in the composition of the oribatid mite community than time changes within a year.

Similarity between oribatid mite genus lists of the same types of sites (SaK/SaT/SaH/XS/Ti-0) differs significantly from the similarity between genus lists originating from the same type of substrate (SaK/SaT/SaH/SaS) if samples were collected with a time lag of maximum 24 weeks in the latter case. At the same time, the distance between genus lists coming from different types of substrates is similar to the distance between oribatid mite genus lists of the same or different habitats in a given topographicum, i.e. the type of substrates plays a similarly important role in the quality of the oribatid mite community as habitat types in a given topographicum.

The distance between genus lists originating from the same or different types of substrate is much smaller than the distance between genus lists of different topographicums (XK/DT/...), independently of the fact if sites in the same or different zoogeographic kingdoms are compared. More remarkable is the fact that the similarity of genus lists coming from the same sites in the same topographicums does not differ significantly from the similarity of genus lists originating from the same sites in different topographicums. It means that oribatid mite communities of the same types of habitats resemble each other nearly in the same way no matter if they originate from the same or different topographicums.

4. Seasonal change of oribatid mite communities (case study)

Several researches have already been done to detect the pattern and composition of oribatid mite communities and their exact causes. However, it is hidden to date by which mechanisms the structure and functioning of the individual communities are affected.

Climatic factors belong to the most determinant ones. The above statement is reflected by the number of researches as well, because most of the studies on oribatid mite communities investigate how the communities react to the meteorological factors changing naturally or artificially, especially to temperature and the amount of precipitation (Gergócs and Hufnagel, 2009). First it is worth investigating the effect of the natural changes of the climatic factors, and the most appropriate way to do this is to study the seasonal changes of the communities. The mistake in most of these studies is that the research on seasonality is conducted over a maximum of a year (e.g. Schenker, 1984), so it cannot be determined whether the observed phenomena occur similarly each year. The most significant research on this topic was conducted by Irmler (2006) in a beech forest in Germany. He investigated monthly changes of ground-dwelling oribatid mite communities over a seven-year period and found that there are no important changes among the communities.

Our research is similar to his one, however, it differs as well because we collected samples every two weeks and not once a month, and we investigated not one but three microhabitats over a one and a half year period in an oak forest in Hungary.

The other difference is that data on oribatid mites were recorded on genus level and not on species level. The decision on the genus level can be justified by several factors. The analysis of Caruso and Migliorini (2006) showed that there were not any important changes in the data when studying human disturbance on oribatid mites and changing from species level to genus level. Podani (1989) found a similar result in case of plants, namely, changing to genus level does not mean a significant difference when comparing the studied habitats. Osler and Beattie (1999), Hammer and Wallwork (1979) and Norton et al. (1993) concluded according to their studies that the many widespread genera and families indicate that the similarity between oribatid mite communities should be studied above species level. In their meta-analyses their assumption was confirmed that taxonomical levels above the species level are more suitable to compare habitats. According to this we also compared the data series recorded with time difference on genus level.

The aim of our study was to explore which seasonal changes occur in oribatid mite communities living in three types of microhabitats in an oak forest and what role the most important climatic factors that is the amount of precipitation and temperature have in these changes.

Samples for our study were collected in 2005 and 2006 in a given quadrate of 100 m² in a dry oak forest near Törökbálint, Hungary (47°25'38" N, 18°54'16" E). In the sampling quadrate the most common tree species is turkey oak (*Quercus cerris*), common trees are field maple (*Acer campestre*), common ash (*Fraxinus excelsior*) and wild service tree (*Sorbus torminalis*). The most important herbaceous plants are broad-leafed Solomon's seal (*Polygonatum latifolium*) and garlic mustard (*Alliaria petiolata*). Samples were collected every two weeks from 15th March till 1st December 2005 and from 26th March till 30th July 2006. Every time samples were collected randomly from three types of substrate within the quadrate: from 500 cm³ leaf litter, 300 cm³ foerna (from under the leaf litter) and 0.5 dm² hypnum moss (*Hypnum cupressiforme*) living on tree trunks. So we obtained altogether 19 moss samples and 18 leaf litter and 18 foerna samples in 2005 (the first time no leaf litter and foerna samples were collected); in 2006 we collected 10 leaf litter, 10 moss and 9 foerna samples (the tenth sample could not be analyzed for technical reasons). Components of the mesofauna were extracted with the help of a modified Berlese-Tullgren funnel (Coleman et al., 2004) and conserved in isopropyl alcohol, then they were sorted into larger groups, and finally oribatid mites were identified on genus level. For the identification the works of Balogh (1965), Balogh and Balogh (1992, 1972), Balogh and Mahunka (1980), Olsanowski (1996) and Hunt et al. (1998) were used.

The samples were analyzed based on oribatid mite communities with the help of multivariate computer aided methods using the software Syn-tax 2000 (Podani, 2001). The size difference between the samples from the three types of substrate was compensated by comparing the ratio and presence-absence data of the oribatid mite genera with the help of non-metric multidimensional scaling and classification method using several distances. Monte Carlo method was used to check these analyses. The means of the genus proportion of the substrate types were calculated, each mean was multiplied by a number created by a random generator twenty times, then these new values were divided by the sum of the random numbers. This way twenty data series were generated per substrate. These were compared using classification and ordination. Correspondence analysis and PCA analysis were used to identify the genera which are responsible for possible differentiations. The difference of the genera in specimen number and proportion was checked using Mann-Whitney test between the substrate types. The average diversity and the genus number were calculated in case of each substrate type.

In order to recognize seasonality patterns, each season was marked (in winter no sampling was conducted), and the changes in genus diversity and total abundance during a period of the study were displayed separately in case of the three substrate types. The relationship between the pattern changes of the communities and meteorological factors was analyzed with the help of data series from the meteorological station in Pestszentlőrinc (47°25'53" N,

19°10′57″ E). This station can be found 21 air km eastwards from our sampling point. In the case of precipitation the total precipitation amount of the sampling day and that of the preceding 5 and 10 days, and the standard deviation of the precipitation amount of the preceding 5, 10, 15, 20, 25 and 30 days were considered. In the case of temperature minimum and maximum values of the given days were available in our database, so in our analyses the means of these two values were considered. As for temperature data, the mean temperature of the sampling day and that of the preceding 5 and 10 days, and the standard deviation of temperature of the preceding 5, 10, 15, 20, 25 and 30 days were involved in our analyses. A redundancy analysis was conducted for all substrates first and then separately. In order to make the relationships more exact, the correlation between certain genera and meteorological factors was studied based on Spearman's analysis.

Seasonal changes could not be detected in the communities of the three substrates. In *Fig. 2* change in the Shannon diversity of the genera in the case of the three substrates, however, a seasonal pattern recurring the following year cannot be detected.

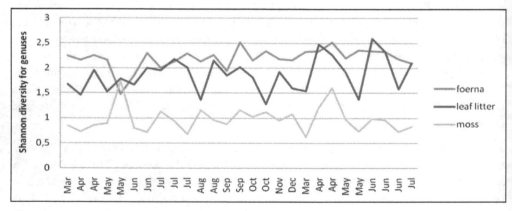

Figure 2. Change in the Shannon diversity of the genera in the case of the three substrates (2005-2006)

There is no relationships between the annual changes in precipitation amount and temperature and the composition and structure of Oribatid mite communities in leaf litter and foerna, however, in the case of moss there is. This can be caused by the genus *Tectocepheus*, which is one of the most frequent genera in moss samples (43% of the adult specimens on average), and its correlation is r=0.38 regarding the five-day mean temperature (Spearman's correlation; p=0.043); and regarding the ten-day mean temperature its correlation is r=0.45 (Spearman's correlation; p=0.014).

According to the research in seasonality during one and a half year we conclude that if any change occurs in the community during the year, it is not seasonal, i.e. neither genus-level diversity, nor abundance, nor the structure of the community have a recurring pattern in leaf litter, foerna and moss microhabitats. These results correspond with several previous data from the literature. Irmler (2004, 2006) observed oribatid mite communities in two different habitats over several years and he did not find any seasonal regularity either, except for some species. Those who interpreted their results in a way that seasonal change was found in oribatid mite communities drew a conclusion based on only one studied year or on the difference between two sampling months (Schenker, 1984; Stamou and Sgardelis, 1989; Askidis and Stamou, 1991; Clapperton *et al.*, 2002). Such a hypothesis was, among others that important changes may occur between two temperate seasons in oribatid mite communities (Stamou and Sgardelis, 1989; Askidis and Stamou, 1991; Clapperton *et al.*, 2002).(kétszer ugyanazok?) Although some of these studies were conducted in other types of habitats, i.e. not in temperate deciduous forests, a study covering a shorter period than a year is considered to be insufficient in the case of all habitats when observing regularities in seasonality. Schenker (1984) carried out a one year study around a turkey oak (*Quercus cerris*), a beech (*Fagus sylvatica*) and a Scots pine (*Pinus sylvestris*) in a moist deciduous forest in Switzerland. He found that annual change in abundance can be observed mostly around the beech and the pine, whereas it can hardly be observed around the oak, and species composition is not affected by seasonality, either. Oribatid mites occurred approximately in the same abundance further away from the oak trunks, than in the soil around it. For our study, data were collected in an oak forest. This may be the reason for the fact that we have not found any seasonal changes or changes in the genus composition, either. However, since Irmler (2006) conducted his studies in a beech forest and could not observe any seasonality either, and Schenker (1984) collected samples for only one year, it cannot be stated that the lack of seasonality is characteristic of oak forests.

If changes in communities do not occur seasonally, this can be brought into connection with micro- and macroclimatic effects. According to the results of Stamou and Sgardelis (1989) it could be concluded that the density of oribatid mites is largely influenced by temperature, although several later studies showed that temperature does not have the power to shape communities (Haimi et al., 2005). Irmler (2006) found that the structure of the community was in connection with the annual mean temperature only, and only some species showed significant correlation with some climatic factors. Webb et al. (1998) showed in the case of oribatid mite species living in polar areas that these species do not depend on seasonal changes, life cycle of the studied oribatid mite species is mainly influenced by temperature fluctuation. In our study proportional change correlating with temperature could be observed in the case of the genus *Tectocepheus* only and only in moss, however, no such connection can be found in the study of Irmler (2006), for example. Based on our results – just like based on those of Irmler (2006) – it can be concluded that the structure of oribatid mite communities is not affected by climatic factors in leaf litter and foerna substrates. In moss samples the connection with temperature was due to the genus *Tectocepheus*.

Seasonality can also be observed in the decomposition of plant material. Quantity and quality of the decomposing plant material change seasonally in the leaf litter and in the soil so it can be assumed that oribatid mite communities may change correspondingly during the year. However, the exact role of oribatid mites in the decomposition of the leaf litter is not completely clear till this day (Lindo and Winchster, 2007). The most important role of oribatid mites in the decomposition is the spreading of microbiota as they feed mainly on fungi and bacteria, and they are not in direct connection with the leaf litter input, accordingly (Maraun *et al.*, 2001). This corresponds with the phenomenon observed by us, that the quantity of leaf litter may not have influenced the compositional changes of the communities.

Our result that leaf litter and foerna substrates differ from moss was not interpreted by other literature yet, however, there are observations regarding other types of substrates. A common result is for example that the oribatid mite community living in the foliage of the trees differs significantly from the one living in the soil under the trees (e.g. Yoshida and Hijji, 2005; Karasawa and Hijii, 2008). Karasawa and Hijii (2004) showed that the substrate of oribatid mite communities living in the soil, in the foliage, on the bark of the tree trunks and on the remnants of algae accumulating on soil significantly differ from each other in seaside forests. In our study the community living in hypnum moss was simpler than the one living in the soil in the forest. Communities of moss and lichen are always relatively simple (Gjelstrup, 1979; Gjelstrup and Søchtig, 1979; Smrz, 1992; Smrz and Kocourková, 1999; Smrz, 2006). The three frequent genera found in moss turned up in the observations of others as well. *Zygoribatula exilis* is assumed to be a species living in moss (Gjelstrup, 1979), however, the *Zygoribatula* species found by us could be found on the forest ground as well. The genus *Tectocepheus* occurs everywhere from drier and more disturbed habitats to intact forests, accordingly it can also be found in moss in great quantities, especially because climatic fluctuation is larger in moss, what only some species can tolerate (Gjelstrup, 1979). A common epiphyte is *Eremaeus oblongus* (Smrz and Kocourková, 1999), in our study the genus *Eremaeus* could be found only in moss (except for only one foerna sample).

Therefore research showed that oribatid mite communities living in soil, leaf litter and hypnum moss, in Hungarian oak forests – similarly to those living in German beech forests – did not show seasonal changes. This result is important on the one hand because according to this, we are not bound to a season regarding sampling. However, besides that it would be important to detect the cause of the still occurring changes and patterns exactly. Furthermore it is unclear as well whether non-woody biocoenoses in the temperate zone or oribatid mite communities living in other climatic zones show seasonal changes.

Communities living in the soil differ from the oribatid mite community of moss living on tree trunks more significantly than from the community of the leaf litter. Oribatid mites in moss, especially the genus *Tectocepheus*, may be influenced by climatic factors to a large degree. In future it would be necessary to study oribatid mite communities of various microhabitats in order to detect exactly by which factors and to what degree their composition, changes and patterns are affected.

5. Spatial similarity pattern of oribatid mite communities (case study)

Oribatid mites of the tropical regions had been almost completely unknown for science before 1958. With the general use of the Berlese-funnel, systematic collections started at that time, which resulted in the collection of hundreds or thousands of species. However, the majority of these samples are unprocessed till these days. Professor J. Balogh and his fellow-workers had to realize, that the description of all the species living there is an impossible task.

And even if the recognizable morphological kinds of the samples will be described with decades of monumental work by the practices of formal describing taxonomy and according to the rules of nomenclature, the biological and coenological information content of these would still remain hidden. However the material extracted with the Berlese-Tullgren device informs us not only about the presence and morphological diversity, but also about the species' abundance and dominance as well. Moreover it has become clear that this material together, as it was brought in front of us, contains a heap of such kind of information, which would be impossible to read from single species or from their constitutional characteristics. The samples collected this way are suitable for zoocoenological examinations. This observation led Humboldt to the recognition of „basic forms" (Grundformen) and later to the revelation of formations, which means the structure that can be found in plant associations without the exact knowledge of species. The emphasis here is on the „visibility" of the vegetation, because the recognition of biological communities began with the sight of the flora: vegetation is a „landscape element". Animal communities – apart from some exceptions – live hidden in the vegetation. But the Berlese-device concentrates and makes them visible.

In this present section of our examinations, the objective is to clarify the methodological possibilities of biological indication and the information content of the coenological data matrices by an appropriately chosen indication case study. We also set the aim to apply univariate indicators and to exploit the possibilities of multivariate coenological pattern analysis.

In this study series, there is need to introduce the main Oribatid sinusia of tropical areas as per climatic, vegetation and elevation zones. According to the holistic approach, we start from the whole and proceed towards the smaller parts. As a first step, we examined the similarities between the Oribatid sinusia of the Neotropical Region and the Notogaea. Stemming from the fact that the vertical stratification of Oribatid fauna follows the vegetation zones, and took samples for examination from 3 elevation zones:

1. tropical rainforest
2. mossforest
3. paramo

In the study, we disregarded the mountain forest zone, which can be found between the zones of tropical rainforest and mossforest, because the determination of its borders is quite uncertain. To avoid transitions, Berlese-samples of tropical rainforest have been selected from 200 m elevation above sea level, close to the forest border. 2 ideal transects were set for representative sampling. The first one crosses Andes at Costa Rica, at the 10. degree of latitude, from coast to coast. The second one starts from Papua New-Guinea, from the valley of Fly River at the 4. degree of Southern latitude, and goes up to the 4000 m high ridge of Mt Wilhelm. As an amendment, samples were also taken near to the 23. degree of Southern latitude in an additional transect, crossing Serra do Mar and Serra do Mantiguera. This transect has been set because 200-300 years ago there's been a belt of dense tropical rainforest – which even exceeded Amazonia in biodiversity – in the most densely populated area of Brazil, along the line marked by Sao Paulo and Rio de Janeiro. Almost 95 % of these forests have been devastated, but it could have been hoped that the rainforest spots reserved the original soil fauna – at least partially. 82 representative samples have been collected on 9 spots of the three transects. The spatial distribution of these samples and the abbreviation of individual sites can be seen in the next table:

	Tropical rainforest	Mossforest	Paramo	
Costa Rica	RC: 10	MC: 10	PC: 10	30
Brazil	RB: 7	MB: 6	PB: 10	23
New-Guinea	RN: 10	MN: 9	PN: 10	29
	27	25	30	82

Table 2.

The base table for our analyses was the 82 individual soil samples from the 9 examined habitat (3 habitat types of 3 areas) containing 111 Oribatid genera. For the various analyses, we created task-oriented assemblies from this base table.

Based on all of the individual samples, paired all samples with all other samples we created the similarity matrix of our data using multiple distance functions. In this current publication, Euclidean distance have been used.

Coenological similarity patterns can be analysed on multiple spatial scales (scale levels). We also analysed the similarity patterns of generic lists of different sites by NMDS with Euclidean distance, and hierarchical cluster-analysis. The two analytical results are shown projected onto each other (*Figure 3*). The multivariate similarity pattern of habitats' Oribatid community gave the expected picture. It can be stated that differences originating from the habitats and continents can also be recognized in the similarity patterns of the generic lists of the examined habitats. However, it can seem surprising that despite the vast geographical distances the pattern generating role of habitats does not disappear, it seems perhaps even more important. In accordance with the real ecological conditions, mossforest plays a transitional role between rainforest and paramo. However, mossforests are the most similar to each other and they are positioned in the middle of the similarity pattern, while the rest of the sites are separated radially. It is clearly visible that Brazilian sites are much more similar to each other than the Costa-Ricans.

Comparison of the sites can be fine-tuned if we also examine the similarity pattern of the individual soil samples considering every sample as different objects independent from the sites. We analysed this similarity pattern also by applying NMDS and Euclidean distance (*Figure 4*). Analysing the collective similarity pattern of every individual sample it can be stated that the groupings of elementary samples reflect their relations to the sites. This justifies the methodological decision by which sites are considered the basic objects of the examination. It can be stated furthermore, that habitat-type is unambiguously more significant pattern-generating factor than geographical attribution. Rainforests, mossforests and paramos lying thousands of kilometers from each other are more similar than sites of other habitats at only a few kilometers distance. Following the results of these pattern analyses, exact examination of the observed phenomena with a regression model seems to be practical.

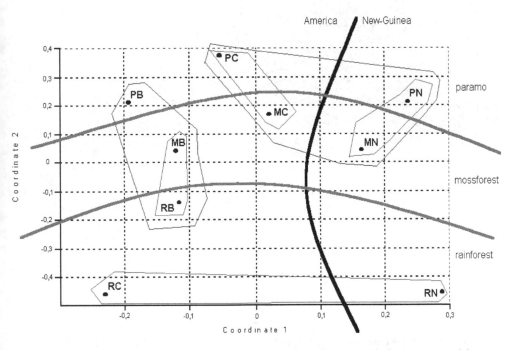

Figure 3. Similarity pattern of the examined sites in an NMDS ordination with the projection of the hierarchical cluster analysis results, applying Euclidean distance.

We pointed out that dissimilarities of habitats caused even by their type and also by the continent they originate from can be recognized in the similarity pattern of genus lists of the examined habitats. But if we analyze the overall similarity pattern of all the individual samples, it is quite conspicuous that the type of habitat is a much more significant pattern-generating factor than the geographical location. Rainforest, mossforests and paramos located many thousand kilometers from each other are more similar to each other than sites of other kind of habitats in only a few kilometers away.

The most important result of our case study is that the list of Oribatid genera as a coenological indicator, primarily characterizes the present ecological effects of the habitat and its climatically determined type of vegetation; and represents the effect of zoocoenological past in a much lesser extent. Thus, it can be concluded that a meritable scientific faunagenetical analysis should not be based upon geohistorical, but climatological grounds. This is why the ecological indication based on Oribatid genus lists provides unique possibilities for the purpose of climate change research.

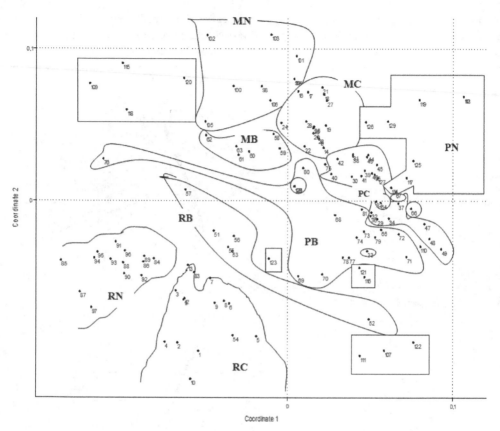

Figure 4. Similarity pattern of elementary samples in an NMDS ordination by applying Euclidean distance.

It became obvious that if we had chosen species and not genera as the basis of our examination, we would not get results that were interpretable from bioindicational aspect. This is because the majority of the described species would have appeared only as local specifica, and they do not provide any meritable information on similarity patterns – unless we are thinking in a very small grade of space. Species-level indication studies would be only rational in the case if we would deal only with cosmopolite species, but their number and detectability would not be sufficient for the majority of examinations for practical tasks. Thus, our important conclusion is that under current circumstances, the recommended taxonomical unit for indicational ecological studies can only be the genus. This statement is also important because many authors (primarily species describing taxonomists) consider a serious problem of indicational research that a number of researchers classify individuals only to genus level (Gulvik 2007). However, based on the work of Caruso (2006), we know that human contamination, intervention and disturbance can be better detected if we examine larger taxonomical units. It is also clear that rapid changes can not be detected on species-, but on community level. From the same work, it is also known that the species data of Oribatid mites can be raised to genus- or family level, and this does not cause loss of data or sensitivity by multivariate methods. According to Osler (1999), it is possible that habitat preference is determined rather at family level. Furthermore it is also known that there is strong relation between species-level and higher taxon level diversity indicators.

At the same time, the standardization and quantification of current Oribatid-collection methods would be obviously necessary for the development of the bioindicational methodology in order to make the data from different authors comparable. The need for this is emphasized more and more in literature (Gulvik 2007). A criteria-system of classifying the collected individuals into genera (or into other optional morphological groups) can be considered as a part of the standardized method. The current identificational and taxonomical practice in oribatodology is almost completely unsuitable for the purposes of biological indication research. Till such methods are unavailable, case studies can be conducted only if the person doing field sample collection and taxonomical processing is identical, and furthermore, this is only true if the coenological matrices have been created with the greatest care.

During our examinations (Balogh et al 2008), we pointed out that by advancing vertically upwards in the tropical high mountains (from rainforests towards the paramo vegetation), the average species number of genera present, and the extent of Holarctic and/or Antarctic relations of present genera grow, which can be well interpreted with the conception system of geographical analogy based on climatic similarities. This indication adequately supports former studies on the role of vegetational similarities and on genera as taxonomical units suitable for indication. Many authors (Andrew 2003, Melamud 2007 state that advancing upwards on the mountains, elevation above sea level and exposure have significant influence on the diversity of Oribatids. Besides, it is known that Oribatid diversity grows from the Boreal region towards warmer climates, but it does not grow further towards the tropics (Maraun 2007).

Furthermore, we pointed out that from the aspect of similarity of individual samples, among the examined habitat types the mossforest is the most homogeneous habitat, and paramo is the most heterogeneous. This latter phenomenon can be well interpreted if we consider the uniform moss cover prevalent in the mossforest, and the role of stable microclimate created by the moss cover. In the paramo, the observed proportion of species with low constancy level and also the heterogeneity of habitats (tussocks with different size) are high, and the role of the resulting microclimatic variability is obvious.

6. Conclusions

The genus lists of oribatid mites provide a unique indication device for the research of climate change, based on their following characteristics:

- Provide uncomparably simple, fast and effective sample collection opportunity in case of nearly any substrate type of nearly any habitat, all over the world.

- Are easy and quick to indentify on a genus- level
- Have well processed, synthetised scientific literature
- Seasonal stability
- Reflect the intrinsic, fundamental ecological characteristics of their biotops, which reflection overwrites the zoogeographical patterns
- Extremely high information content of the distance functions of genus lists.

Category orders as results of the analyses mostly met our expectations, first of all in case of genus lists, i.e. larger distances between genus lists correspond to larger spatial scales. However, time difference within a year could not be pointed out to a considerable extent either on genus or on family level. The most important differences in the orders are the followings:

1. the difference between samples originating from the same and different types of substrate;
2. the difference between the distance of lists coming from the same type of substrate in a given site and the distance of genus lists originating from the same topographicum but from different or the same types of site;
3. the difference between the distance of lists coming from the same or different types of substrate in a given site and the distance of lists originating from different topographicums;
4. the difference between the distance of lists coming from the same types of sites in a given topographicum and the distance of lists originating from different topographicums.

The analysis on family level differs from that of genus level that family lists of the zoogeographic kingdoms resemble each other as much as family lists originating from the same types of substrate.

Thus, compiled a reference list with the help of which it can be expressed to what spatial distances the similarity – shown with distance functions – of the genus or family lists of two oribatid mite communities originating from samples of unknown quality corresponds. Disturbed and transformed habitats can be compared with the help of oribatid mite communities based on an existing reference list.

Author details

V. Gergócs
Department of Plant Taxonomy, Ecology and Theoretical Biology, Eötvös Loránd University, Faculty of Science, Budapest, Hungary

R. Homoródi*
Department of Mathematics and Informatics, Corvinus University of Budapest, Faculty of Horticulture, Budapest, Hungary

---

* Corresponding Author

L. Hufnagel
Department of Mathematics and Informatics, Corvinus University of Budapest,
Faculty of Horticulture, Budapest, Hungary
„Adaptation to Climate Change" Research Group of the Hungarian Academy of Sciences,
Budapest, Hungary

Acknowledgement

We have to thank Prof. János Podani for his kind methodological help, Dr. Péter Balogh for his valuable professional help with the research of oribatid mites and for making the manuscripts from the Balogh-legacy available. We thank the "Adaptation to Climate Change" Research Group of the Hungarian Academy of Sciences and particularly the late Zsolt Harnos, who ensured the professional prerequisites of the research. Our research was supported by the Research Assistant Fellowship Support (Corvinus University of Budapest), „ALÖKI" Applied Ecological Research and Forensic Institute Ltd., the Landscape Architecture and Landscape Ecology PhD School of the Corvinus University of Budapest and the "Bolyai János" Research Fellowship (Hungarian Academy of Sciences, Council of Doctors). This work was supported by the research project of the National Development Agency TÁMOP 4.2.1.B-09/1/KMR-2010-0005.

7. References

[1] Altesor, A., Pineiro, G., Lezama, F., Jackson, R.B., Sarasola, M., Paruelo, J.M. (2006) Ecosystem changes associated with grazing in subhumid South American grasslands. – Journal of Vegetation Science 17:323-332.

[2] Andres, P. (1999) Ecological risks of the use of sewage sludge as fertilizer in soil restoration effects on the soil microarthropod populations. – Land Degradation & Development 10 (1): 67-77.

[3] Andres, P. (1999) Ecological risks of the use of sewage sludge as fertilizer in soil restoration effects on the soil microarthropod populations. Land Degradation & Development 10 1:67-77.

[4] Andrew, N.R., Rodgerson, L., Dunlop, M. (2003) Variation in invertebrate-bryophyte community structure at different spatial scales along altitudinal gradients. – Journal of Biogeography 30 (5): 731-746.

[5] Arroyo, J., Iturrondobeitia, J.C. (2006) Differences in the diversity of oribatid mite communities in forests and agrosystems lands. – European Journal of Soil Biology 42: 259-269.

[6] Askidis, M.D., Stamou, G.P. (1991) Spatial and temporal patterns of an oribatid mite community in an evergreen-sclerophyllous formation (Hortiatis, Greece). – Pedobiologia 35: 53-63.

[7] Badejo, M.A., Akinwole, P.O. (2006) Microenvironmental preferences of oribatid mite species on the floor of a tropical rainforest. – Experimental and Applied Acarology 40: 145-156.

[8] Badejo, M.A., Espindola, J.A.A., Guerra, J.G.M., De Aquino, A.M., Correa, M.E.F. (2002) Soil oribatid mite communities under three species of legumes in an ultisol in Brazil. – Experimental and Applied Acarology 27(4): 283–296.

[9] Balogh, J. (1965) A synopsis of the World Oribatid (Acari) Genera. – Acta Zoologica Academiae Scientiarum Hungaricae 9(1-2): 5-99.

[10] Balogh, J. (1972) The Oribatid Genera of the World. – Akadémiai Kiadó, Budapest.

[11] Balogh, J., Balogh, P. (1992) The Oribatid Mites Genera of the World. – The Hungarian National Museum Press, Budapest.

[12] Balogh, J., Mahunka, S. (1980) Atkák XV - Acari XV. – Magyarország Állatvilága, Fauna Hungariae, XVIII.(19).

[13] Balogh, J., Mahunka, S. (1983) Primitive Oribatids of the Palaearctic Region. – Akadémiai Kiadó, Budapest.

[14] Balogh, P., Gergócs, V., Farkas , E., Farkas, P., Kocsis, M., Hufnagel, L. (2008) Oribatid assemblies of tropical high mountains on some points of the „Gondwana-bridge" – a case study. – Applied Ecology and Environmental Research 6(3) 127-158.

[15] Bargagli, R. (1998) Trace Elements in Terrestrial Plants – Springer and Landes Company, Berlin.

[16] Battigelli, J.P., Spence, J.R., Langor, D.W., Berch, S.M. (2004) Short-term impact of forest soil compaction and organic matter removal on soil mesofauna density and oribatid mite diversity. – Canadian Journal of Forest Research-Revue Canadienne de Recherche Forestiere 34(5) 1136-1149.

[17] Bedano, J.C., Cantu, M.P., Doucet, M.E. (2005) Abundance of soil mites (Arachnida : Acari) in a natural soil of central Argentina. – Zoological Studies 44(4) 505-512.

[18] Behan-Pelletier, V., Winchester, N. (1998) Arboreal oribatid mite diversity: Colonizing the canopy. – Applied Soil Ecology 9: 45-51.

[19] Behan-Pelletier, V.M. (1999) Oribatid mite biodiversity in agroecosystems: role for bioindication. – Agriculture, Ecosystems and Environment 74(1-3) 411-423.

[20] Behan-Pelletier, V.M. (1999) Oribatid mite biodiversity in agroecosystems: role for bioindication. – Agriculture, Ecosystems and Environment 74: 411-423.

[21] Bengtsson, G., Tranvik, L. (1989) Critical metal concentrations for forest soil invertebrates. – Water, Air and Soil Pollution 47: 381-417.

[22] Berch, S.M., Battigelli, J.P., Hope, G.D. (2007) Responses of soil mesofauna communities and oribatid mite species to site preparation treatments in high-elevation cutblocks in southern British Columbia. – Pedobiologia 51(1) 23-32.

[23] Berch, SM; Battigelli, JP; Hope, GD. 2007. Responses of soil mesofauna communities and oribatid mite species to site preparation treatments in high-elevation cutblocks in southern British Columbia. Pedobiologia, 51:23 – 32

[24] Berg, N.W., Pawluk, S. (1984) Soil mesofauna studies under different vegetative regimes in north central Alberta. – Can. J. Soil Sci. 64: 209-223.

[25] Berthelsen, B., Olsen, R., Steinnes, E. (1995) Ectomycorrhizal heavy metal accumulation as a contributing factor to heavy metal levels in organic surface soils. – The Science of the Total Environment 170: 141-149.

[26] Black, H.I.J., Parekh, N.R., Chaplow, J.S., Monson, F., Watkins, J., Creamer, R., Potter, E.D., Poskitt, J.M., Rowland, P., Ainsworth, G., Hornung, M. (2003) Assessing soil biodiversity across Great Britain: national trends in the occurrence of heterotrophic bacteria and invertebrates in soil. – Journal of Environmental Management 67(3) 255-266.

[27] Bolger, T., Curry, J.P. (1984) Influences of pig slurry on soil microarthropods in grasslands. Rev. – Écol. Biol. Sol 21: 269-281.

[28] Caruso T, Migliorini M (2006a) Micro-arthropod communities under human disturbance: is taxonomic aggregation a valuable tool for detecting multivariate change? Evidence from Mediterranean soil oribatid coenoses. Acta Oecol 30:46–53

[29] Caruso T, Migliorini M (2006b) A new formulation of the geometric series with applications to oribatid (Acari, Oribatida) species assemblages from human-disturbed mediterranean areas. Ecol Mod 95:402–406

[30] Caruso, T., Migliorini, M. (2006) Micro-arthropod communities under human disturbance: is taxonomic aggregation a valuable tool for detecting multivariate change? Evidence from Mediterranean soil oribatid coenoses. – Acta Eocologica 30(1) 46-53.

[31] Cepeda-Pizarro, J.G., Gutierrez, J.R., Valderrama, L., Vasquez, H. (1996) Phenology of the edaphic microarthropods in a Chilean coastal desert site and their response to water and nutrient amendments to the soil. – Pedobiologia 40(4) 352-363.

[32] Clapperton, M.J., Kanashiro, D.A., Behan-Pelletier, V.M. (2002) Changes in abundance and diversity of microarthropods associated with Fescue Prairie grazing regimes. – Pedobiologia, 46(5) 496-511.

[33] Cole L., Buckland S.M., Bardgett R.G. (2008) Influence of disturbance and nitrogen addition on plant and soil animal diversity in grassland. – Soil Biology & Biochemistry 40(2) 505–514.

[34] Coleman, D., Fu, S., Hendrix, P., Crossley Jr., D. (2002) Soil foodwebs in agroecosystems: impacts of herbivory and tillage management. – Eur. J. Soil Biol. 38(1) 21-28.

[35] Coleman, D.C. (2008) From peds to paradoxes: Linkages between soil biota and their influences on ecological processes. – Soil Biology & Biochemistry 40(2) 271-289.

[36] Coleman, D.C., Crossley, Jr. D.A., Hendrix, P.F. (2004) Fundamentals of Soil Ecology. Chapter 9: Laboratory and Field Axercises in Soil Ecology. – Elsevier Academic Press, Burlington.

[37] Cortet, J., Gomot-De Vauflery, A., Poinsot-Balaguer, N., Gomot, L., Texier, C., Cluzeau, D. (1999) The use of invertebrate soil fauna in monitoring pollutant effects. – Eur. J. Soil Biol. 35(3) 115-134.

[38] Coulson, S.J., Hodkinson, I.D., Webb, N.R., Block, W., Bale, J.S., Strathdee, A.T., Worland, M.R., Wooley, C. (1996) Effects of experimental temperature elevation on high-arctic soil microarthropod populations. – Polar Biology 16(2) 147-153.

[39] Coulson, S.J., Leinaas, H.P., Ims, R.A., Sovik, G. (2000) Experimental manipulation of the winter surface ice layer: the effects on a High Arctic soil microarthropod community. – Ecography 23(3) 299-306.

[40] Coulson, SJ; Leinaas, HP; Ims, RA; Sovik, G. Experimental manipulation of the winter surface ice layer: the effects on a High Arctic soil microarthropod community Ecography 23 3:299-306.

[41] Denneman, C.A.J., Van Straalen, N.M. (1991) The toxicity of lead and copper in reproduction tests using the oribatid mite Platynothrus peltifer. – Pedobiologia 35: 305-311.

[42] Edwards, C.A., Lofty, J.R. (1975) The influence of cultivations on soil animal populations. – In: Vanek, J. (Ed.), Progress in Soil Zoology Academia Publishing House, Prague, pp. 399-406.

[43] Enami, Y., Shiraishi, H., Nakamura, Y. (1999) Use of soil animals as bioindicators of various kinds of soil management in northern Japan. – Jarq-Japan Agricultural Research Quarterly 33(2) 85-89.

[44] Erdmann, G., Floren, A., Linsenmair, K.E., Scheu, S., Maraun, M. (2006) Little effect of forest age on oribatid mites on the bark of trees. – Pedobiologia 50: 433-441.

[45] Fagan, L.L., Didham, R.K., Winchester, N., Behan-Pelletier, V., Clayton, M., Lindquist, E., Ring, R.A. (2006) An experimental assessment of biodiversity and species turnover in terrestrial vs. canopy leaf litter. – Eocologia 147: 335-347.

[46] Fox, C.A., Fonseca, E.J.A., Miller, J.J., Tomlin, A.D. (1999) The influence of row position and selected soil atributes on Acarina and Collembola in no-till and conventional continuous corn on a clay loam soil. – Appl. Soil Ecol. 13(1) 1-8.

[47] Franklin, E., Magnusson, W.E., Luizao, F.J. (2005) Relative effects of biotic and abiotic factors on the composition of soil invertebrate communities in an Amazonian savanna. – Applied Soil Ecology 29(3) 259-273.

[48] Franklin, E; Magnusson, WE; Luizao, FJ. 2005 Relative effects of biotic and abiotic factors on the composition of soil invertebrate communities in an Amazonian savanna Applied Soil Ecology 29 3:259 273

[49] Gackowski, G., Seniczak, S., Klimek, A., Zalewski, W. (1997) Soil mites (Acari) of young Scots pine forests in the region polluted by a copper smelting works at G1ogo'w (in Polish) – Zeszyty Naukowe ATR, Bydgoszcz, Ochrona S' rodowiska 208: 27-35.

[50] Gergócs, V., Garamvölcsi, Á., Homoródi, R., Hufnagel, L. (2011) Seasonal change of oribatid mite communities (Acari, Oribatida) in three different types of microhabitats in an oak forest. – Applied Ecology and Environmental Research

[51] Gergócs, V., Garamvölcsi, Á., Hufnagel, L. (2010) Indication strength of coenological similarity patterns based on genus-level taxon lists. – Applied Ecology and Environmental Research

[52] Gergócs, V., Hufnagel, L. (2009) Application of Oribatid Mites as Indicators. – Applied Ecology and Environmental Research 7(1) 79-98.

[53] Gill, R.W. (1969) Soil microarthropod abundance following old-field litter manipulation. – Ecology 50: 805-816.

[54] Gjelstrup, P. (1979) Epiphytic cryptostigmatic mites on some beechand birch-trees in Denmark. – Pedobiologia 19: 1-8.

[55] Gjelstrup, P., Søchting, U. (1979) Cryptostigmatic mites (Acarina) associated with Ramalina siliquosa (Lichenes) on Bornholm in the Baltic. – Pedobiologia 19: 237-245.

[56] Gulvik, M.E. (2007) Mites (Acari) As Indicators of Soil Biodiversity and Land Use Monitoring: a Review. – Polish Journal of Ecology 55(3) 415-440.

[57] Hågvar, S. (1998) Mites (Acari) developing inside decomposing spruce needles: biology and effect on decomposition rate. – Pedobiologia 42: 358-377.

[58] Hågvar, S., Abrahamsen, G. (1990) Microarthropods and Enchytraeidae (Oligochaeta) in naturally lead-contaminated soils: a gradient study. – Environmental Entomology 19: 1263-1277.

[59] Haimi, J., Laamanen, J., Penttinen, R., Räty, M., Koponen, S., Kellomäki, S., Niemelä, P. (2005) Impacts of elevated CO_2 and temperature on the soil fauna of boreal forests. – Applied Soil Ecology 30(2) 104-112.

[60] Hammer, M. (1952) Investigation On The Microfauna Of Nothern Canada, Part I, Oribatidae. – Acta Arctica.

[61] Hammer, M. (1958) Investigation On The Oribatid Fauna Of The Andes Mountains, I. Argentine and Bolivia.

[62] Hammer, M. (1961) Investigations On The Oribatid Fauna Of The Andes Mountains, II. Peru.

[63] Hammer, M. (1962) Investigations On The Oribatid Fauna Of The Andes Mountains, III. Chile.

[64] Hammer, M. (1962) Investigations On The Oribatid Fauna Of The Andes Mountains, IV. Patagonia.

[65] Hammer, M. (1966) Investigations On The Oribatid Fauna Of New Zealand, Part I-III.

[66] Hammer, M. (1972) Investigations On The Oribatid Fauna Of Tahiti And On Some Oribatids Found On The Atoll Rangiroa.

[67] Hammer, M., Wallwork., J.A. (1979) A review of tbe world distribution of oribatid mites (Acari: Cryptostigmata) in relation to continental drift. – Biol. Skrift. 22: 1-31.

[68] Hammer, Ř., Harper, D.A.T., Ryan, P.D. (2001) PAST: Paleontological Statistics Software Package for Education and Data Analysis. – Palaeontologia Electronica 4(1) 9. http://palaeo-electronica.org/2001_1/past/issue1_01.htm

[69] Hodkinson, I.D., Coulson, S.J., Webb, N.R., Block, W. (1996) Can high Arctic soil microarthropods survive elevated summer temperatures? – Functional Ecology 10(3) 314-321.

[70] Hopkin, S.P., Watson, K., Martin, M.H., Mould, M.L. (1985) The assimilation of heavy metals by Lithobius variegatus and Glomeris marginata (Chilopoda, Diplopoda) – Bijdragen tot de Dierkunde 55: 88-94.

[71] Hulsmann, A., Wolters, V., (1998) The effects of different tillage practices on soil mites, with particular reference to Oribatida. - Appl. Soil Ecol. 9: 327-332.

[72] Hunt, G.S., Colloff, M.J., Dallwitz, M.J, and Walter, D.E. (1998) Oribatid mites: an interactive key to Oribatid mites of Australia. CD-ROM for MS-Windows. – CSIRO Publishing, Melbourne.

[73] Hunt, G.S., Norton, R.A., Kelly, J.P.H., Collof, M.J., Lindsay, S.M., Dallwitz, M.J., Walter, D.E. (1998) Oribatid mites: an interactive glossary to oribatid mites. CD-ROM for MS-Windows. – CSIRO Publishing, Melbourne.

[74] Irmler, U. (2004) Long-term fluctuation of the soil fauna (Collembola and Oribatida) at groundwater-near sites in an alder wood. – Pedobiologia 48(4) 349-363.

[75] Irmler, U. (2006) Climatic and litter fall effects on collembolan and oribatid mite species and communities in a beech wood based on a 7 years investigation. – European Journal of Soil Ecology 42(1) 51-62.

[76] Jing, S., Solhoy, T., Wang, H.F., Vollan, T.I., Xu, R.M. (2005) Differences in soil arthropod communities along a high altitude gradient at Shergyla Mountain, Tibet, China. – Arctic, Antarctic and Alpine Research 37(2) 261-266.

[77] Johnston, J.M., Crossley Jr. D.A. (2002) Forest ecosystem recovery in a southeast US: soil ecology as an essential component of ecosystem management. – Forest Ecology and Management 155(1-3) 187-203.

[78] Kaneko, N., Salamanca, E.F. (1999) Mixed leaf litter effects on decomposition rates and soil microarthropod communities in an oak-pine stand in Japan. – Ecological Research 14(2) 131-138.

[79] Kaneko, N., Sugawara, Y., Miyamoto, T., Hasegawa, M., Hiura, T. (2005) Oribatid mite community structure and tree species diversity: A link? – Pedobiologia 49(6) 521-528.

[80] Kaneko, N; Sugawara, Y; Miyamoto, T; Hasegawa, M; Hiura, T.2005 Oribatid mite community structure and tree species diversity: A link? Pedobiologia 49 6:521- 528.

[81] Karasawa, S., Hijii, N. (2004) Effects of microhabitat diversity and geographical isolation on oribatid mite (Acari: Oribatida) communities in mangrove forests. – Pedobiologia, 48(3) 245-255.

[82] Karasawa, S., Hijii, N. (2008) Vertical stratification of oribatid (Acari: Oribatida) communities in relation to their morphological and life-history traits and tree structures in a subtropical forest in Japan. – Ecological Research 23(1) 57-69.

[83] Khan, A.G., Kuek, C., Chaudhry, T.M., Khoo, C.S., Hayes, W.J. (2000) Role of plants, mycorrhizae and phytochelators in heavy metal contaminated land remediation. – Chemosphere 41(1-2) 197-207.

[84] Koehler, H., Born, H. (1989) The influence of vegetation structure on the development of soil mesofauna. – Agriculture Ecosystems and Environment 27(1-4) 253-269.

[85] Korentajer, L. (1999) A review of the agricultural use of sewage sludge: benefits and potential hazards. – Water Air Soil Pollut. 17: 189-196.

[86] Koukoura, Z., Mamolos, A.P., Kalburtji, K.L. (2003) Decomposition of dominant plant species litter in a semi-arid grassland. – Appl. Soil Ecol. 23(1) 13-23.

[87] Kovác, L., L'uptácik, P., Miklisová, D, Mati, R. (2001) Soil Oribatida and Collembola communities across a land depression in an arable field. – Eur. J. Soil Biol. 37: 285-289.

[88] Krogh, P.H., Pedersen, M.B. (1997) Ecological effects assessment of industrial sludge for microarthropods and decomposition in a spruce plantation. – Ecotoxicol. Environ 36(2) 162-168.

[89] Lebrun, Ph., van Straalen, N.M. (1995) Oribatid mites: prospects for their use in ecotoxicology. – Experimental & Applied Acarology 19: 361-379.

[90] Liiri, M., Haimi, J., Setala, H. (2002) Community composition of soil microarthropods of acid forest soils as affected by wood ash application. – Pedobiologia 46(2) 108-124.

[91] Lindberg, N., Bengtsson, J. (2005) Population responses of oribatid mites and collembolans after drought. – Applied Soil Ecology 28(2) 163-174.

[92] Lindberg, N., Bengtsson, J. (2006) Recovery of forest soil fauna diversity and composition after repeated summer droughts. – Oikos 114: 494-506.

[93] Lindberg, N., Engtsson, J.B., Persson, T. (2002) Effects of experimental irrigation and drought on the composition and diversity of soil fauna in a coniferous stand. – Journal of Applied Ecology 39(6) 924-936.

[94] Lindo, Z., Visser, S. (2004) Forest floor microarthropod abundance and oribatid mite (Acari: Oribatida) composition following partial and clear-cut harvesting in the mixedwood boreal forest. – Canadian Journal of Forest Research-Revue Canadienne de Recherche Forestiere 34(5) 998-1006.

[95] Lindo, Z., Winchester, N. (2006) A comparison of microarthropod assemblages with emphasis on oribatid mite in canopy suspended soils and forest floors associated with ancient western redcedar trees. – Pedobiologia 50: 31-41.

[96] Lindo, Z., Winchester, N. (2007) Oribatid mite communities and foliar litter decomposition in canopy suspended soils and forest floor habitats of western redcedar forests, Vancouver Island, Canada. – Soil Biology and Biochemistry 39(11) 2957-2966.

[97] Lindo, Z; Visser, S. 2004 Forest floor microarthropod abundance and oribatid mite (Acari : Oribatida) composition following partial and clear-cut harvesting in the mixedwood boreal forest Canadian Journal of Forest Research-Revue Canadienne De Recherche Forestiere. 34 5:998-1006.

[98] Luxton, M. (1981) Studies on the Oribatid mites of a Danish Beech wood soil IV. Developmental biology. – Pedobiologia 21: 312-340.

[99] Maraun, M., Alphei, J., Beste, Ph., Bonkowski, M., Buryn,R., Migge, S., Peter, M., Schaefer, M., Scheu, S. (2001) Indirect effects of carbon and nutrient amendments on the soil meso- and microfauna of a beechwood. – Biology and Fertility of Soils 34(4) 222-229.

[100] Maraun, M., Salamon, J.A., Schneider, K., Schaefer, M., Scheu, S. (2003) Oribatid mite and collembolan diversity, density and community structure in a moder beech forest (Fagus sylvatica) effects of mechanical perturbations. – Soil Biology & Biochemistry 35(10) 1387-1394.

[101] Maraun, M; Schatz, H; Scheu, S. 2007. Awesome or ordinary? Global diversity patterns of oribatid mites. ECOGRAPHY, 30:209 – 216- (benne van, de formátum??)

[102] Marra, J.L., Edmonds, R.L. (1998) Effects of coarse woody debris and soil depth on the density and diversity of soil invertebrates on clearcut and forested sites on the Olympic Peninsula, Washington. – Environ. Entomol. 27(5) 1111-1124.

[103] Melamud, V., Beharav, A., Pavlicek, T., Nevo, E. (2007) Biodiversity interslope divergence of Oribatid Mites at „Evolution Canyon", Mount Carmel, Israel. – Acta Zoologica Academiae Scientiarum Hungaricae 53(4) 381-396.

[104] Migliorini, M., Bernini, F. (1999) Oribatid mite coenoses in the Nebrodi Mountains (Northern Sicily). – Pedobiologia 43(4) 372-383.

[105] Migliorini, M., Fanciulli, P.P., Bernini, F., (2003) Comparative analysis of two edaphic zoocoenoses (Acari Oribatida, Hexapoda Collembola) in the area of Orio al Serio Airport (Bergamo, northern Italy). – Pedobiologia 47(1) 9-18.

[106] Migliorini, M., Petrioli, A., Bernini, F. (2002) Comparative analysis of two edaphic zoocoenoses (Oribatid mites and Carabid beetles) in five habitats of the 'Pietraporciana' and 'Lucciolabella' Nature Reserves (Orcia Valley, central Italy). – Acta Eocologica 23: 361-374.

[107] Migliorini, M., Pigino, G., Caruso, T., Fanciulli, P.P., Leonzio, C., Bernini, F. (2005) Soil communities (Acari Oribatida, Hexapoda Collembola) in a clay pigeon shooting range. – Pedobiologia 49(1) 1-13.

[108] Minor, M.A., Norton, R.A. (2004) Effects of soil amendments on assemblages of soil mites (Acari : Oribatida, Mesostigmata) in short-rotation willow plantings in central New York Canadian Journal of Forest Research-Revue (Canadienne de Recherche Forestiere) 34: 1417-1425.

[109] Mitchell, R.J., Campbell, C.D., Chapman, S.J., Osler, G.H.R., Vanbergen, A.J., Ross, L.C., Cameron, C.M., Cole, L. (2007) The cascading effects of birch on heather moorland: a test for the top-down control of an ecosystem engineer. – Journal of Ecology 95(3) 540-554.

[110] Moldenke, A.R., Thies, W.G. (1996) Application of chloropicrin to control laminated root rot: Research design and seasonal dynamics of control populations of soil arthropods. – Environmental Entomology 25(5) 925-932.

[111] Moore, J.C., Snider, R.J., Robertson, L.S. (1984) Effects of different management practices on Collembola and Acarina in corn production systems. 1. The effects of no-tillage and Atrazine. – Pedobiologia, 26(2) 143-152.

[112] Müller, P. (1980) Arealsysteme und Biogeographie. – Ulmer, Stuttgart.

[113] Neave, P., Fox, C.A. (1998) Response of soil invertebrates to reduced tillage systems established on a clay loam soil. – Appl. Soil Ecol. 9(1-3) 423-428.

[114] Norton, R.A. (1994) Evolutionary aspects of oribatid mites life histories and consequences for the origin of the Astigmata. – In: M.A. Houck (Ed.), Mites: Ecological and Evolutionary Analyses of Life-History Patterns, Chapman & Hall, New York, pp. 99-135.

[115] Norton, R.A., Behan-Pelletier, V.M. (1991) Calcium carbonate and calcium oxalate as cuticular hardening agents in oribatid mites (Acari: Oribatida). – Canadian Journal Of Zoology 69(6) 1504-1511.

[116] Norton, R.A., Kethley, J.B., Johnston, B.E., Oconnor, B.M. (1993) Phylogenetic perspectives on genetic systems and reproductive modes of mites. – In: Wrensch, D, and Ebbert, M. (eds) Evolution and diversity of sex ratio in insects and mites, Chapman and Hall.

[117] Noti, M.I., Andre, H.M., Ducarme, X., Lebrun, P. (2003) Diversity of soil oribatid mites (Acari : Oribatida) from High Katanga (Democratic Republic of Congo) a multiscale and multifactor approach – Biodiversity And Conservation 12(4) 767-785.

[118] Noti, M.I., André, H.M., DufrSne. M. (1996) Soil oribatid mite communities (Acari: Oribatidaj from high Shaba (Zäire) in relation to vegetation. – Applied Soil Ecology, 5(1) 81-96.

[119] O'Lear, H.A., Blair, J.M. (1999) Responses of soil microarthropods to changes in soil wateravailability in tallgrass prairie. – Biology and Fertility of Soils 29(2) 207-217.

[120] Olejniczak, I. (2004) Communities of soil microarthropods with special reference to collembola in midfield shelterbelts. – Polish Journal of Ecology 52(2) 123-133.

[121] Olsanowski, Z. (1996) A monograph of the Nothridae and Camisiidae of Poland (Acari: Oribatida: Crotonoidea). – Genus International Journal of Invertebrate Taxonomy (Supplement), Wrocław.

[122] Osler, G.H.R., Beattie, A.J. (1999) Taxonomic and structural similarities in soil oribatid communities. – Ecography 22(5) 567-574.

[123] Osler, G.H.R., Cole, L., Keith, A.M. (2006) Changes in oribatid mite community structure associated with the succession from heather (Calluna vulgaris) moorland to birch (Betula pubescens) woodland. – Pedobiologia 50(4) 323-330.

[124] Osler, G.H.R., Harrison, L., Kanashiro, D.K., Clapperton, M.J. (2008) Soil microarthropod assemblages under different arable crop rotations in Alberta, Canada. – Applied Soil Ecology, 38(1) 71-78.

[125] Osler, G.H.R., Korycinska, A., Cole, L. (2006) Differences in litter mass change mite assemblage structure on a deciduous forest floor. – Ecography 29: 811-818.

[126] Osler, G.H.R., Murphy, D.V. (2005) Oribatid mite species richness and soil organic matter fractions in agricultural and native vegetation soils in Western Australia. – Applied Soil Ecology 29(1) 93-98.

[127] Osler, GHR; Beattie, AJ. 2001 Contribution of oribatid and mesostigmatid soil mites in ecologically based estimates of global species richness Austral Ecology 26 1:70-79.

[128] Podani, J. (1989) Comparison of ordinations and classifications of vegetation data. – Vegetatio 83: 111-128.

[129] Podani, J. (1997) Bevezetés a többváltozós biológiai adatfeltárás rejtelmeibe. – Scientia Kiadó, Budapest.

[130] Podani, J. (2001) SYN-TAX 2000 user's manual. – Scientia Kiadó, Budapest.

[131] Prinzing A., Kretzler S., Badejo A., Beck L., 2002 Traits of oribatid mite species that tolerate habitat disturbance due to pesticide application, Soil Biology & Biochemistry 34 1655–1661

[132] Prinzing A., Kretzler S., Badejo A., Beck L. (2002) Traits of oribatid mite species thattolerate habitat disturbance due to pesticide application. – Soil Biology & Biochemistry 34(11) 1655-1661.

[133] Reynolds, B.C., Crossley, D.A., Hunter, M.D. (2003) Response of soil invertebrates to forest canopy inputs along a productivity gradient. – Pedobiologia 47(2) 127- 139.

[134] Salmon, S., Mantel, J., Frizzera, L., Zanella, A. (2006) Changes in humus forms and soil animal communities in two developmental phases of Norway spruce on an acidic substrate. – Forest Ecology and Management 237(1-3) 47-56.

[135] Salmon, S; Mantel, J; Frizzera, L; Zanella, A. 2006. Changes in humus forms and soil animal communities in two developmental phases of Norway spruce on an acidic substrate. Forest Ecology and Management. 237: 47-56.

[136] Schenker, R. (1984) Spatial and seasonal distribution patterns of oribatid mites (Acari: Oribatei) in a forest soil ecosystem. – Pedobiologia 27(22) 133-149.

[137] Scheu, S., Schulz, E. (1996) Secondary succession, soil formation and development of a diverse community of oribatids and saprophagous soil macro-invertebrates. – Biodiversity and Conservation 5(2) 235-250.

[138] Seastedt, T.R. (1984) The role of microarthropods in decomposition processes. – Annual Review of Entomology 29: 25–46.

[139] Seniczak, S., Bukowski, G., Seniczak, A., Bukowska, H. (2006) Soil Oribatida (Acari) of ecotones between Scots pine forest and lakes in the National Park „Bory Tucholskie". – Biological Lett 43(2) 221-225.

[140] Seniczak, S., Dabrowski, J., Dlugosz, J. (1995) Effect Of Copper Smelting Air Pollution On The Mites (Acari) Associated With Young Scots Pine Forests Polluted By A CopperSmelting Works At Giogow, Poland. I. Arboreal Mites. – Water, Air, and Soil Pollution 94(3-4) 71-84.

[141] Siepel H, 1996, Biodiversity of soil microarthropods: The filtering of species Biodiversity and Conservation 5, 2: 251-260

[142] Siepel, H. (1995) Are some mites more ecologically exposed to pollution with lead than others? – Experimental and Applied Acarology 19(7) 391-398.

[143] Siepel, H. (1996) Biodiversity of soil microarthropods: The filtering of species. – Biodiversity and Conservation 5(2) 251-260.

[144] Skubala, P., Gulvik, M. (2005) Pioneer Oribatid Mite Communities (Acari, Oribatida) in Newly Exposed Natural (Glacier Foreland) and Anthropogenic (Post-Industrial Dump) Habitats. – Polish Journal of Ecology 53(3) 395-407.

[145] Skubala, P., Kafel, A. (2004) Oribatid mite communities and metal bioaccumulation in oribatid species (Acari, Oribatida) along the heavy metal gradient in forest ecosystems. – Environmental Pollution 132(1) 51-60.

[146] Smrž, J. (1992) The ecology of the microarthropod community inhabiting the moss cover of roofs. – Pedobiologia 36(6) 331–340.

[147] Smrž, J. (2006) Microhabitat selection in the simple oribatid community dwelling in epilithic moss cover (Acari: Oribatida). – Naturwissenschaften 93(11) 570-576.

[148] Smrž, J., Kocourková, J. (1999) Mite communities of two epiphytic lichen species (Hypogymnia physodes and Parmelia sulcata) in the Czech Republic. – Pedobiologia 43: 385–390.

[149] Stamou G.P.; Asikidis M.D.; Argyropoulou M.D.; Iatrou G.D., 1995 Respiratory Responses of Oribatid Mites to Temperature Changes , Journal of Insect Physiology, 41 3:229-233(5)

[150] Stamou, G.P., Asikidis, M.D., Argyropoulou, M.D., Iatrou, G.D. (1995) Respiratory Responses of Oribatid Mites to Temperature Changes, Journal of Insect Physiology 41(3) 229-233(5).

[151] Stamou, G.P., Sgardelis, S.P. (1989) Seasonal distribution patterns of oribatid mites (Acari: Cryptostigmata) in a forest ecosystem. – Journal of Animal Ecology 58(3) 893-904.

[152] Steiner, W.A. (1995) Influence of air pollution on moss-dwelling animals: 3. Terrestrial fauna, with emphasis on Oribatida and Collembola. – Acarologia 36(2) 149-176.

[153] Straalen N.M., Butovsky R.O., Pokarzhevskii A.D., Zaitsev A.S., Verhoef S.C., 2001, Metal concentrations in soil and invertebrates in the vicinity of a metallurgical factory near Tula (Russia) Pedobiologia 45, 451–466

[154] Straalen, N.M., Butovsky, R.O., Pokarzhevskii, A.D., Zaitsev, A.S., Verhoef, S.C. (2001) Metal concentrations in soil and invertebrates in the vicinity of a metallurgical factory near Tula (Russia). –Pedobiologia 45(5) 451-466.

[155] Sulkava, P., Huhta, V. (2003) Effects of hard frost and freeze-thaw cycles on decomposer communities and N mineralisation in boreal forest soil. – Applied Soil Ecology 22(3) 225-239.

[156] Taylor, A.R., Schröter, D., Pflug, A., Wolters, V. (2002) Response of different decomposer communities to the manipulation of moisture availability: potential effects of changing precipitation patterns. – Global Change Biology 10(8) 1314-1324.

[157] Taylor, A.R., Wolters, V. (2005) Responses of oribatid mite communities to summer drought: The influence of litter type and quality. – Soil Biology & Biochemistry 37(11) 2117-2130.

[158] Trueba, D.P., Gonzalez, M.M.V., Aragones, C.R. (1999) Soil mesofaunal communities from a periodically flooded lowland forest in the Si'an Kaan, Biosfere Reserve, Quintana Roo, Mexico. – Revista De Biologia Tropical 47(3) 489-492

[159] Tsiafouli, M.A., Kallimanis, A.S., Katana, E., Stamou, G.P., Sgardelis, S.P. (2005) Responses of soil microarthropods to experimental short-term manipulations of soil moisture – Applied Soil Ecology 29(1)17 26

[160] Uvarov, A.V. (2003) Effects of diurnal temperature fluctuations on population responses of forest floor mites. – Pedobiologia 47(4) 331-339.

[161] Valix, M., Tang, J.Y., Malik, R. (2001) Heavy metal tolerance of fungi. – Minerals Engineering 14(5) 499-505.

[162] Wallwork, J.A. (1983) Oribatids in forest ecosystems. – Annual Review of Entomology 28: 109–130.

[163] Walter, D.E., Proctor, H.C. (1999) Mites: Ecology, Evolution and Behaviour. – CABI Publishing, Wallingford. 322pp.

[164] Webb, N.R., Coulson, S.J., Hodkinson, I.D., Block, W., Bale, J.S., Strathdee, A.T. (1998) The effects of experimental temperature elevation on populations of cryptostigmatic mites in high Arctic soils. – Pedobiologia 42(4) 298-308.

[165] Yoshida, T., Hijii, N. (2005) The composition and abundance of microarthropod communities on arboreal litter in the canopy of Cryptomeria japonica trees. – Journal of Forest Research 10(1) 35-42.

[166] Zaitsev, A.S., Chauvat, M., Pflug, A., Wolters, V. (2002) Oribatid mite diversity and community dynamics in a spruce chronosequence. – Soil Biology & Biochemistry 34(12) 1919-1927.

[167] Zaitsev, A.S., van Straalen, N.M. (2001) Species diversity and metal accumulation in oribatid mites (Acari, Oribatida) of forests affected by a metallurgical plant. – Pedobiologia 45(5) 467-479.

[168] Zaitsev, AS; van Straalen, NM. 2001 Species diversity and metal accumulation in oribatid mites (Acari, Oribatida) of forests affected by a metallurgical plant Pedobiologia 45 5:467-479.

Chapter 10

The Role that Diastrophism and Climatic Change Have Played in Determining Biodiversity in Continental North America

Stuart A. Harris

Additional information is available at the end of the chapter

1. Introduction

Diastrophism includes both tectonic movements and plate tectonics. Together with changes in climate, they largely explain the high number of endemic species and biodiversity of both plants and animals on the North American continent. Each continent has had a different climatic and geological history, which explains why each continent differs in its biota from the others, although there are considerable similarities in the evolution of the biota. Thus in Europe, there are relatively few endemic species of plants (Birks, 2008) in contrast to 70% endemic vascular plants in the North American Tundra and Boreal Forest of the western Cordillera of Canada (Harris, 2008), and close to 100% endemic species in some groups of insects. This chapter provides an outline of the evolution of the biota of North America from the time that it was part of the massive continent (Pangaea) occupying the area around the South Pole, until the present day.

2. Palaeozoic era

At the beginning of the Palaeozoic Era, Laurasia was near the equator on the periphery of Pangaea. The latter was a large land mass consisting of the progenitors of the existing continents that were located around the South Pole at that time. In the Cambrian Period (600-500 million years ago), North America was orientated so that the equator ran from western Mexico northwards to the Arctic Ocean along the Yukon-NWT border (Briden & Irving, 1964). During the Ordovician Period (500-425 million years ago), the continent rotated so that the equator lay in a line from just south of Baja California northeastwards to the east coast of Hudson Bay. The rotation of North America continued and by the Carboniferous Period (345-280 million years ago), it passed through San Diego east-northeastwards to Cape Breton. Thus the North American plate was rotating in a clock-wise direction as well as moving south relative to the Earth's magnetic poles, which are assumed to approximate the poles of rotation. Throughout this time, it lay in the Tropics.

It was during the Silurian Period (425-405 million years ago), that plants started to move from the sea on to the tropical parts of the land. Centipedes and spiders appeared amongst the vegetation. In the succeeding Devonian Period (405-345 million years ago), forests were spreading across the hot, humid land areas and the first amphibians appeared on land. By the Carboniferous Period (425-380 million years ago), the extensive coal beds in both Europe and North America indicate that parts of the area had a hot, continuously humid climate supporting a dense equatorial-type forest (Schwarzbach, 1961). Absence of evidence of glacial deposits but the presence of sand dune deposits suggests that there were also areas of hot deserts in the southwest of the United States. It is sometimes called the "Age of the Amphibians" since they were most numerous during the early part of this Period. Off-shore, extensive limestone deposits were being formed together with extensive coral reefs. By the end of this era, many of the genera of horsetails, conifers, reptiles and insects had already been evolved and were present on land. The tropical forests that produced the coal beds had a diverse fauna and flora, the remains of which are found entombed as fossils. The animals were of predominantly different species and genera than those present today, though the main groups of primitive plants were already well represented. The ancestral club-mosses (*Lycopodium*) that had first appeared in the Silurian Period in what is now Australia, grew to over 30m in height in these forests, as did the trees of the class *Cordiatales*. The first conifers subsequently evolved from these in the Permian Period. Giant horsetails (*Equisetum*) were also abundant. These plants were, of course, essential to support the terrestrial food chain, and the enormous production of coal indicates that there was a spectacular increase in release of oxygen and an enormous reduction in carbon dioxide in the atmosphere (Harris 2010), though some of the carbon dioxide was replaced by volcanic activity during the Permian Period. The swampy areas in the forests also supported a wide range of primitive animals including small amphibians, eel-like creatures and the first small reptiles. The latter were a reasonable match for the amphibians but were not capable of extirpating them. In the air, giant insects flitted about. However, these animals were adapted to a very specific microenvironment and when that ceased to exist due to erosion in the west and mountain building in the east, most were extirpated, e.g., the enormous flying insects. A few key species were able to adapt to the altered environment, and of these, some genera of mycchoriza fungi, *Lycopodium* and *Equisetum* are still to be found today.

By the Permian Period (280-230 million years ago), the equator lay in a line from the California-Oregon border to just north of Newfoundland. Much of the land area was now reduced to a low plain, although mountain chains were rising in the Appalachian region due to collisions of plates (the Appalachian Orogeny). Present-day North America continued to experience a tropical climate, but the mountain chains cut off the rain-bearing winds coming inland from the east coast. This dry environment was ideal for the evolution of a wide variety of dinosaurs from the survivors of the small Paleozoic reptiles, and extensive sand dunes occurred in the southwest of the United States. In the wet areas, the first true reptiles and the ancestors of the crocodiles, alligators and caimen appeared.

Amongst the new genera and species that evolved on this new landscape were the first Laurasian stoneflies (Plecoptera) and several other groups of insects (Illies, 1965; Zwick and Teslenko, 2000). The best record of these insects comes from the Baltic amber of Eocene age (38-54 million years ago) but by then, the fauna resembled the modern one (Hynes, 1988). The Mecoptera which consist of only about 500 species worldwide today (Penny & Byers, 1979) were particularly conspicuous across Laurasia in the Lower Permian sediments (Carpenter, 1930). In general, the genera and families present at that time are also found in the Mesozoic sediments, but died out during the Tertiary Era as the mean annual air temperatures decreased.

The jostling of the plates also resulted in tension, causing faulting in many parts of the continents, and this resulted in widespread out-pouring of basaltic magma. This released considerable quantities of carbon dioxide into the atmosphere which would provide sustenance for the tropical forests on the fans below the mountains. In these forests, the first Conifers became abundant, and those insects and amphibians that survived the environmental changes became more widespread. Laurasia began moving northwards away from Gondwanaland just before the end of this period. Since mammals were later to become important species on all the continents, it is probable that the first ancestors of the mammals evolved at this time, prior to the split, although the first known mammalian fossils are dated to about 200 million years ago in the early Jurassic Period.

3. Mesozoic era

The northward movement of the Laurasian plate during the Jurassic period (230-180 million years ago) began about 220 million years before present. The plate consisted of North America, northern Europe and northern Asia. Tropical forests grew in the wetter areas, while sand dunes existed in the drier regions. Laurasia had a varied topography with mountains, plains and shallow seas. The equator now lay across the continent from southern Baja California to New York, so that north-west Alaska was at 50°N (Briden & Irving, 1964, Figure 8). Since the land mass of North America was surrounded by warm oceans, the climate remained hot.

South of the land mass, the Tethys Sea was developing between Laurasia and Gondwanaland. It stretched from central and southern China westwards to the Atlantic Ocean and separated Laurasia from South America, Africa and India. The result was a circum-equatorial ocean. This had an enormous effect on the temperature of the Earth since water absorbs approximately five times as much solar radiation as soil. This ocean provided warm surface and thermohaline currents that carried heat northwards to the Arctic Ocean via the North Pacific, as well as hot tropical air masses (Harris, 2002a). The ice cap over Antarctica melted, leaving a series of large islands where a tropical forest evolved (Francis et al., 2008). A mega warm event had begun (Harris, 2012) that was to last from about 200 million years before present until 44 million years ago. Summer temperatures averaged 20° C during this global thermal maximum. The genera present in this Antarctic flora were the ancestors of the present-day tropical flora.

In the Northern Hemisphere, a tropical biota evolved that was adapted to much higher temperatures than the present-day tropical flora. This extended north to the polar areas, except perhaps on the highest mountains. A tremendous variety of dinosaurs evolved, some reaching gigantic proportions. The first mammals consisted of small rat-like creatures that managed to survive running around the feet of the large, dominant dinosaurs. The oldest known specimen of the Asilidae (robber flies) dates to about 110 million years ago (Grimaldi, 1990), though that author put the evolution of this group at about 144 million years before present. However, they could have originated before Laurasia parted from the southern continents (Yeates & Grimaldi, 1993, Yeates & Irwin, 1996). By Late-Cretaceous times, small primitive marsupials and insectivores similar to shrews and hedgehogs were fairly abundant, and would survive the Cretaceous/Tertiary die-off of the previously dominant Dinosaurs. Cycads and conifers were the main components of the forests, while crocodiles, turtles and lamellibranchs, e.g., oysters, were common in freshwater lakes and swamps. It was during this time that the first birds appeared, apparently evolving from certain groups of dinosaurs.

Eurasia was not a single land mass (Cox, 1974). Shallow epicontinental seas covered part of the plate, so that the Turgai Straits separated Scandinavia and Britain (which was part of North America at that time) from the main Asian land mass. Thus although Britain and Scandinavia were part of the same plate, movement of the terrestrial biota across the land areas to Asia was restricted in time and space.

About 150 Ma before present, North America and Eurasia started to separate at their southern margin, and the North Atlantic Ocean began to form. Initially, the biota of both continents remained the same and the opening proceeded slowly, but as the plates continued to move northwestwards, the biota that required very hot tropical conditions could no longer pass back and forth between Eurasia and North America. The first angiosperms appeared at 144 million years ago.

In early Cretaceous times (135-63 million years ago), an epicontinental sea (the Mid-Continental Seaway) encroached onto the land that is now along the Mackenzie valley, and gradually extended south until it reached the ocean in what is now the Gulf of Mexico. This divided the floras and faunas into eastern and western populations, but by Late Cretaceous times the sea had largely dried up. During this time, the two sides had developed different insect faunas since they could not cross the wide expanse of water, but these became homogenized when the sea no longer acted as a barrier. Combined with the Turgai Straits, the Mid-Continental Seaway also resulted in there being two distinct faunas of land animals including the dinosaurs (Cox, 1974, Noonan, 1988, Wolfe, 1975). One was called Asia-America (Asia plus western North America) with a connection between them at high latitude across present-day Alaska and Siberia, and the other, Euroamerica (eastern North America plus northwest Europe). These faunas remained distinct after the disappearance of the Mid-Continental Seaway. During the early phases of the opening of the North Atlantic in the Middle Cretaceous (90-95 million years ago), a rift is believed to have formed between Greenland and Labrador, resulting in Greenland, the Rockall Plateau and Europe being a single land mass separate from North America (Heirtzler, 1973).

About 80 million years before present, the Laurasian plate no longer had room to continue moving northwards. Eurasia became stationary, but North America started its western movement that continues until today. This gradually extended the Atlantic Ocean northwards as two separate plates were formed from the former Laurasian plate. It was also during the Mesozoic Era that South America, Africa and India separated from the land masses around the South Pole. Africa and India headed towards Eurasia, but South America moved northwestwards towards the eastern part of Asia-America.

4. Cenozoic era

At the beginning of the Cenozoic era (63 million years ago), the climate was still tropical, though North America was moving to much higher latitudes. The Pacific Plate was moving north-northwest, but about 43 million years before present, its direction changed to west-northwest, as indicated by the change in direction of the Emperor and Hawaiian sea-mount chains (Clague & Jarrard, 1973). At the same time, the Aleutian Trench and Chain started to form (Worrall, 1991). This altered the geometry of the Beringian Gateway for warm currents carrying heat to the Arctic Ocean. About 38 million years ago, the Turgai Strait Gateway to the Arctic Ocean closed (Marincovich et al., 1990, McKenna, 1975). At the end of the Paleocene (c. 60 million years ago) in the Southern Hemisphere, the tropical forests were gradually displaced by floras dominated by the *Araucaria* conifers and the southern beech *Notofagus* which could survive freezing winter temperatures. Australia and Antarctica were still joined, but Australia started moving north around 55 million years ago [Kamarovitch & Geoph, 2009). Oxygen isotope data from the Atlantic seas of South America show that there was a marked cooling of the ocean where the planktonic foraminifera lived from about 19°C during the late Paleocene, resulting in sea temperatures of about 6°C by the early Oligocene (Shackleton & Kennett, 1995). A shallow water connection developed between the southern Indian and Pacific Oceans over the South Tasman Rise by about 39 million years ago, and at 38 million years before present, a large area of sea ice had developed over Antarctica. Bottom ocean temperatures plummeted and the thermohaline ocean circulation was initiated. The Drake Passage between Patagonia and Antarctica probably opened up about this time. Meanwhile in the Arctic Ocean, ice-rafted sediments were deposited on the sea floor from about 44 million years ago (Tripati et al., 2008). The mega cold event that we live in was beginning.

Soon after (before 29 million years ago), the Tethys Sea became fragmented into the Mediterranean Sea and an eastern portion including the Indian Ocean, as the Arabian plate collided with the almost stationary Eurasian plate. This resulted in the crumpling of the marine sediments in the former Tethys, which were uplifted into the mountains ranges of Iraq, Persia and Afghanistan (Nomura et al., 1997). Gone was a substantial part of the heat source for the Arctic, and the climate of the northern land areas was cooling (Harris, 2002a). The gradual closing of the gap between Asia and North America reduced the flow of warm ocean currents into the Arctic Basin by about 23.5 million years before present, though this was ultimately offset by currents flowing along the sea connection by the opening of the North Atlantic Ocean about 20 million years ago.

These events appear to have resulted in a dramatic drop in mean annual air temperatures around 35 million years ago (Frakes, 1979). This heralded the end for many of the tropical species of biota inhabiting the northern regions. They were replaced by species that could tolerate the cooler temperatures, and these changes in the biota have continued until today. The loss of the warm ocean currents entering the Arctic Basin would appear to be the most likely cause of the abrupt change in mean annual air temperature around the basin and the consequent extinction of so many species and genera that had survived for so long.

The Cenozoic was when the Mammals became the dominant land animals, aided by the presence of fur and by being warm-blooded. Within 10 million years, they had become greatly diversified and lived in almost all micro-environments. They included herbivores and carnivores, e.g., whales and bats. Those weighing more than about 45 kg are referred to as the Megafauna, and first appeared in Eocene times (55-30 million years ago). These became abundant in the early and middle Cenozoic, but largely died out in the Pliocene and Pleistocene. They included herbivores such as *Coryphodon*, *Uintatherium* and *Arsinoitherium*, together with carnivores such as *Andrewsarchus* and smaller wolf-like predators and sabre-toothed cats. About 40 million years ago, the first camels evolved (Harrington, 1978). At first, they were rabbit-sized with four toes, but they became much larger about 24 million years ago. By about 5 million years before present, some were substantially larger than the present-day camels of Africa and Asia (Harrison, 1985). They were common on the dry scrub grasslands of central North America from 600,000 years ago until about 10,000 years before today. A larger species of camel entered Alaska and crossed the Bering land bridge to Asia about 5 million years before present, and evolved into the species still surviving in Africa and Asia. About 20,000 years ago, they died out in North America. Among the other mammals that originated in North America are the horses, mastodons and mammoths. These also migrated across the Bering Strait, but subsequently died out in North America. Our ancestors are believed to have helped kill them off, though some mammoths survived in relatively inaccessible areas of Alaska (Haile et al., 2009) and Northern Siberia (Boeskorov, 2004, 2006, Vartanyan et al., 1993) well into the Holocene. The last wild horse carcass in Siberia dates from as late as 2,150 years ago, but horses were reintroduced into North America during the second invasion by the Spanish Conquistadors.

Stewart & Stark (2002, Figure 3.2) also conclude that a considerable number of genera of Plecoptera (stoneflies) moved across the Bering Land Bridge in both Miocene and early Pliocene times, since today part of their distribution extends south of Alaska to California. Before the closure of the North Atlantic Land Bridge, there was an exchange of at least 5 genera of stoneflies with Europe that now exhibit an Amphi-Atlantic pattern of distribution.

There was also a tremendous explosion of species in the other terrestrial groups including amphibians, birds, fish, insects and reptiles. Many are endemics, often with a very limited distribution. Weber (1965) discussed the plant geography of the southern Rocky Mountains, and determined that there were also a number of species now living in Colorado and California that are also found in sub-tropical Asia. He concluded that they must have crossed the Bering Land Bridge while the climate was still sub-tropical in late Cenozoic times.

As the Atlantic Ocean extended northwards, the remaining land connection was limited to higher and higher latitudes. The history of its extension is quite complex with several phases taking place (Hallam, 1981, 1994). In the latter part of the life of the land bridge, only warm temperate and subtropical biota could pass across from one continent to the other. Molecular studies suggest that some interchange in flora has continued until very recently, but the mechanism has yet to be determined. No similar evidence has been found for interchange of animals.

Northward movement of the Indian plate resulted in the uplift of the Himalayan mountain range about 21-17 million years before present, further reducing the size of the tropical seas. The northward movement of the African plate further reduced the remnants of the Tethys, producing the mountains of southern Europe including the Alps and Carpathians.

Cooling of the Northern Hemisphere continued, so that by 6 million years ago, an open Boreal mixed forest had become established in southern Alaska. By then, the species present had largely been replaced by the ancestors of the present-day biota that were adapted to the much colder environments. A land connection across the Bering Strait briefly allowed warm temperate species to be exchanged between Asia and North America. Subsequently, only Arctic and Subarctic biota could cross the Land Bridge due to the marked cooling. Matthews (1980) has argued that most of the modern genera and species of insects in most of North America date from about 5 million years ago and are endemic. Relatively few insects live in the Arctic, most being found further south in warmer climates.

5. Onset of major cold events in North America

About 3.5 million years ago, the first major cold event (extensive glaciations and development of permafrost) affected the Cordillera of western North America. Cold conditions also extended across northern Canada, and there must have been a substantial southward movement of the climatic zones. Another brief connection with Asia across the Bering Land Bridge occurred when the sea level dropped during this first major cold event. There were to be 5 subsequent occasions when the land bridge was open, the last one being during the late Wisconsin cold event about 15,000-20,000 years ago. Altogether there were about 13 major cold events (Harris, 1994; 2000; 2005) separated by shorter interglacials.

The biota that do not live in the Arctic or Subarctic could not move across the Bering Land Bridge during the few times it was open during the last 3.5 million years. These include the Hispine beetles (Staines, 2006). Only one species of Lasiopogon (L. hinei) is found on both sides of the Bering Strait (Cannings, 2002), but that species is exceptional in its wide range throughout Eurasia.

The second major cold event at 3 million years ago produced the most extensive glaciation in Alaska and the Yukon Territory. The moisture came primarily from the open Arctic Ocean, but by the time the third major cold event occurred (2.58 million years before today), the Arctic Ocean had frozen over and permafrost with tundra was present along the Arctic coast. This split the temperate humid vegetation into separate eastern and western populations. These were forced to move south along the respective coasts as cooling continued, and any components of the biota that could not adjust to the changing environment were extirpated. Since the climate, topography and micro-environments were different on the two sides of North America, different Temperate and Subtropical species that lived in the more humid areas evolved on the two sides of the Continent. The eastern populations have significantly more species than the western populations. Likewise, the animal populations in the two areas show distinctive species adapted to the local environment. Only the Boreal Forest and Tundra biomes are distributed across the northern part of the continent. Further south, the dry central steppe (Prairies) separates the two populations of biota that are adapted to wetter climates. The Tropical conditions moved south into Central America, while Subtropical climates were limited to the extreme southern United States, even during the Interglacial periods.

Initially, the later glaciations only affected relatively small areas in the north, but subsequently, the ice caps have become far more extensive. Permafrost was particularly widespread across the northern parts of North America, and facilitated the spreading of the Tundra flora south along the western Cordillera (Harris, 2007a). Only in the last 100ka have the Milankovitch cycles become significantly correlated with the onset of glaciations (Harris, 2012, Imbrie & Imbrie, 1980). During the last cold event and probably during the earlier events of the last million years, permafrost extended down to the southern part of Arizona and New Mexico, so the biota of all the climatic zones had to migrate long distances to survive or else find local refugia. The ice sheets wiped out all the vegetation in their path as they advanced, and when they retreated, the biota had to rapidly migrate north again over distances in excess of 1500 km (Dynesius & Jansson, 2000; Harris, 2010a). The result was that only those species of plants and animals that could migrate, adapt or find a suitable refugium could survive.

Refugia were present throughout these climatic changes, providing suitable habitat for the biota, whether the changes involved mean annual air temperature, precipitation, or both. In the more northerly mountain valleys, a combination of cold air drainage, temperature inversions and steam fog provide a buffering of the mean annual air temperature, so that the effects of cooling events were greatly reduced (Harris, 2007b, 2010b). This undoubtedly helped the biota in the eastern part of Beringia survive during the major cold events of the last 3 million years.

It should be noted that there are multiple kinds of refugia. Until now, most of the literature only discusses the effects of variations in temperature (e.g., Willis and Whittaker, 2000; Stewart et al., 2010). As these authors point out, refugia exist for species during both warmer and colder conditions. However, the vegetation is actually controlled by a number of factors, of which the moisture regime is undoubtedly equally important. Since the vegetation cover is a critical part of the ecosystem, it is also a major factor in providing a suitable microenvironment for animals.

The colder climate of the Late Wisconsin event would have resulted in an expansion of the ranges of species adapted to the cold conditions southwards and also on to lands becoming exposed by the eustatic drop in sea level along the Grand Banks area off Newfoundland, the Gulf coast of the southern United States and along the Beringian land bridge. Undoubtedly in the north, these provided expanded ranges for the arctic mammalian fauna, as well as the limited number of arctic insects such as butterflies (Layberry et al., 1998). Many Arctic mammalian species were destined to become extirpated by a combination of hunting and climate change, but 6 species of butterflies still live in both Alaska and the adjacent part of Siberia. The existence of closely related but different species of butterflies in eastern Beringia (Alaska and the Yukon Territory) and in western Beringia (East Siberia) is apparent evidence of Holocene speciation there.

In the southwest United States, the climatic changes were accompanied by widespread tectonic movements and volcanism (Wahrhaftig & Birman, 1965). This orogenic activity started in middle Miocene times and is continuing today. Pluvial lakes developed in the inland drainage basins (Morrison, 1965) and the sediments in these basins contain scattered vertebrate fossils, freshwater mollusks and diatoms, some dating back to 3.4 million years ago. Fossils in the marine terraces provide further evidence of the climatic changes and their effects on the local biota. The isolated volcanic mountains tend to have local endemic biotas that have evolved to cope with the local microenvironments. During pluvial events, animals such as voles are believed to have descended from the mountains and became widespread in the Great Basin and Mohave Deserts (Findley & Jones, 1962, Norris, 1958). Southward movement of sage voles and other animals resulted in their presence as fossils in the Isleta Caves of New Mexico (Harris & Findley, 1964). Around Lake Bonneville, speciation resulted in the appearance of a new species of *Oxyloma* (*O. missoula,* Succinidae) that is limited to that particular drainage area (Harris & Hubricht, 1982).

In the southeast United States, a narrow zone of subtropical and tropical vegetation may have persisted along the coast during the glaciations, but permafrost conditions were present along the higher and more northern parts of the Appalachian Mountains. Once again there is clear evidence of the southward migration of the biota during the glaciations. Remains of mammoths and other cold arctic fauna are well known from along the eastern seaboard of the United States, and have even been found in the sediments on the shallow sea floor that would have been exposed as dry land due to eustatic sea level changes during the glaciations. There was a gradual change in the mammalian fauna throughout the sequence of glaciations, indicating that speciation was fostered by the climatic changes (see Hibbard et al., 1965, for a summary). This contrasts with only slow speciation being reported in the insect world, since speciation in insects seems to have been slower than the speed of the major climatic changes (Matthews, 1980). Other animals, e.g., the Mollusca, tended to evolve in a similar way to the mammals in response to the environmental changes during the last 3.5 million years.

The aquatic biota also had to adjust their ranges. The north-south Mississippi River was very important, since it facilitated the migration south of fish and aquatic mollusks from the interior of the continental United States. The biota had to move into the warmer waters close to the Caribbean Sea which also diminished in size as sea levels dropped due to the accumulation of ice on land. During the last few glaciations, the Mississippi River acted as a spillway for the melting ice to the north, so the biota would have had to find refugia in smaller tributary streams. During deglaciation, they would migrate upstream to reclaim their previous habitats, or find new ones. Migratory birds presumably altered their migration patterns as they do today, to dwell in suitable habitats. When a cold event ended, they would adjust their migrations to make use of the new environments as they became available.

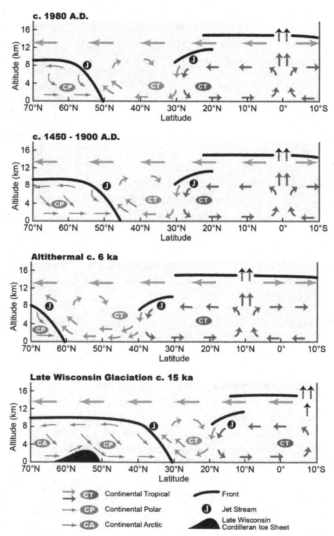

Figure 1. Changes in distribution of the main air masses during the main climatic extremes of the last 15,000 years along a north-south transect from the Arctic Ocean to Brazil along the Cordillera of the Americas (modified from Harris, 2010a). Note firstly that the zone of intertropical convergence moved south into northern Brazil during the Late Wisconsin cold event, and secondly, that the latitudinal range of the cold cA/cT air extended some 20° south of the ice sheets almost to the Mexican border along the Cordillera.

Origin	Common Name
North America	Rabbits
	Squirrel
	Field mice
	Cats
	Skunks and otters
	Foxes
	Bears
	Horses
	Camels and llamas
	Deer
South America	Opossums
	Armadillos
	Giant ground sloths
	Three-toed sloth
	Anteaters
	Monkeys
	Porcupines
	Guinea pigs

Table 1. Some mammal families that took part in the Great American Faunal Interchange (after Webb, 1997 and MacDonald, 2003).

6. Closing of the Panamanian Seaway

About 3.5 million years ago, the Panamanian Seaway between North and South America began a step-wise closure. The Isthmus of Panama was finally dry land by about 2.4 million years before today. South America had been moving northwestwards and had finally run into the southern margin of the North American plate. The northwards motion of the South America plate is still resulting in volcanism and the formation of the islands in the Caribbean Sea. Both plants and animals have taken part in exchanges between North and South America since then. During Interglacials, Tropical Rain Forest occurred along the Isthmus and permitted the exchange of that biota, but during major cold events, savannah conditions occurred there, allowing an exchange of the biota found in much drier areas (Harris, 2010a). As a result, many of the species of grasses and other vascular plants of

Guyana and northeast Brazil are the same as in Costa Rica, Honduras, sub-humid portions of Mexico and the south coast of the United States of America (FNA, 1993 - ?).

There was also a spectacular interchange of mammalian faunas between the two continents (Table 1, modified from Webb, 1997) that has been called "the Great American Interchange" (Marshall et al., 1982). The mammals from North America have tended to displace the marsupials of South America, whereas the marsupials could only move a limited distance north into the United States due to the cold winters and competition from the indigenous mammalian species. A small camel species crossed the Panamanian Land Bridge into South America and evolved into the present-day Llamas, Alpacas, Guanacos, and Vicunas. However only two genera (*Anacroneuria* and *Amphinemura*) of the Plecoptera which occur as a distinct Austral-American group in South America and two Nearctic groups in North America have managed to cross the Land Bridge (Stewart & Stark, 2002, p.16-17).

7. The Late Wisconsin Glaciation

This took place between about 25,000 years and 10,000 years ago in North America, although the timing of the glacial maximum is highly diachronous. Harris (2010) describes the evidence for the spectacular southward shift of the climatic belts along the western Cordillera during the main event (Figure 1). Permafrost extended down to Arizona and New Mexico, and the arid areas of Mexico moved south into Venezuela and the Guianas. The intertropical convergence zone moved at least 6 degrees to the south, so that sand dunes developed on the slopes just inland of the present-day coast of the Guianas (Harris, 2010a) and in northern Venezuela (Rabassa et al., 2005, Rabassa, 2008).

The ice sheets covered most of Canada (Figure 2), bull-dozing the landscape and destroying the vegetation in their paths. The cold events also allowed the migration of the biota adapted to the drier conditions occurring today in the area from the Mid-West down to Mexico to move south into South America across the Isthmus of Panama, since the climatic zones had moved south (Harris, 2010a). The rest of the biota had to adapt very quickly, or for those species living south of the ice sheets, migrate down the mountain sides and find a suitable refugium. Those that could not adapt perished. The Butterfly species, dependent on specific plants for survival, would have had to follow the changes in distribution of those plants. The species that migrate with seasonal changes would have needed to modify their migration patterns, probably shortening the distance of travel to adapt to the changing climatic zones.

Refugia around the ice sheets included Eastern Beringia (Harris, 2004), various islands along the northwest Pacific coast, the chief of which were the Queen Charlotte Islands (Calder & Taylor, 1968), unglaciated areas in southwest Alberta around Plateau Mountain (Harris, 2007a, 2008), the "Driftless Area" of southwest Wisconsin (Nekola & Coles, 2001, 2010), the Grand Banks east of New England, postulated areas in the far north of the Arctic Islands, and the area south of the ice sheets in the United States (Rogers et al., 1991). This was a time of rapid speciation of plants in the northern refugia (Harris, 2007a, 2008), the number of new species increasing with severity of the climatic change in the refugium. At least two species of land snails (*Gastrocopta rogersensis* and *Vertigo meramecensis*) are regarded as having evolved in the Driftless Area (Nekola and Coles, 2001, 2010), while about 40 species of Arctic-Alpine vascular plants evolved in Eastern Beringia (Harris, 2007a; 2008). This represents far more speciation in a given sized population than during the same period in most environments south of the ice sheets.

8. Deglaciation

About 14,000 years ago, there was a marked change in the relative strengths of the air masses affecting North America (Figure 1). This resulted in the northward movement of the climatic zones and the zone of intratropical convergence. It was not a continuous process; the climate fluctuated with both warmer and colder periods, thus complicating the revegetation process. The exact timing of the fluctuations and their areal extent is still being examined. Localized readvances of glaciers provide evidence of these fluctuations, as do variations in pollen, diatom and finger clam distribution at the base of the oldest post-glacial sediments at the bottom of lakes in the formerly glaciated areas. During the early part of the deglaciation process, the Cordilleran ice sheet had rapid local readvances, but in general, it down-wasted in situ with widespread ice stagnation in the valleys in British Columbia and in the Prairies from south-central Alberta across to North Dakota (Alley, 1976, Harris, 1985, Prest et al., 1968). In the Cordillera, the mountain tops appeared first from under the ice and there were numerous lakes in the valleys, with ice blocking the centre of the valley. The former water levels are marked by gravel terraces and hanging deltas along the valley walls. In the main part of central and eastern Canada, the ice persisted as a single entity centered on Hudson Bay (Figure 2) until about 9,000 years ago (Prest et al., 1968). It then split into two parts, one centered on the highlands of northern Québec and the other on the highlands west of Hudson Bay. Complete deglaciation did not occur until 6-7ka in these centers and about 4ka on Baffin Island.

The warming first commenced in the south, and slowly and jerkily moved north (Wright, 1970). Thus the prairie vegetation at 20,000 years before present was limited to a narrow north-south zone extending from central Oklahoma to the east coast of Mexico (Ross, 1970). Lehmkuhl (1980) discusses the movement of insects into the evolving Prairies with their harsh temperature regime and unpredictable precipitation. In the case of the grasshoppers that are found today across the Prairies, Ross reports that only 3 out of 82 species are endemics, 31 species have moved in from the west, 34 from Mexico, 10 from the Caribbean coast, and 7 from the eastern part of the United States. In the case of aquatic Mollusca, Clarke (1973) estimates that only about 21 species out of a potential 103 species now populate the western interior of Canada, probably due to the vast distances, poor dispersal mechanisms, and harsh climate. In contrast, mammals and birds could migrate readily across the region.

The flora and fauna adapted to the cold permafrost land would have had to contract their ranges with some cold-adapted butterflies and rock crawlers (Gylloblattidae) still surviving in small areas on mountain summits along the Cordillera of western North America. The mountain sheep and goats survive in this way, whereas the larger herbivores such as Mammoths could not survive in the southern parts of their former ranges.

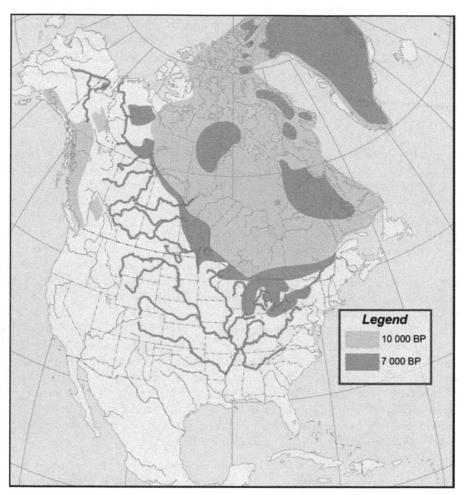

Figure 2. Position of the ice sheets in North America at 10,000 and 7,000 years before present (after Denton and Hughes, 1981). Also shown is the distribution of the proglacial drainage and the proglacial lakes that permitted the fishes and other aquatic biota to migrate northwards across present-day drainage divides. They were followed by many terrestrial snails that require moist microenvironments.

The flora and associated fauna of the more humid regions moved north through Mexico, divided into two coastal biotas by the semi-desert of the interior. The western group moved north into California and the Southern Rocky Mountain States, while the eastern biota moved east along the Gulf coast of the United States before moving further north. The results of this can be seen in the present distribution of the Monarch Butterfly (Urquhart, 1960). It feeds on various species of Milkweed that are found in southeast and southwest Canada, where it spends its summers (Figure 3). In winter, the two populations migrate south, the western population over-wintering on trees in the sheltered coastal canyons near Goleta, while the eastern populations over-winter in Florida or along the eastern coast of Mexico.

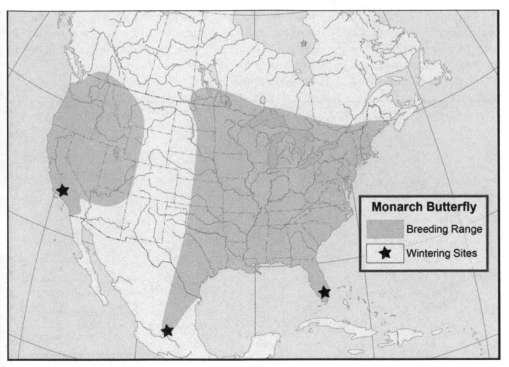

Figure 3. Present-day wintering sites and summer breeding range of the Monarch Butterfly (Partly after Urquhart, 1960). In winter, they migrate south, the western population forming giant clusters on groups of trees in California (Brower & Malcolm, 1991), e.g., in a gulley above the Pacific Ocean at Goleta, with the eastern population overwintering at several locations along the Caribbean coast of Mexico and near Mexico City (Wassenaar & Hobson, 1998). The Floridian group is believed to overwinter on the Yucatan Peninsula.

In the Western Cordillera of the United States, the basins dried up and the biota moved up the mountains, forming isolated ecosystems on the various mountain ranges. This accounts for the considerable endemism found in this region in the plants, reptiles and birds. In the case of the Kangaroo rats, the sharp boundaries between species distribution may be partly caused by competition (Brown & Lomolino, 1998). However in general, many species of birds and 80% of the butterflies rely on one or two families or even species of plants for food, so their distribution is limited by the distribution of the food plant. Additional tropical and subtropical species moved north, adding to the species richness of the region. Cacti, creosote bush and the associated biota moved into the semi-deserts from the south.

In the case of the trans-continental snake *Diadophis punctatus* (Figure 4), phylogeographic studies of the present–day populations in the United States indicate that there are 13 populations with distinctive DNA that are apparently adapted to specific environmental conditions (Fontanella et al., 2008). The nine southern populations have stable distribution boundaries and appear to represent a pre-glacial distribution that was relatively unaffected by the Late Wisconsin cold event (Figure 4). In the case of the 4 northern populations, they are currently migrating northwards into the areas with suitable climates following deglaciation. The preglacial populations in these four areas would have been extirpated in the northern part of these ranges during the cold event. Since the snake has limited ability to migrate over long distances, the adjustment of its northern boundaries of distribution is still taking place as it expands northwards into suitable environments.

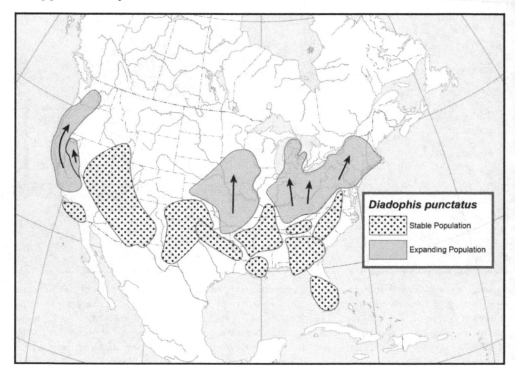

Figure 4. The distribution of the present-day distinct haplotypes (populations that have distinctive DNA) of the snake, *Diadophis punctatus* (after Fontanella et al., 2008). The distributions of the southern populations are stable, and may have lived in the same area since before the Late Wisconsin glaciations because of an acceptable microclimate. Also shown are the populations that are slowly expanding northwards to occupy areas that are now suitable for its existence, but were too cold for the snake during the Late Wisconsin Glaciation.

The formerly glaciated areas are another matter. The migration of species into the areas vacated by the glaciers was highly dependent on glacial history, distribution of refugia, the climate which limited the possible migrations, and the waterways in the case of aquatic and associated biota.

The downwasting of glaciers in western Canada and North Dakota slowed the recolonization process of trees (Figure 5), even though deglaciation commenced earlier at about 15,000 years ago. The Jack Pine, which survived the glaciations in the northeast United States, was able to move west along the Laurentide ice front, whereas the Lodgepole Pine that used the west coast as a refugium was largely blocked off from eastward movement by the ice in the mountain valleys and the dry area on the Prairies (MacDonald & Cwynar, 1985). Western Red Cedar only penetrated the Coast Ranges into the Interior Plateau of British Columbia about 3,000 years ago.

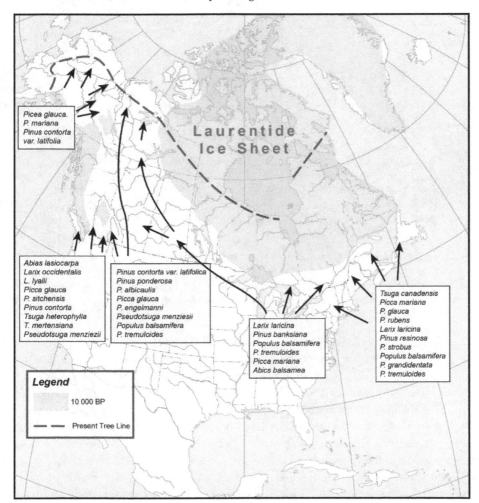

Figure 5. The influence of the residual ice masses on the migration of some of the common tree species into the deglaciated areas about 10,000 years ago.

In the case of the vascular plants of the present-day alpine tundra and boreal forest, revegetation in western Canada came primarily from Eastern Beringia (Figure 6), with only limited northward movement from the southern refugium (Harris, 2007a, 2008). The species in the SW Alberta refugium spread west across the limestone mountains, while those that survived the glaciation around the Queen Charlotte Islands moved north and south along the west side of the Coast Ranges. In the east, the vegetation gradually moved north following the retreating ice front in the period between 10,000 and 7,000 years ago.

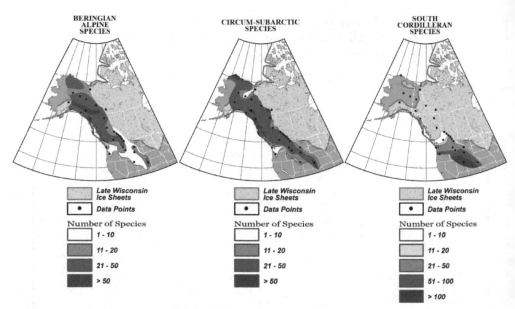

Figure 6. Migration of selected alpine tundra species from the Beringian refugium along the western Cordillera of North America contrasted with the limited northward migration from south of the maximum extent of the Cordilleran ice sheet (modified from Harris, 2007a). This migration had to occur in the short period when it was cold enough for their movement after the down-wasting of the ice sheet and before the mountain valleys became warm enough for invasion by the Montane Boreal Forest.

Various species of Succinidae (Mollusca) survived the glaciations in refugia (Figure 7), either in Beringia (e.g., *Succinea strigata*) or south of the ice sheets in the Rocky Mountains (*Oxyloma nuttalliana*), on the Central Plains (e.g., *S. indiana*), south of the Great Lakes (*Catinella avara, O. retusum,* and *S. ovalis*), on the exposed Grand Banks (now inundated by the sea) or in the unglaciated parts of New England (e.g., *O. groenlandicum* and *O. verrilli*). A new species (*O. missoula*) evolved in the Lake Bonneville drainage, but it did not migrate elsewhere during the Holocene. Most of the species of the Succineids need to be close to water (c. 1-2 m) and the bodies of terrestrial snails consist of about 50% water. They must lay down a thin layer of mucus as they move, so they are particularly dependent on moisture for survival. As a result, they had difficulty moving on to the rocky and sandy substrates of the Laurentian Shield following deglaciation, except where the waters of Glacial Lake Agassiz drained northwards into Hudson Bay.

Climate certainly limited many northward migrations, e.g., many vascular plants and the freshwater bivalves. The latter only entered the previously glaciated Cordilleran areas in northern Washington, northern Montana and in Southern British Columbia. A few species were able to move further north in the Prairie Provinces, but many only just entered Southern Ontario and the Red River Valley in Manitoba. They only occur along the margins of the Shield on the calcareous rocks and sediments in southern Québec, though the late deglaciation and colder climate there undoubtedly inhibited the northward migration.

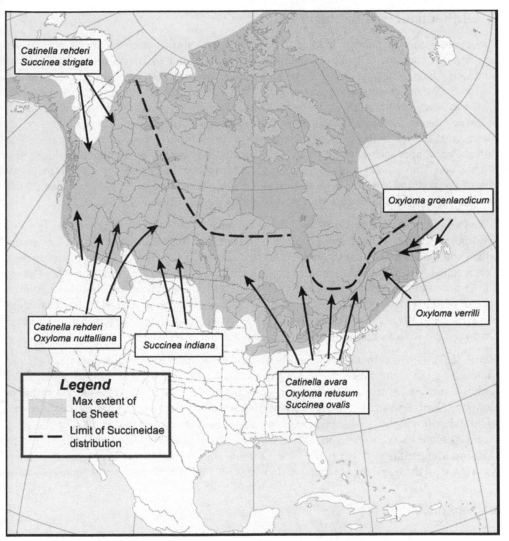

Figure 7. Migration routes of the various species of Succinedae that survived the Late Wisconsin glaciations south of the ice sheets, and moved north about 10,000 years ago following the proglacial drainage shown in Figure 2 (partly after Harris and Hubricht, 1975). Most of these species must live within one or two meters of water, and so they subsequently followed the water draining north from Lake Agassiz into the south end of Hudson Bay about 8700 years ago.

The biota that live all or part of their life in fresh water and could tolerate cold climates, e.g., fish and some insects (Lehmkuhl, 1980, Morgan & Morgan, 1980), moved along the main drainage-ways and proglacial lakes, e.g., Glacial Lake Agassiz, crossing into other watersheds using the temporary spillways. Many came from the unglaciated areas to the south, but others came from Beringia.

9. Altithermal/Hypsithermal event

The post-glacial warming culminated in a period of time when the mean annual air temperature was about 2 degrees warmer than at present in the Eastern Cordillera of Southern Alberta (Harris, 2002b). It reached its peak about 9,000 years ago in the southern United States, but the peak only affected the Prairie Provinces of Canada from about 6,500-4,500 years before today. It was the result of a weakening of the cP/cA air masses, resulting in the northward movement of the Arctic front (Figure 1). The vegetation zones migrated up the mountains, while those species that were at the mountain tops were extirpated. The Boreal Forest approached the Arctic Ocean about 5,000 years ago (Anderson et al., 1989, Ritchie and Hare, 1971) with White Spruce having been reported on the Tuktoyaktuk Peninsula at about the same time. It was at this time when the last remnants of the mammalian Megafauna disappeared in the isolated parts of Alaska and Siberia.

In the southern Canadian Cordillera, it began after 6,830 years ago. The westerly rain-bearing mP air mass as well as the cold cA/cP air mass weakened relative to the cT air mass, which was therefore able to move north into the southern Yukon Territory before turning east. This resulted in drought conditions across the Prairies with drying up of the ponds and lakes. The Prairie vegetation moved north of latitude 60° (Strong & Hills, 2003), and these authors concluded that it may have migrated westwards into the interior plateau of the Cordillera along the Liard River valley. Remnants of prairie-type vegetation may be found today around Carmacks and Kluane, Yukon Territory, e.g., *Krascheninnikovia lanata* (Douglas et al., 2001).

Harris & Pip (1973) demonstrated that there was considerable northward migration of land snails along the main river valleys. Tiny land snails that are currently limited to the east-central United States of America migrated along the major river flood plains to the Cordillera along the North Saskatchewan and Missouri rivers. Today remnants of these populations can be found surviving at sheltered places, but are slowly being extirpated (Figure 9). Thus *Gastrocopta armifera, G. similis* and *G. pentodon* were present in the vicinity of Lake Louise in 1880 but were extirpated when the CPR railroad was built. There are still some isolated occurrences of these species remaining at sheltered isolated sites along the river valleys.

10. Neoglacial events

A series of three cold Neoglacial events began about 4,500 years ago during which the MAAT was significantly cooler, and periodic localized increases in precipitation caused the glaciers to advance a short distance down-valley. The balance of the air masses north of about 48 degrees latitude changed (Figure 1), but this did not affect the area further south. There was also no change in the position of the zone of inter-tropical convergence. It was during the first event (3500 years ago) that the western red cedar (*Thuja plicata*) finally migrated up the Skeena River valley from the coast and colonized the wetter portions of the interior of southern British Columbia (MacDonald & Cwynar, 1975). The Prairies exhibited a large area of sand dunes during the last Neoglacial event (Wolfe et al., 2001; Wolfe & Hugenholtz, 2009). That is why the fur traders chose to paddle up the North Saskatchewan River rather than cross the sand dunes to reach the mountains. Similar sand seas existed in Nebraska. In the north, the Boreal Forest retreated 5 degrees to the south on the adjacent

Prairie by 4,000 years before today, while the Aspen Parkland moved about 1 degree south. These cooler events were separated by warmer intervals at least as warm as today, with that just before the last Neoglacial being even warmer. These changes favoured some species, but caused others to migrate latitudinally or altitudinally, or become extirpated. Similar changes took place throughout the continent.

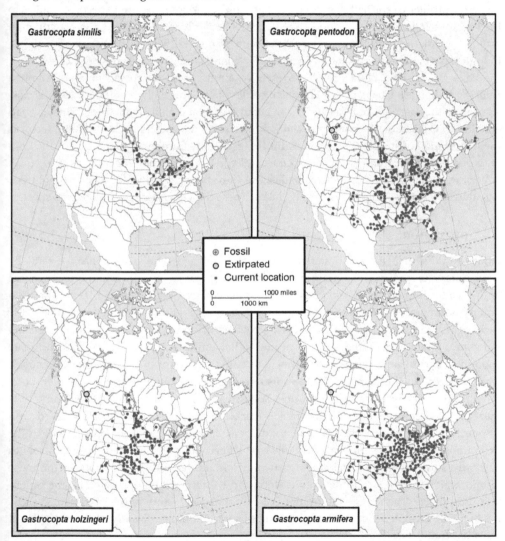

Figure 8. Distribution of 4 species of the genus *Gastropoda* that exhibit evidence of having extended their ranges westwards and northwards during the Altithermal/Hypsithermal warm event from their main area of present-day distribution. Today, remnants of the extended populations are found in favourable microenvironments along the river floodplains that they followed during the migration process. They are known to have been extirpated in the vicinity of Lake Louise, Alberta, since the late 1800s.

11. Post-Neoglacial changes

The last Neoglacial event ended about 110 years ago with the development of the present-day distribution of air masses (Figure 1), resulting in a minor change in temperature, but important changes in precipitation. There was a marked reduction in precipitation in the Eastern Cordillera, causing the vegetation that required more moisture to become limited to the high precipitation areas around Lake Louise (Harris, 2012). After 1943, the precipitation on the Canadian Prairies increased, and the sand dune field has largely become stabilized, though it is now used for irrigation farming. Grassland species such as the Prairie Swift Fox, the Sage Grouse and Black-Footed Ferret have almost been extirpated on the Canadian Prairies. In the last 500 years, the European settlers have gradually modified the landscape, starting in the east and south as well as at isolated coastal regions in the west. This has had an enormous effect on the biota, especially by destruction of habitat. An additional factor is the importation of species from other countries, especially Europe. Forty percent of the flora of Nova Scotia is from elsewhere (Zinck, 1998). Ships discharging water into North American waterways that is used as ballast are introducing fresh water fish and mollusks, e.g., the Zebra Mussel into the rivers and lakes. These then devastate the indigenous species.

12. Conclusion

Until the advent of European settlement, climatic changes and diastrophism essentially determined the biodiversity of the biota of North America. The species found today evolved in the last 6 million years in response to the marked cooling of the continent. There had been limited immigration of present-day species from Asia and South America, and little exchange with Europe. The alternating major warm and cold events caused repeated massive migrations latitudinally and altitudinally, unless a given species was fortunate enough to survive in a refugium. Species that could not adapt or migrate quickly enough were extirpated. The climatic changes also resulted in speciation in the vascular flora, though not in many of the insect groups. Speciation in most of the latter takes more time than the duration of most climatic major warm or cold events. This has resulted in a primarily endemic biota that is able to disperse into new environments rapidly. The exceptions are mainly found in the southwest United States on isolated mountain ranges currently surrounded by deserts. Of particular note is the split in the biota of the more humid regions at lower latitudes into eastern and western groups separated by the central semi-arid plains. This split is the result of the early glacial history of the continent. Clearly, the biota of North America has had a unique history that is significantly different to that of the other continents.

Author details

Stuart A. Harris
Department of Geography, University of Calgary, Canada

Acknowledgement

Professor G. M. MacDonald (UCLA) kindly commented on an earlier version of part of the subject matter and Robin Poitras drew the Figures.

13. References

Alley, N. F., 1976. The palynology and paleoclimatic significance of a dated core of Holocene peat, Okanagan Valley, British Columbia. *Canadian Journal of Earth Sciences*, 13: 1133-1144.

Anderson, T. W., Mathewes, R. W. & Schweger, C. E., 1989. Holocene climatic trends in Canada with special reference to the Hypsithermal interval. In Fulton, R. J., Ed.. *Quaternary geology of Canada and Greenland, the Geology of North America*, K-1: 520-528. Geological Survey of Canada, Ottawa.

Birks, H. H., 2008. The Late-Quaternary history of arctic and alpine plants. *Plant Ecology and Diversity*, 1: 135-146.

Boeskorov, G. G., 2004. The North of Eastern Siberia: Refuge of Mammoth Fauna in the Holocene. *Gondwanaland Research*, 7(2): 451-455.

Boeskorov, G. G.., 2006. Arctic Siberia: refuge of the Mammoth fauna in the Holocene. *Quaternary International*, 142- 143: 119-123.

Briden, J. C. & Irving, E., 1964. Palaeolatitude Spectra of sedimentary palaeoclimatic indicators. In: Nairn, A. E. M., Ed.. *Problems in Palaeoclimatology*. Interscience Publishers, London: 199-224.

Brower, L. P. & Malcolm, 1991. Animal migrations: endangered phenomena. *American Zoologist*, 31: 265-276.

Brown, J. H. & Lomolino, M. V., 1998. *Biogeography*. 2nd Edition. Sinauer Associates, Sunderland, MA.

Byers, G. W., 1988. Geographic affinities of the North American Mecoptera. *Memoires of the Entomological Society of Canada*, #144: 25-30.

Calder , J. A. & Taylor, R. L., 1968. *Systematics of the vascular plants. Flora of the Queen Charlotte Islands* volume 1. Monograph #4, Research Branch, Agriculture Canada. Ottawa.

Cannings, R. A., 2002. The Systematics of *Lasiopogon* (Diptera: Asilidae). Royal British Columbia Museum, Victoria. 354p.

Carpenter, F. M., 1930. The Lower Permian insects of Kansas. Part 1. Introduction and the order Mecoptera. *Bulletin of the Museum of Comparative Zoology, Harvard*, 60: 69-101.

Clague, D. A. & Jarrard, R. D., 1973. Tertiary Pacific plate motion deduced from Hawaiian-Emperor chain. *Geological Society of America Bulletin*, 84: 1135-1154.

Clarke, A. H., 1973. The Freshwater mollusks of the Canadian Interior Basin. *Malacologia*, 13(1-2): 509p.

Cox, C. C., 1974. Vertebrate Palaeodistributional Patterns and Continental Drift. *Journal of Biogeography*, 1: 75-94.

Denton, G. H. and Hughes, T. J., 1981. *The Last Great Ice Sheets*. New York. John Wiley and Sons,494p.

Douglas, G. W., Argus, G. W., Dickson, H. L. & Brunton, D. F., 1981. The Rare Vascular Plants of the Yukon. *Syllogeus*, 28: 1-96.

Dynesius, M. & Jansson, R., 2000. Evolutionary consequences of changes in species' geographical distributionsdriven by Milankovitch climate oscillations. *PNAS*, 97: 9115-9120.

Findley, J. S. & Jones, C. J., 1962. Distribution and variation of the voles of genus *Microtus* in New Mexico and adjacent areas. *Journal of Mammology*, 43: 154-166.

Fontanella, F. M., Feldman, C. R., Siddall, M. E. & Burbrink, F. T., 2008. Phylogeography of *Diadophis punctatus*: Extension Lineage Diversity and repeated patterns of historical demography of a trans-continental snake. *Molecular Phylogenetics and Evolution*, 46: 1049-1070.

FNA, 1993 - ?. *Flora of North America north of Mexico*. 1-30. Cambridge University Press, Cambridge.

Frakes, L. A., 1979. *Climates throughout geologic time*. Elsevier Scientific Publishing Co., New York. 310p.

Francis, J. E., Ashworth, A., Cantrill, D. J., Crame, J. A., Howe, J., Stephens, R., Tosolini, A.-M. & Thorn, V., 2008. 100 million years of Antarctic Climate Evolution: Evidence from fossil plants. In: Cooper, A. K., Barrett, P. J., Stagg, H., Storey, B., Stump, E., Wise, W. & the 10th ISAES editorial team, Eds.. *Antarctica; A Keystonein a Changing World*. Proceedings of the 10th International Symposium on Anarctic Earth Scienes, Washington,

D.C. The National Academies Press. Grimaldi, D., 1990. Diptera. In: Grimaldi, D., Ed.. *Insects from the Santana Formation, Lower Cretaceous, of Brazil*. Bulletin of the American Museum of Natural History 195: 1-191.

Haile, J., Froese, D. G., MacPhee, R. D. E., Roberts, R. G., Arnold, L. J., Reyes, A. V., Rasmussen, M., Nielsen, R., Brook, B. W., Robinson, S., Demuro, D., Gilbert, M. T. P., Munch, Austin, J. J., Cooper, A., Barnes, I., Möller, P. & Willerslev, E., 2009. Ancient DNA reveals late survival of mammoth and horse in interior Alaska. *Proceedings of the National Academy of Sciences*, 106(52): 22363-22368.
www.pnas.org/cgi/doi/10.1073/pnas.0912510106

Hallam, A., 1981. Relative importance of plate movements, eustasy, and climate on controlling major biogeographical changes since the early Mesozoic. In: Nelson, G. and Rosen, D. E., Eds.. *Vicariance Biogeography: A Critique*. New York, Columbia University Press.

Hallam, A., 1994. *An outline of Phanerozoic Biogeography*. Oxford University Press, Oxford.

Harrington, C. R., 1978. Quaternary vertebrate faunas of Canada and Alaska and their suggested chronological sequence. *Syllogeus*, 15:1-105.

Harris, A. H. & Findley, J. S., 1964. Pleistocene-Recent fauna of the Isleta Caves, Bernaillo County, New Mexico. *American Journal of Science*, 262: 114-120.

Harris, S. A., 1985. Evidence for the nature of the early Holocene climate and palaeogeography, High Plains, Alberta, Canada. *Arctic and Alpine Research*, 17: 49-67.

Harris, S. A., 1994. Chronostratigraphy of glaciations and permafrost episodes in the Cordillera of western North America. *Progress in Physical Geography*, 18: 366-395.

Harris, S. A., 2000. Pliocene and Pleistocene glaciations and permafrost events proven to date in the Cordillera of western North America. *Earth Cryology*, 4: 24-43. [In Russian].

Harris, S. A., 2002a. Global Heat Budget, Plate Tectonics and Climatic Change. *Geografiska Annaler*, 84A (1): 1-9.

Harris, S. A., 2002b. Biodiversity of the Vascular Timberline Flora of the Rocky Mountains of Alberta, Canada. In: Köerner, C. and Spehn, E., Eds.. *Mountain Biodiversity: A global assessment*. Parthenon Publishing Group, Lancashire, U.K.: pp. 49-57.

Harris, S. A., 2004. Source areas of north Cordilleran endemic flora: Evidence from Sheep and Outpost Mountains, Kluane National Park, Yukon Territory. *Erdkunde*, 58: 62-81.

Harris, S. A., 2005. Thermal history of the Arctic Ocean environs adjacent to North America during the last 3.5 Ma and a possible mechanism for the cause of the cold events (major glaciations and permafrost events). *Progress in Physical Geography*, 29:1-19.

Harris, S. A., 2007a. Biodiversity of the alpine vascular flora of the N.W. North American Cordillera: The Evidence from phytogeography. *Erdkunde*, 61(4): 344-357.

Harris, S. A., 2007b. Reaction of continental mountain climates to the postulated "global warming": Evidence from Alaska and the Yukon Territory. *Earth Cryosphere*, 11(3): 78-84. [In Russian.]

Harris, S. A., 2008. Diversity of Vascular Plant Species in the Montane Boreal Forest of western North America. *Erdkunde*, 62 (1): 59-73.

Harris, S. A., 2010a. Climatic change in Western North America during the last 15,000 years: The role of changes in the relative strengths of air masses in producing changing climate. *Sciences in Cold and Arid Regions*, 2 (5): 371-383.

Harris, S. A., 2010b. Evidence for increased stability of temperatures in areas of mountain permafrost in interior valleys and closed basins in wide Cordilleras in North America. In: *Cryospheric change and its influences*. Program and Abstracts, Lijiang, China, 52-54.

Harris, S. A., 2012. Climatic Change: Comparison of some of the Causes, and a Theory of how they operate together. .*Advances in Meteorology*. In Press.

Harris, S. A. & Hubricht, L., 1982. Distribution of the species of the genus *Oxyloma* (Succinidae) in southern Canada and the Adjacent Portions of the United States. *Canadian Journal of Zoology*, 60: 1607-1611.

Harris, S. A. & Pip, E., 1973. Molluscs as indicators of Late and Post-Glacial history in Alberta. *Canadian Journal of Zoology*, 51: 209-215.

Harrison, J. A., 1985. Giant Camels from the Cenozoic of North America. *Smithsonian Contributions to Paleobiology*, 57: 1-29.

Heirtzler, J. R., 1973. The Evolution of the North Atlantic Ocean. In: Tarling, D. H. and Runcorn, S. K., Eds.. Implications of Continental Drift to the Earth Sciences, Volume1, Academic Press, New York, pp. 191-196.

Hibbard, C. W., Ray, D. E., Savage, D. E., Taylor, D. W. & Guilday, J. E., 1965. Quarternary Mammals in North America. In: Wright, H. E., Jr. and Frey, D. G., Eds.. *The Quaternary of the United States*. Princeton University Press, Princeton, New Jersey, 509-525.

Hynes, H. B. N., 1988. Biogeography and Origins of the North American stoneflies (Plecoptera). *Memoires of the Entomological Society of Canada*, #144: 31-37.

Illies, H. B. N., 1965. Phylogeny and Zoogeography of the Plecoptera. *Annual Review of Entomology,* 10: 117-140.

Imbrie, J. & Imbrie, J. Z., 1980. Modeling Climatic Response to Orbital Variations. *Science,* 207: 943-953.

Karmarovitch, N. & Geoph, P., 2009. Hansen Mars Challenge – a challenge to Hansen et al., 1988. *Australian Institute of Geophysicists* (AIG), NEWS 96, May, 2009. http://aig.org.au/assets/194/AIGnews_May09.pdf

Layberry, R. A., Hall, P. W. & Lafontaine, J. D., 1998. *The Butterflies of Canada.* Toronto. University of Toronto Press.

Lehmkuhl, D. M., 1980. Temporal and spatial changes in the Canadian insect fauna: Patterns and explanation. *The Canadian Entomologist,* 112(11): 1145-1159.

MacDonald, G. M., 2003. *Biogeography: Introduction to Space Time and Life.* New York. J. Wiley and Sons: 518.

MacDonald, G. M. & Cwynar, L. C., 1985. A fossil pollen based reconstruction of the Late Quaternary historyof lodgepole pine (*Pinus contorta* var. *latifolia*) in the western interior of Canada. *Canadian Journal of Forest Research,* 15: 1039-1044.

Marincovich, L., Jr., Brouwers, E. M., Hopkins, D. M. & McKenna, M. C., 1990. Late Mesozoic and Cenozoic paleogeographic and paleoclimatic history of the Arctic Ocean Basin, based on shallow-water marine faunas and terrestrial vertebrates. *Geological Society of America, The Geology of North America,* L: 403-426.

Marshall, L. G., Webb, S. D., Sepkoski, J. J., Jr. & Raup, D. M., 1982. Mammalian evolution and the great American Interchange. *Science,* 215: 1351-1357.

Matthews, J. V., 1980. Tertiary land bridges and their climate: Backdrop for development of the present Canadian insect fauna. *The Canadian Entomologist,* 112(11): 1089-1103.

McKenna, M. C., 1975. Fossil mammals and Early Eocene North Atlantic land continuity. *Annals of the Missouri Botanic Gardens,* 62: 335-353.

Morgan, A. V. & Morgan, A., 1980. Faunal assemblages and distributional shifts of Coleoptera during the Late Pleistocene in Canada and the Northern United States. *The Canadian Entomologist,* 112(11): 1105-1144.

Morrison, R. B., 1965. Quaternary Geology of the Great Basin. In: Wright, H. E., Jr. and Frey, D. G., Eds.,*The Quaternary of the United States.* Princeton University Press, Princeton, New Jersey, 265-285.

Nekola J. C. & Coles, B. F., 2001. Systematics and ecology of *Gastrocopta (Gastrocopta) rogersensis* (Gastropoda: Pupillidae), a new species of land snail from the Midwest of the United States of America. *The Nautilus,* 115(3): 105-114.

Nekola J. C. & Coles, B. F., 2010. Pupillid land snails of eastern North America. *American Malacological Bulletin* 28: 29-57.

Nomura, R., Seto, K., Nishi, H., Takemura, A., Iwai, M., Motoyama, I., & Maruyama, T., 1997. Cenozoic paleoceanography in the Indian Ocean; Paleoceanographic biotic and abiotic changes before the development of monsoon system. *Journal of the Geological Society of Japan,* 103(3): 280-303.

Noonan, G. R., 1988. Faunal relationships of Eastern North America and Europe as shown by Insects. *Memoires of the Entomological Society of Canada,* #144: 39-53.

Norris, K. S., 1958. The evolution and systematic of the iguanid genus *Uma* and its relation to the evolution of other North American desert reptiles. *American Museum of Natural History Bulletin*, 114: 247-326.

Penny, N. D. & Byers, G. W., 1979. A Check-list of the Mecoptera of the World. *Acta Amazonica*, 9: 365-388.

Prest, V. K., Grant, D. R. & Rampton, V. N., 1968. Glacial Map of Canada. *Geological Survey of Canada*, Map 1253A.

Rabassa, J., Coronato, A. & Salemme, M., 2005. Chronology of the Late Cenozoic Patagonian glaciations and their correlation with biostratigraphic units in the Pampean region(Argentina). *Journal of South American Earth Sciences*, 20(1-2): 363-379.

Rabassa, J., 2008. The Late Cenozoic of Patagonia and Tierra del Fuego. *Developments in Quaternary Science*, 11: 151-204.

Ritchie, J. C. & Hare, F. K., 1971. Late-Quaternary vegetation and climate near the arctic tree line in Northwestern North America. *Quaternary Research*, 1: 331-342.

Rogers, R. A., Rogers, L. A., Hoffmann, R. S. & Martin, L. D., 1991. Native American biological diversity and the biogeographic influence of Ice Age refugia. *Journal of Biogeography*, 18: 623-630.

Ross, H. H., 1970. The ecological history of the Great Plains: Evidence from Grassland Insects. In: Dort, W. and Jones, J. K., Jr., Eds., *Pleistocene and Recent Environments of the Central Great Plains*. Special Publication of the University of Kansas Department of Geology, 3: 225-240.

Schwarzbach, M., 1961. The Climatic History of Europe and North America. In: Nairn, A. E. M., Ed.. *Descriptive Palaeoclimatology*. London, InterScience Publishers. Chapter 11: 255-291.

Shackleton, N. J. & Kennett, J. P., 1995. Late Cenozoic oxygen and carbon isotopic changes at DSDP site 284: Implications for Glacial History of the Northern Hemisphere and Antarctica. In: *Initial Reports of the Deep Sea Drilling Project* 29: 801. U.S. Government Printing Office, Washington, D.C..

Staines, C. L., 2006. *The Hispine Beetles of America North of Mexico (Chrysomelidae: Cassidinae.* Virginia Museum of Natural History, Special Publication #5, Martinsville, Virginia. 178p.

Stewart, J. R., Lister, A. M., Barnes, I. & Dalén, L., 2010. Refugia revisited: individualistic responses of species in space and time. *Proceedings of the Royal Society, B,* 277: 661-671.

Stewart, K. W. & Stark, B. P., 2002. *Nymphs of North American Stonefly Genera.* The Caddis Press, Columbus, Ohio. 510p.

Strong, W. L. & Hills, L. V., 2005. Late-glacial and Holocene palaeovegetation zonal reconstruction for central and north-central America. *Journal of Biogeography*, 32: 1043-1062.

Tripati, A.K, Backman, J., Elderfield, H. & Ferretti, P., 2008. Eocene bipolar glaciations associated with global carbon cycle changes. *Nature,* 463: 341-346.

Urquhart, F. A., 1960. *The Monarch Butterfly.* University of Toronto Press, Toronto.

Varanyan, S. L., Garutt, V. E. & Sher, A. V., 1993. Holocene dwarf mammoths from Wrangel Island in the Siberian Arctic. *Nature,* 362: 337-340.

Wahrhaftig, C. & Birman, J. H., 1965. The Quaternary of the Pacific Mountain System in California. In: Wright, H. E.,Jr.and Frey, D. G., Eds., *The Quaternary of the United States.* Princeton University Press, Princeton, New Jersey, 299-340.

Wassenaar, L. J. & Hobson, K. A., 1998. Natal origins of migratory monarch butterflies at wintering colonies in Mexico: New isotopic evidence. *Proceedings of the National Academy of Sciences,* 95: 15436-15439.

Webb, S. D., 1997. The Great American Faunal Interchange. In: Coates, A. G. (Ed)., *Central America*: 97-122. New Haven, Connecticut, Yale University Press.

Weber, W. A., 1965. Plant Geography in the Southern Rocky Mountains. In: Wright, H. E., Jr. and Frey, D. G., Eds.. *The Quaternary of the United States, 453-468.* Princeton, New Jersey, Princeton University Press. 922p.

Willis, K. J. & Whittaker, R. J., 2000. The Refugial Debate. *Science,* 287: 1406-1407.

Wolfe, J. A., 1975. Some aspects of plant geography of the northern Hemisphere during the Late Cretaceous and Tertiary. *Annals of the Missouri Botanical Garden,* 62: 264-279.

Wolfe, S. A., Huntley, D. J., , David, P. P., Ollerhead, J., Sauchin, D. J., & MacDonald, G. M., 2001. Late 18th century drought induced sand dune activity, Great Sand Hills, Saskatchewan. *Canadian Journal of Earth Sciences,* 38: 105-117.

Wolfe, S. A. & Hugenholtz, C. H., 2009. Barchan dunes stabilized under recent climate warming on the southern Great Plains. *Geology,* 37(11): 1039-1042.

Worrall, D. M., 1991. Tectonic history of the Bering Sea and evolution of Tertiary strike-slip basins of the Bering shelf. *Geological Society of America Special Paper,* 257: 1-120.

Wright, H. E., Jr., 1970. Vegetational history of the Central Plains. In: Dort, W. and Jones, J. K., Jr., Eds., *Pleistocene and Recent Environments of the Central Great Plains.* Special Publication of the University of Kansas Department of Geology, 3: 152-172.

Yeates, D. K. & Grimaldi, D., 1993. A new *Metatrichia* window fly (Diptera: Scenopinidae) in Dominican amber, with a review if the systematic of the genus. *American Museum Novtates,* 3078: 1-8.

Yeates, D. K. & Irvin, M. E., 1996. Apioceridae (Insecta: Diptera): cladistic reappraisal and biogeography. *Zoological Journal of the Linnean Society,* 166: 247-301.

Zinck, M., 1998. *Roland's Flora of Nova Scotia.* Halifax, Nova Scotia Museum and Nimbus Publishing. 3rd Edition: 2 volumes.

Zwick, P. & Teslenko, V., 2000. Phylogenetic system and zoogeography of the Plecoptera. *Annual Review of Entomology,* 45: 709-746.

Health and Humanity

Chapter 11

Marine Environment and Public Health

Jailson Fulgencio de Moura, Emily Moraes Roges, Roberta Laine de Souza, Salvatore Siciliano and Dalia dos Prazeres Rodrigues

Additional information is available at the end of the chapter

1. Introduction

The oceans represent a significant source of biological diversity, water, biomass, oxygen, and other important aspects to human health [1-3]. The quality of the ocean is essential for maintaining the planet, and thus to public health. However, the complex and fragile evolutionary stabilization of the ocean and coastal regions has been disrupted by human activities in a short time scale [4]. The vast majority of waste produced by human activities for centuries has reached the oceans, even over long distances and in inhospitable places [3, 5]. In recent decades there have been evident the vast scope of the changes of the marine environment caused by anthropogenic activities, as well as the many responses to these changes that tend to impact ecological processes, putting endangered species susceptible and producing various diseases in the human population [3, 6]. These changes are not restricted to oceanic scale, but are strongly associated with the continents, consequently, strong pressure on the health of terrestrial ecosystems, with impacts on socioeconomic and cultural activities and, finally, to public health. Recently, the trend has grown to incorporate the term health within the definitions of environmental health. The term health of the oceans, the second definition of the Panel on Health of the Oceans (HOTO/GOOS), refers to the condition of the marine environment from a perspective of adverse effects caused by anthropogenic activities, in particular: habitat destruction, changes in the proportion of sedimentation, mobilization of contaminants and climate changes [7, 8].

Indeed, the human utilization of ocean environment has negatively and extensively impacted the ecological system that people are connected. The human activities in coastal zones, such as agricultures, urban development, fisheries, coastal industries and aquacultures, have contributed to chemical, physical and ecological impacts that may be interconnected [8, 9]. For example, the human activities cited generate a significant input of chemical pollutants (e.g. metals, persistent organic pollutants, nanoparticles, radionuclides and nutrients) that is known to impact the biodiversity and the marine ecological system [3, 10].

Marine microbiological pollution represents an expressive impact on biodiversity and human health. The microbiological activity in coastal environment can result in direct impact in human health, but can trigger the biodiversity loss, degradation of ecosystem function and impact in recreation, tourism and human wellbeing [1, 2, 6, 11, 12]. Marine pollution, such as nutrients input, runoffs, and regional and international navigation by ships can load new pathogens to the environment, and the climate change may exacerbate their effects and establishment in an area. For example, the oceans have been identified as the source of introduction of *Vibrio cholerae* that resulted in outbreaks in South America [11, 13]. Potential pathogens from the Family Vibrionaceae and Aeromonadaceae have been frequently identified in coastal humans and marine top predators. It is important to highlight that pathogens of these families are not associated with fecal contamination [1, 3].

The current scenario on the conservation of the oceans has been reflected in numerous human diseases related to marine life. The relationships of the oceans to human activities and public health is already consensus; however, its mechanisms are not well understood due to its complexity. These relationships include the focus on climate change, toxic algae poisoning and chemical and microbial contamination of marine waters and fish (Figure 1) [4, 14, 15].

The marine environment provides valuable benefits for human activities, including protein sources and economic activity through fisheries, aquaculture and navigation. Furthermore, there are the economic benefits from tourism, culture, biomedicine, recreation activities and renewable energy [4, 16]. The oceans represent a great source of biodiversity and play a vital role in water and biogeochemistry cycle. Other human benefits from the oceans are clear, and important for human wellbeing, such as artistic inspiration, increased physical activity and therefore fitness, reduced levels of stress and simply the harmony as a result of healthy oceans and their stable biodiversity [3].

The relationship between public health and the health of the oceans are also growing due to increasing number of people living in coastal areas, mainly in tropical and subtropical regions [1, 3, 4]. In these regions, increases vulnerability to social and environmental stability resulting from natural disasters that involve the ocean and health. It is estimated that world population has reached 6.6 billion in 2007, with a projected growth to 9.3 billion by 2050, developing countries are primarily responsible for this increase [17]. Approximately 65% of the human population lives within 159 km of shoreline with growth estimated at 75% for 2025. In coastal regions the oceans remain an important source of protein, quality of life, recreation, and are an integral part of economic activities in various localities [6, 18].

Coastal residents are highly vulnerable to climate variability and extreme events. As an example, the event of a tsunami in Indonesia has caused at least 175,000 deaths in 2005. In addition to the physical impacts on the health effects of these events, epidemics occur frequently due to the favorable conditions that follow extreme phenomena, and that end up being magnified by the conditions of social and environmental vulnerability of affected populations [4]. Various infectious agents found in marine hosts including bacterial, viral and protozoan result in infectious diseases in humans [19]. The effects of climate and temperature on disease vectors, such as the growing prevalence of malaria following El Niño events also have been suggested [20].

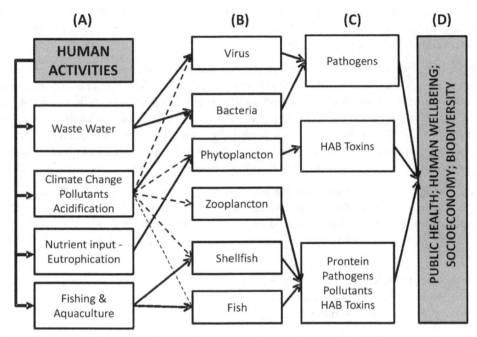

Figure 1. Schematic illustration of the anthropogenic pressures impacts (A) of the marine environment (B) and the subsequent result of the marine biodiversity alterations (e.g. pathogens, HABs) (C) on human health and biodiversity, wellbeing and socioeconomic relationships.

2. Climate variations

The oceans play an extreme important role in the climate by the storage and transportation of heat around the globe. The interaction of the ocean currents and atmospheric winds operate regulating the climate. The marine ecological processes are dependent of the variation of the temperature, as the availability of nutrients that is associated with this factor, and tend to maintain the ecologic stability [4]. An example of an extreme inter-annual variability is the El Niño Southern Oscillation (ENSO).

ENSO is a semi-periodic variability of the inter-annual climate cycle that occurs in intervals of 2-7 years as a result of the discontinuity of the up-welling system in the eastern Equatorial Pacific, forced by the change in wind pattern [20, 21]. The ENSO results in changes in the oceanic temperature and in the atmospheric pressure in the Pacific basin. However, the impacts of the ENSO are not limited to the Pacific Basin, but can influence many continental and marine regions around the globe by changing the atmospheric circulation that disturb temperature and precipitation pattern, resulting in extreme periods and intensity of drought and heavy rains in different areas [20].

Climatic variations triggered by El Niño events are associated with ecosystem changes that results in impact on public health. These climate variations influence the population density and dispersal pattern of vectors, for example, mosquitoes and rodents, which tend to cause infectious diseases in epidemic proportions, such as malaria, dengue and hantavirus [20, 22, 23]. In addition, other diseases such as leishmaniasis and cholera outbreaks have often been associated with this climatic event [23].

Besides the problems related to ENSO events, other extremes events such as drought, have more insidious effects on health for the loss in agricultural production and, consequently, for severe nutritional disorders [4]. Therefore, it is not only direct impacts, but also because it tends to aggravate the socioeconomic structure of the societies affected, causing an amplification of the impacts on public health. In cases of drought triggered by climate variations associated with ENSO forests tend to become more vulnerable to fires, resulting in the massive loss of biodiversity and respiratory diseases linked to poor air quality [22].

The occurrence of El Niño in 1997-1998 resulted in the deaths of more than 21,000 people in 27 countries around the world. Altogether, 117 million people were affected. The occurrence of morbidities as a result of the pressures of these phenomena have affected around 540,000 people, while 4.9 million people were displaced from their homes, becoming homeless [24].

Like the variations, understood as an intrinsic property of the climate system, responsible for natural variations in the patterns observed in geographical scales, global climate changes occur due to temperature rise caused by anthropogenic emissions of greenhouse gases during decades.

Global climate change may have both direct and indirect effects on public health [1, 25]. The greenhouse gases, naturally present in low concentrations in the atmosphere keep the earth's average temperature around 15°C. Without this mechanism of regulation of the global atmospheric temperature, the Earth's average could be -18°C and the planet would freeze, preventing the extensive existing biodiversity [26]. However, the anthropogenic release of greenhouse gases has increased the global temperature resulting in catastrophic effects on human and environmental health, while causing socioeconomic and cultural upheavals [27].

Focusing on disorders caused by the effects of climate change in the long term in the oceans, the Intergovernmental Panel on Climate Change (IPCC) points as main influences the level rise of the oceans, the global temperature increase, the varying levels of salinity, the changes in the circulation of water masses, the decreasing concentration of oxygen, the sea level rise, and probable increase in intensity and frequency of hurricanes and cyclones [28].

One of the most discussed of global warming on the oceans is increasing the sea level. This can have catastrophic effect of introducing salt water into fresh water systems in the continent, affecting the quality and availability of this for consumption [1]. Moreover, according to the fourth IPCC report [28], there is observational evidence that an increase in the number of tropical cyclones in the North Atlantic, which began around 1970, is associated with increased surface temperature of the sea. Global warming may also promote changes in the general pattern of fecal-oral infections and foodborne illness. It is hoped that the wide geographic distribution (by both the altitude and latitude) of organisms that transmit disease (vectors) not only increase the potential for transmission, but also change the dynamics of the life cycle (e g, reproduction, survival and potential of infection) of vectors of parasitic infectious organisms [23, 25, 26, 29].

The imbalance in ecological relationships, due to climate change may alter the natural mechanisms of control of vectors and their host organisms, and populations of parasites. In addition, more frequent droughts and rising sea levels may force human populations to migrate to areas where infectious organisms are located, but that currently produce little impact on people. Additional effects include impacts of global change on agriculture, reductions in the ozone layer, economic impacts and increased vulnerability to disease and malnutrition. The many effects of climate change will affect all life forms on Earth, including all its biodiversity and ecological processes.

3. Extreme events

Due to the increasingly human populations residing on coastal regions, extreme events such as tsunamis, tornadoes, cyclones, storms and floods tend to mobilize international public attention due to increased social vulnerability [18, 30, 31]. Extreme events of the same magnitude and similar characteristics, impact differentially the different population groups depending on their level of vulnerability [1, 3, 4, 32]. While the rich industrialized nations suffer most from economic loss as a consequence to natural disasters, the poor and developing countries often suffer from extensive loss of life, incidence of diseases, and loss of social and physical structures [30]. An example is the Indian Ocean tsunami event in 2004, which triggered a series of tsunamis responsible for approximately 220,000 deaths, Indonesia being one of the countries most affected with more than 400,000 homeless.

Natural disasters force a temporary condition of people living in crowded conditions with poor sanitation, poor management of human waste, impoverished nutrition, and incidence of waterborne diseases, low immunity and susceptibility to infectious diseases such as pneumonia, cholera, dengue, malaria, addition of trauma resulting from the magnitude of the events [32]. In addition, extreme events can also interfere in the continuity of health services due to impacts on infrastructure, or force changes of priority in health policies. Some infectious diseases may be aggravated by malnutrition or hunger-related as a result of human migration. Recent studies show that the destructive power of hurricanes has grown around the world, dramatically raising its frequency in the last two decades in the Atlantic [33, 34]. Often the ability to anticipate and respond to natural disasters is based on understanding of climate systems, which depend on the complex interaction of the atmosphere, the continents and oceans. However, usually the main importance is focused on developing and improving measures to prevent population to environmental extremes, there is a need to improve the socioeconomic conditions in order to reduce these impacts.

The Brazilian coast present 8,698 kilometers long of extension, covering about 514,000 square kilometers. The heterogeneity and vulnerability of this coastal region is obstacle for environmental management, principally due to the proportion of the population living in this environment (18%). As an example, 16 out of 28 metropolitan regions in Brazil are located along the coast. Coastal erosion is particularly a phenomenon that results in an elevated risk to the large number of people inhabiting coastal areas along the Brazilian coast [35, 36]. Despite of the widespread range or coastal eroded regions, the configuration of the magnitude of the disasters are not equally distributed. Environmental influences (e.g. wind, wave and wave partners and trends) have been identified as the developer to seashore erosion, but human intervention in the morphodynamic of river mouth or sedimentary flux has influenced such disasters. In Atafona beach, São João da Barra (northern coast of Rio de Janeiro state) the coastal erosion has dramatically impacted the region [37]. The landward advance of the sea has already caused several consequences for local residents, including habitation loss, economical impacts, and historic and touristic impairments. In places where before there were houses and streets, and an established local commerce, is now part of coastal water or shows a scenario of destruction: about 400 houses in 16 blocks away have been demolished by the power of the waves (Figure 2). Atafona is located at the south side of the Paraíba do Sul River, the main river of the Rio the Janeiro state. The environmental variables and anthropogenic influences are thought to trigger the disasters that have been observed since 1950 [37]. The reduction of the fluvial discharge, as result of the human activities along Paraíba does Sul River, has contributed to the degradation of the coastal zone in Atafona. The sea level rise triggered by the climate change probably may influence increasing the impact in the coastal.

Figure 2. Images showing the coastal erosion caused by the sea energy in Atafona, São João da Barra, northern Rio de Janeiro state, Brazil. Downloaded from: http://viafanzine.jor.br/site_vf/pag/1/na_terra_fotos.htm

4. Harmful algal bloom

The toxins produced by toxic algal blooms (HAB - Harmful Algal Bloom) have the ability to bioconcentrate through the food chain. Therefore, humans, like many other animals that occupy the highest scales of this chain are vulnerable to the adverse effects of these toxins [3, 38]. The greatest risk of poisoning and gastrointestinal infections are linked to sea food consumption, especially of bivalve mollusks (mussels and oysters), because they are filter feeders, which makes these organisms accumulate large amounts of HABs. Bathers are also exposed to the effects of blooms of toxic algae by ingestion and inhalation of "spray" produced by the action of breaking waves containing HABs [39].

Worldwide, sea weed toxins have been associated with cases of human poisoning and animals fatalities [38, 40]. Moreover, massive blooms of toxic and nontoxic algae can cause sharp decrease of oxygen (hypoxia) in place of occurrence, resulting in massive death of marine life and affecting recreation, fish commerce, tourism and public health [38]. From 5000 species of phytoplankton, about 300 occur in massive blooms and slightly more than 80 are known to be toxic [3, 41]. The HAB species are classified as toxin producers (can contaminate sea food or kill fish) and as high biomass producers (can cause hypoxia or anoxia and die off of marine life, when reach high concentrations) [42]. The "toxic producers" HAB species can cause shellfish poisonings and potential impacts on public health, and the "high biomass produces" are thought to promote massive mortalities of fish and reductions in yields in deteriorated environments.

Some blooms of toxic algae can persist in the environment due to the inhibiting power of the toxins on the growth of other phytoplankton species, or reduce the predation of zooplankton. The human poisonings caused by exposure to HABs cause serious problems to human health, which can lead to death or produce sequels. However, it is not uncommon physicians in coastal regions, where most cases occur, erroneously diagnose the symptoms of poisoning, or attributed other factors to them [43]. In addition, there is evidence that colorectal cancer is strongly associated with the ingestion of biotoxins produced by marine microalgae through the consumption of bivalve mollusks [44]. There are five recognized types of poisoning caused by ingestion of HAB: paralytic shellfish poisoning (PSP), neurotoxic shellfish poisoning (NSP), diarrheic shellfish poisoning (DSP), amnesic shellfish poisoning (ASP) and ciguatera poisoning (CFP) [1, 3, 4, 6, 38, 39]. Different from the other four types of HAB poisoning the CFP is caused by the ingestion of reef fishes contaminated by toxins produced by dinoflagellates. Therefore the toxin can enter in the food chain and impact top predators, such as humans [42].

Although there is record of HABs before the transformation of coastal ecosystems by anthropogenic activities, in recent decades has increased dramatically the number of problems associated with HABs around the globe. However, part of this growth is associated with the growth of environmental monitoring. A potential route of spread of these organisms lies in the transport of ballast water in ships. In addition, bivalve mollusks commercially introduced for aquaculture in the countries can also carry the organism in various ways [45]. Global environmental changes such as the destruction of reefs, nutrient enrichment of coastal waters by nitrogen and phosphorus, as well as global climate change, may serve to explain the increase of red tides reported worldwide, as well as the growth of human diseases related with exposure to marine toxins or associated with the events. Also, cholera outbreaks have been associated with HABs from the knowledge that marine copepods are capable of carrying the bacteria *Vibrio cholerae*, feed of algal blooms. Therefore, these blooms can lead to spread of cholera and outbreaks associated with the frequency of flooding and extreme events [46]. Shuval [40] estimated that marine biotoxins associated mainly with blooms of toxic algae cause an estimated 100,000 to 200,000 cases of severe poisoning annually worldwide and approximately 10,000 to 20,000 deaths and a similar number of very severe cases with neurological sequel, such as paralysis. Furthermore, HABs events can produce mass deaths of marine organisms and cause heavy economic losses, mainly in the extractive fishing, aquaculture and tourism [1, 4, 43].

In Brazilian coastal water, toxic algae bloom has caused impacts on biodiversity and resulting economic impairment, principally on aquaculture and fishery activities. In Baía de Todos os Santos, Bahia state (Brazilian coast), a massive mortality of fishes and shellfishes was registered in 2007. About 50 tons of fishes and shellfishes were killed, which resulted in negative consequences for fisheries and aquaculture, due to the prohibition to commercialize organisms for consumption from this contaminated area. Similarly, in Florianópolis, Santa Catarina state, southern Brazil, an harmful algal bloom of pseudo-nitzschia were identified, resulting in a preventing official measure to protect the population against the effect caused by the event. Despite the identification of harmful algal blooms in coastal regions of Brazil and their association with impacts on biodiversity, few studies have focused to understand the impacts on human health. However, cases of human death have been registered associated with consumption of water from reservoirs with bloom of cyanobacteria potentially hazardous.

A drastic epidemic gastroenteritis outbreak was registered in Itaparica, region of Bahia State associated with the flooding of a Dam reservoir in 1988. From about 2,000 gastroenteritis cases identified, 88 resulted in death along a period of 42 days. Bacterial, toxicological and virological analises were conducted in fecal and blood samples from the patients, and drinking water was examined for microorganisms and metals. Clinical results were also reviewed to understand and identify the etiologic agent. The laboratory analyses indicated that the source of the epidemiology was water from the Dam, which revealed the presence of high concentrations of toxin produced by cyanobacteria (genus *Anabaena* and *Microcystis*). The cases of infectious disease were restricted to the areas supplied by drinking water from the dam. Also in 1996, 54 fatalities were recorded in Caruaru (Pernambuco state, northeastern Brazil) in hospitalized patients with chronic renal failure. During hemodialysis sessions, the patients received untreated water contaminated with cyanobacteria.

5. Microbial pathogenic pollution

The microbiological activities are of great importance for many ecological processes in the marine ecosystem. Their functions are essential for the maintenance of biogeochemical cycles required for the maintenance of life [48]. Marine ecosystem provides a natural habitat for a range of microbial pathogens such as bacteria, viruses and parasites. Some pathogens inhabit the water, while other can live attached to particles or inside of marine organisms.

High concentrations of these microbes within coastal waters should indicate that the water or even seafood may be contaminated by human waste [1, 49]. However, the use of indicator microbes to test the quality coast waters for recreation and sea food consumption has been questioned, particularly in the subtropical and tropical marine environments, mainly in areas with no point source of contamination identified [1, 50]. An example is bacteria species from the families Aeromonadaceae and Vibrionaceae that are naturally inhabitants of the marine environments. Many species of this family are not related with fecal contamination of coastal waters; therefore, the use of enteric bacteria as indicators of microbiological water quality is strongly limited [1, 50]. *Vibrio* species, especially *V. cholerae, V. parahaemolyticus,* and *V. vulnificus,* are frequently associated with infectious diseases through the ingestion of shellfish or even fish [42, 50, 51]. *Vibrio* and *Aeromonas* species are clinically important for humans and biodiversity health, causing gastroenteritis of infections through the open wounds resulting in septicemia. A large number of people worldwide have been impacted by the infections of pathogenic microbes in coastal waters. More than 170 million cases of respiratory and enteric impairments associated with recreation and seafood consumption coastal waters contaminated with infectious microbes have been reported [52]. In the United States (USA) 33% of shellfish harvesting waters are impacted by micropathogens. Currently, 62% of the coastal beaches of the Rio de Janeiro state (Brazil) are classified as inappropriate for recreation, principally due to contamination with fecal bacteria. In addition, 20,300 recreational beach warnings were reported in USA in 2008 due to the fecal microbe presence in coastal waters [53].

In Table 1 we present some results of bacteriological surveys (Aeromonadaceae and Vibrionacea species) that have been carried out with many specimens of marine mammals, seabirds and sea turtles from Brazilian coast. The microbiological samples were collected during a long term beach monitoring program for research and conservation of marine mammals, seabirds and sea turtles. This monitoring program has been conducted since 1999 by the GEMM-Lagos from the National School of Public Health (ENSP/FIOCRUZ). The bacteriological analyses were conducted in the *National Reference Laboratory for Bacterial Enteroinfections* (LRNEB) from the Oswaldo Cruz Institute (IOC/FIOCRUZ).The samples were collected through the sterilized swabs introduced carefully in the mouth, eyes, nostrils, genital slit, anus and open wounds of sick or recently dead animals. Twenty species of bacteria were detected in the animals sampled, five and 15 belonging to the family Aeromonadaceae and Vibrionaceae respectively. The most prevalent microbial species at the marine animals sampled were *Vibrio alginolyticus* and *Aeromonas caviae*, both representing 69% of detection. Green sea turtle (*Chelonia mydas*), Guiana dolphins (*Sotalia guianensis*) and Kelp gull (*Larus dominicanus*) presented 65, 45 and 35% of the 20 species of bacteria found in the analyses, respectively.

Interestingly, the three more affected species share the same habitat preferences. Both marine species are commonly observed in coastal waters of Rio de Janeiro state and prey on coastal marine food. Kelp gull are commonly found in high density consuming rest of human food at the beach, and coastal dead fishes. It is important to highlight that humans are exposed to the feces of this bird during recreation on the beach. All green sea turtles sampled were juveniles and use coastal waters mainly to eat sea algae. Many specimens have been found sick and associated with the ingestion of marine debris [54, 55]. Guiana dolphins use coastal estuarine waters where they prey on fishes, squids and shrimps. The most important conservation problems of the species is the accidental mortality in fishing nets, but persistent pollutants seems to be also a problem for the conservation of this species [10, 56, 57]. Simultaneous occurrence of different species isolated were observed in the species sampled what could revels the possibility of synergic actions. Considering that these bacteria are recognized as emergent pathogens, and the relevance of the findings for public health in light of the growing area of "ocean and human health" we would like to emphasize the importance of this investigation, which indicates the aquatic environment as a possible route of transmission among marine biota, which includes humans. Aquatic animals are prone to bacterial infections in the same way as land animals, especially when they are under stress condition. Disease may occur systemically or be confined to external surfaces such as the skin or gills specially by pathogenic bacteria which are ubiquitous in the environment, or may form part of the normal internal bacterial flora of an aquatic animal [51].

In the marine ecosystem, the distribution of a viral or bacterial pathogen is directly determined by its virulence, as well as the number of susceptible hosts available. This balance between pathogen and host generates and maintains the variety of both groups. In some occasions, this delicate and normal relationship breaks, mainly due to the forces of aggression on the environment or environmental imbalances, resulting in the abundance of pathogens and increased vulnerability on marine biodiversity and public health [48]. Physical, chemical and biological marine environment may influence the number and diversity of marine microbes. However, Wang et al. [58] observed that high levels of organochlorine pollutants have been found in the tissues of Hong Kong's cetaceans, this class of chemical can cause immunosuppression, with an increased vulnerability to bacterial infections. Aquatic mammals are animals sensitive to changes in their habitat and for that reason considered excellent health indicators in environmental monitoring programs.

Ingestion of inadequately cooked seafood exposes people to parasitic infections, particularly with anisakids and cestodes, which reports increase of parasitic contamination in shellfish from polluted waters [1, 59]. In addition, many studies have shown human pathogens emerging in the marine environment and associated with infectious diseases in marine mammals exposed to polluted waters including: giardiasis, papillomavirus, brucellosis, lobomycosis, toxoplasmosis, etc.[60, 61].

Shuval [40] estimated that each year about 2.5 million clinical cases of hepatitis infections occur globally, with about 25,000 deaths and 25,000 cases of liver deficiencies associated with consumption of contaminated seafood, especially mussels. Moreover, this author estimated an overall economic impact of 7.2 billion per year associated with these conditions. Iwamoto et al. [62] showed the report to CDC during 1973 to 2006, 188 outbreaks of seafood-associated infections, causing 4,020 illnesses, 161 hospitalizations, and 11 deaths, were reported to the Food-Borne Disease Outbreak Surveillance System. Most of these seafood-associated outbreaks (n=43; 76.1%) were due to a bacterial agent; 40 (21.3%)

outbreaks had a viral etiology, and 5 (2.6%) had a parasitic cause. Therefore it is necessary to adress appropriate studies to characterize the impact over the ocean's capacity to maintain environmental quality important to the health of marine population and the microbiological hazards present in marine ecosystems to prevent outbreaks by seafood consumption and recreational use of these waters.

BACTERIA FAMILY (BELOW)	BACTERIA SPECIES(BELOW)	WHALES			DOLPHINS				SEABIRDS			SEA TURTLES		
		Megaptera novaeangliae	*Eubalaena australis*	*Balaenoptera acutorostrata*	*Stenella frontalis*	*Sotalia guianensis*	*Pontoporia blainvillei*	*Delphinus sp.*	*Sula Leucogaster*	*Larus dominicanus*	*Spheniscus magellanicus*	*Chelonia mydas*	*Lepidochelys olivacea*	*Caretta caretta*
AEROMONADACEA	A. veronii	■									■			
	A. caviae		■		■	■	■	■	■			■		
	A. hydrophila		■											
	A. media													
	A. trota						■							
VIBRIONACEAE	V. alginolyticus	■	■					■	■		■			■
	V. vulnificus	■	■						■					
	V. parahaemolyticus	■			■							■		
	V. cincinnatiensis	■	■									■		
	V. fluvialis											■		■
	V. furnisii				■			■						
	V. mimicus							■						
	V. harveyi		■					■						
	V. mediterranei							■			■			
	V. aestuarinus						■							
	V. pelagius													
	V. campbelii											■		
	V. hepatarius													
	V. coralliitycus	■												
	V. fischeri									■		■		

Table 1. Vibrio and Aeromonas species isolated form marine mammals, seabirds found sic on the beaches alond the coast of Rio de Janeiro state, southeastern Brazil.

6. Bioinvasion in coastal systems

The bioinvasion, refers to some exotic species introduced into a new environment, and which for the absence of natural controls such as parasites and diseases, become extremely harmful to local biodiversity, especially in disturbed habitats [11, 63]. When a species introduced into a new environment has success in establishing itself and its population increases, it tends to compete and eliminate native species, or cause damage to local ecology and affect socio-economic pattern and public health [45]. The bioinvasion is considered one of the most important threats to biodiversity and integrity of marine ecosystems, especially in coastal regions. However, this question had deserved attention only after the signing of the Convention on Biological Diversity in June 1992. Bioinvasions have occurred in all regions of the world, and the largest carrier of exotic species to new areas is navigation, where the ballast water of ships acting as a "vector" for introduction of species [63].

Climate change, nitrogen deposition and contaminants in the marine environment appear to help the successful accommodation of invasive species in a new habitat, especially microorganisms [11, 64]. Several marine species have caused heavy economic and ecologic impacts in a habitat invaded. Once established, the elimination of exotic species in the new habitat is very costly or even impossible, therefore, the policies related to bioinvasion have been linked to measures to prevent introduction of exotic species [65]. The exchange of ballast water of ships in coastal areas of a new marine ecosystem is considered the main introduction factor for alien species [11, 63]. One of the main problems of bioinvasion related to public health is the introduction of toxic algae that cause poisoning and other pathogens such as *Vibrio cholerae*, which causes of infection [13, 63, 64].

In 1991, cholera appeared in Latin America, and until recently caused more than 1.2 million of infections and 12,000 deaths. It is believed that Peru served as an entry in the South American continent [63]. However, Brazil has achieved the highest number of cases across the continent in 1993 and 1994, most recently in 1999 on the coast of Paraná, with 467 confirmed cases [13]. There is scientific evidence showing that the first cases of cholera occurred in the coastal ports, which suggests that outbreaks or epidemics could have been caused by the ballast water of ships arriving from endemic areas [63]. In a study conducted by the National Health Surveillance Agency (ANVISA) in 2002 detected the presence of *Vibrio cholerae* and *Escherichia coli* in high proportions in samples collected from ballast water of ships in various ports of Brazil, supporting the hypothesis of ships as carriers of the pathogen [13].

7. Environmental contaminants

Chemical contamination is one of the main challenges for the conservation status of the marine environment. Environmental contaminants have compromised the quality of water and air, affecting biodiversity in ecosystems, contaminating food and endangering human health. The vast majority of waste produced by anthropogenic activities inevitably reaches the oceans and is widely dispersed and may even reach free regions of the release of pollutants, such as the Antarctic region [3, 5, 12, 66, 67].

Approximately 80% of the contamination that reaches the oceans has their emission sources on the continents, via air routes, direct discharges into the oceans by effluents, industrial, agricultural and other sources [68]. The ocean contamination in associated with the concentration of people living in coastal regions around the world [18]. The contaminants of highest concern are those that have environmental persistence, are capable of long-range transport, can biomagnify in food chain and bioaccumulate in humans and animal tissues and have potentially significant impacts on humans and environmental health [66, 69-71]. The sources and amount of emissions are also extremely important. Persistent organic pollutants (POPs), polycyclic aromatic hydrocarbons (PAHs) and some metals present the chemical characteristics mentioned above.

The human activities have considerably altered the geochemical and biogeochemical cycles of the metals in nature, especially during the last and current century. Once the environment, the metallic elements can occur in various chemical forms and thus may increase or decrease its toxic properties [1, 72]. Mercury, which has been associated with various human health problems, is used in wide range of industrial processes. When released into the environment, bacteria can quickly transform its inorganic form in inorganic mercury (methyl-mercury). Methyl-mercury can concentrate in the marine food chain, and may cause cytotoxic effects, kidney and brain of those exposed [10, 57, 72]. Concentrations 1-2 mg / kg brain tissue may cause neurological damage. Furthermore, due to its ability to cross the placental barrier, methyl-mercury becomes extremely harmful to fetuses exposed [1]. Due to the extensive contamination with mercury, individuals consuming fish (principally predator species) frequently exhibit the highest levels of methyl-mercury in their tissues. Top predator species such as marine mammals, sharks and seabirds present extremely high concentrations or mercury in their tissues, and people that consume meat of organisms from these groups generally are exposed to high concentrations.

The human vulnerability for persistent contaminants in the ocean is strongly linked to the origin and trophic position of the marine food consumed. An example of this is people in Iraq and Japan, which may have higher levels of 50-100 ppm of methyl-mercury in hair samples, when the average concentration of this compound in humans is less than 1 ppm [73]. Cadmium also has the ability to bioaccumulate in the marine environment and is often found in biological samples taken from this environment. Cadmium is recognized as a human carcinogen; however, the increased risk is related to human exposure to this element can lead to proteinuria and renal failure.

Arsenic and lead are also potentially harmful to human and environmental health. These are usually found in living organisms and marine sediments, industrial discharges being a major source of environmental emissions. Several related contaminants have been found in tissues of marine organisms, and in some cases these have been associated with adverse effects on the exposed organisms [3]. One of the variables which can cause confusion and lack of causal association studies is the presence of mixtures of a considerable range of these specific contaminants present in the oceans. This mixture could cause adverse effects acting in concert, and perhaps at low levels, which could obscure associations in studies using only specific contaminants.

Persistent organic pollutants (POPs) pose potential risks to human health and the environment. Exposure to POPs can cause serious human and environmental health impacts including certain cancers, birth defects, dysfunctional immune and reproductive systems and greater susceptibility to disease [1, 70].

The main human exposure to POPs in the oceans is through fish consumption [74]. One of the most relevant POPs even today is the pesticide DDT, which despite its commercialization and application banned in most countries, is still used in some tropical and subtropical nations for vector control, such as malaria [69, 75, 76]. According to the International Agency for Research on Cancer (IARC), DDT is possibly carcinogenic and sub-acute exposures may cause problems in the central nervous system and also impair the immunological integrity. Similarly, PCBs (polychlorinated biphenyls) have caused severe impacts on the exposed organisms and public health, mainly through fish consumption [74].

PAHs are pollutants of great environmental persistence, and together with its derivatives have important carcinogenic, mutagenic and genotoxic [71]. PAHs are formed by thermal transformation of fossil fuels. Thus, forest fires, industrial processes and petrochemical activities are major contributors to environmental contamination by PAHs [1, 71]. These can also be formed naturally, but anthropogenic is that is causing concern. PAHs are highly soluble and rapidly absorbed through the lungs, the intestines and the skin of experimental animals, regardless of route of administration. The carcinogenic effects of some PAHs Of crucial importance to environmental and public health, fish consumption is the main source of human exposure relating to ocean pollution.

8. Oceans for public health and well-being

The oceans have a valuable relationship with human wellbeing through ecosystem services, the source of discoveries for pharmacology and biomedicine, cultural values, and simply the satisfaction of people, which stems from the harmony of healthy oceans and their stable biodiversity. The marine ecosystem services include the stabilization of the coast, the regulation of nutrients and climate, and the management of pollutants, energy resources, and natural products of values for biomedicine, tourism and recreation. Therefore, besides the importance of the quality of the oceans to maintain the integrity of biodiversity residing in this biome, oceans also produce beneficial effects and essential for the maintenance and stability of terrestrial ecosystems to the welfare and human health [2, 3, 42].

The coastal regions provide an important natural place for human leisure, which contribute for both physical and psychological benefits. There is medical evidence showing that the access to natural environments improves health and wellbeing, prevents disease and helps the development of recover from illness. Coastal environments stimulate fitness and leisure activities (e.g. swimming, surfing and coastal walking, beach sports) [42]. These physical and mental exercises can prevent cardiovascular diseases and help to reduce obesity and cancer [42]. In addition, the leisure activities may help to prevent or improve many mental health issues, such as reduction of stress.

Great efforts have been made to evaluate the complex economic values of environmental services and natural resources. Generally, the conservation of the ecosystem is considered more economically profitable than the economic values arising from the acquisition and use of its resources, which often leave severe environmental liabilities [30, 77]. Constanza et al. [77] showed that while the coastal areas cover only 8% of global land surface, the services and benefits from this area are responsible for approximately 43% of the total value of global ecosystem services valued at 12.6 trillion dollars.

In the last six decades there has been a growing interest in bioactive substances with properties derived from marine organisms [1, 4, 16, 78]. Already in the 1950s Bergman and Feeney [79] discovered two drugs of importance to medicine (ARA-C and ARA-A), based on nucleoside present in marine sponges (*Tectitethya crypta* and *Streptomyces antibiotics*). Formulated synthetically from the discovery of these researchers, the Ara-C is indicated for the treatment of non-lymphocytic leukemia, the leukemia meninges and chronic myelocytic leukemia, whereas the Ara-A is indicated for the treatment of viral infections caused by *Herpes simplex* and *Herpes zoster* [4, 43, 80]. Another valuable contribution of importance to medicine was the discovery of azidothymidine, AZT. This synthetic derivative, originating from marine sponges, is currently still one of the most effective drugs in the treatment of acquired immunodeficiency syndrome (AIDS) [43, 80]. From the work of these researchers, scientists began to explore marine biodiversity and its potential for the discovery of new bioactive compounds, aimed at advancement of pharmacology and biomedicine in the treatment of diseases known to cause severe damage on the population. The success of the discovery of new bioactive compounds and their pharmacological effects, extracted from marine organisms has been demonstrated from formulations of new anticancer treatments, and infectious diseases and inflammation [81]. However, much emphasis has been attributed to the discovery of anti-cancer compounds derived from marine organisms due in large part to the availability of funds for supporting studies aiming to find new compounds. The oceans are rich source of chemical and biological diversity, with hundreds of thousands, maybe even millions of new species are still unknown, especially micro-organisms that represent a great opportunity for the discovery of new species and new chemicals. Another approach of extreme importance is the study of marine organisms as a basis for discovery in biomedicine. Research on the natural history, taxonomy, physiology and biochemistry of marine organisms has served as a model for biomedical research to elucidate issues relevant to the physiology, biochemistry and human disease.

9. Conclusions

The pressure of human activities on marine environment generates ecosystem modifications that affect the people depending on the vulnerability of the population exposed. The past and current human development needs great modification to ensure the stabilization and homeostasis of the ocean. In addition, it is important to better understand the dynamic of the marine processes which can contribute to prevent the risks associated with human exposure. It includes the development of a system capable to generate information of a wide range of complex environment processes that should be used to prevent human and biodiversity impacts. Climate change and other anthropogenic pressures have the ability to influence many environmental factors, important for human health, such as fisheries, HABs, pathogens and contaminants. Environmental model are required to better understand the ocean and ecosystem dynamics their role on climate change, as well as to prevent impacts resulting from the modifying ecosystems. The development of indicators is needed to establish measures to study and prevent the impacts of the oceans changes on human health. The conservation of the marine environments, principally those with no apparent alterations, are greatly encouraged to avoid human and biodiversity.

Author details

Jailson Fulgencio de Moura and Salvatore Siciliano
National School of Public Health - ENSP; Oswaldo Cruz Foundation – FIOCRUZ,
Brazil

Emily Moraes Roges, Roberta Laine de Souza and Dalia dos Prazeres Rodrigues
National Reference Laboratory for Bacterial Enteroinfections – LRNEB;
Oswaldo Cruz Institute – IOC; Oswaldo Cruz Foundation – FIOCRUZ,
Brazil

Acknowledgement

We thank the laboratory team from the National *Reference Laboratory for Bacterial Enteroinfections* (LRNEB) from the Oswaldo Cruz Institute (IOC/FIOCRUZ). We would like to thank the PhDs Rosalina Koifman, Sérgio Koifman and Aldo Pacheco Ferreira from the National School of Public Health (ENSP/FIOCRUZ) for the encouragement to prepare this work. The first author J. F. Moura is funded by the Fiocruz Foundation (FIOCRUZ). In addition, we thank the invite of the In Tech to publish this book chapter.

10. References

[1] Fleming LE, Broad K, Clement A, Dewailly E, Elmir S, Knap A, et al. Oceans and human health: Emerging public health risks in the marine environment. Marine pollution bulletin 2006;53(10-12);545-60.

[2] Fleming LE, Laws E. Overview of the oceans and human health. Oceanography 2006;19(2);18-23.

[3] Moura JF, Cardozo M, Belo MS, Hacon S, Siciliano S. A interface da saude publica com a saude dos oceanos: producao de doencas, impactos socioeconomicos e relacoes beneficas. Ciencia & saude coletiva 2011;16(8);3469-80.

[4] NRC NRC. From Monsoons to Microbes: Understanding the Ocean's Role in Human Health: National Academy Press; 1999.

[5]   Aono S, Tanabe S, Fujise Y, Kato H, Tatsukawa R. Persistent organochlorines in minke whale (Balaenoptera acutorostrata) and their prey species from the Antarctic and the North Pacific. Environmental Pollution 1997;98(1);81-89.

[6]   Dewailly E, Knap A, Oceanography AFftoahhbrab, 19(2):84-93. Food from the oceans and human health: balancing risks and benefits. Oceanography 2006;19(2);84-93.

[7]   IOC. The final design plan for the HOTO module of GOOS. Paris: Intergovernmental Oceanographic Commission; 2002.

[8]   Andersen NR. An early warning system for the health of the oceans. Oceanography 1997;10(1);14-23.

[9]   Mora C, Metzger R, Rollo A, Myers RA. Experimental simulations about the effects of overexploitation and habitat fragmentation on populations facing environmental warming. Proceedings Biological sciences / The Royal Society 2007;274(1613);1023-8.

[10] Moura JF, Hacon Sde S, Vega CM, Hauser-Davis RA, de Campos RC, Siciliano S. Guiana dolphins (Sotalia guianensis, Van Beneden 1864) as indicators of the bioaccumulation of total mercury along the coast of Rio de Janeiro state, Southeastern Brazil. Bulletin of environmental contamination and toxicology 2012;88(1);54-9.

[11] Drake LA, Doblin MA, Dobbs FC. Potential microbial bioinvasions via ships' ballast water, sediment, and biofilm. Marine pollution bulletin 2007;55(7–9);333-41.

[12] Franco T, Druck G. Padrões de industrialização, riscos e meio ambiente. Ciência e Saúde Coletiva 1998;3(2);61-72.

[13] ANVISA. Brasil água de lastro ANVISA. Projeto GGPAF 2002. Brasília: Agência Nacional de Vigilância Sanitária - ANVISA-MS; 2003.

[14] Laws EA, Fleming LE, Stegeman JJ. Centers for Oceans and Human Health: contributions to an emerging discipline. Introduction. Environmental health : a global access science source 2008;7 Suppl 2S1.

[15] Malone TC. The coastal module of the Global Ocean Observing System (GOOS): an assessment of current capabilities to detect change. Marine Policy 2003;27(4);295-302.

[16] Grossel M, Walsh PJ. Benefits from the sea: sentinel species and animal models of human health. Oceanography 2006;19(2);126-33.

[17] PRB. World population highlights: key findings from PRB's 2007 world population data sheet: Population Reference Bureau; 2007.

[18] Cohen JE. Population growth and earth's human carrying capacity. Science 1995;269(5222);341-46.

[19] Sogin ML, Morrison HG, Huber JA, Mark Welch D, Huse SM, Neal PR, et al. Microbial diversity in the deep sea and the underexplored "rare biosphere". Proceedings of the National Academy of Sciences of the United States of America 2006;103(32);12115-20.

[20] Kovats RS, Bouma MJ, Hajat S, Worrall E, Haines A. El Nino and health. Lancet 2003;362(9394);1481-9.

[21] Kovats RS. El Nino and human health. Bulletin of the World Health Organization 2000;78(9);1127-35.

[22] Confalonieri UEC, Chame M, Najar A, Chaves SAM, Krug T, Nobre C, et al. Mudanças globais e desenvolvimento: importância para a saúderme Epidemiológico do SUS 2002;11(3);139-54.

[23] Confalonieri UEC. Variabilidade climática, vulnerabilidade social e saúde no Brasil. Terra Livre 2003;1(20);193-204.

[24] GESAMP. Land-based sources and activities affecting the quality and uses of the marine, coastal and associated freshwater environment Protecting the oceans from land-based activities Group of Experts on the Scientific Aspects of Marine Environmental Protection (GESAMP); 2001.

[25] Few R. Health and climatic hazards: Framing social research on vulnerability, response and adaptation. Global Environmental Change 2007;17(2);281-95.

[26] Haines A, McMichael AJ, Epstein PR. Environment and health: 2. Global climate change and health. Canadian Medical Association Journal 2000;163(6);729-34.

[27] Wilkinson P, Campbell-Lendrum DH, Bartlett CL. Monitoring the health effects of climate change. In: McMichael AJ, Campbell-Lendrum DH, Corvalan CF, Ebi KL, Githeko A, Scheraga JD, et al., eds. Climate Change and Human Health: risks and responses. Geneva: World Health Organization (WHO); 2003. p204-19.

[28] Bindoff NL, Willebrand J, Artale V, Cazenave A, Gregory J, Gulev S, et al. Observations: Oceanic Climate Change and Sea Level. In: Solomon S, Qin D, Manning M, Chen Z, Marquis M, Averyt KB, et al., eds. Climate Change 2007: The Physical Science Basis Contribution of Working Group I to the Fourth Assessment Report of the Intergovernmental Panel on Climate Change. Cambridge/United Kingdom/New York: Cambridge University Press; 2007. P Available from http://www.ipcc.ch/ publications_and_data/ ar4/wg1/en/ch5.html.

[29] Patz JA. A human disease indicator for the effects of recent global climate change. Proceedings of the National Academy of Sciences of the United States of America 2002;99(20);12506-8.

[30] Costanza R, Farley J. Ecological economics of coastal disasters: Introduction to the special issue. Ecological Economics 2007;63(2–3);249-53.

[31] Brown C. Marine and coastal ecosystems and human well-being. Kenya: UNEP; 2006. Kenya: United Nations Environment Programme/Millennium Ecosystem Assessment; 2006.

[32] Waring SC, Brown BJ. The Threat of Communicable Diseases Following Natural Disasters: A Public Health Response. Disaster Management & Response 2005;3(2);41-47.

[33] Webster PJ, Holland GJ, Curry JA, Chang H-R. Changes in Tropical Cyclone Number, Duration, and Intensity in a Warming Environment. Science 2005;309(5742);1844-46.

[34] Emanuel K. Increasing destructiveness of tropical cyclones over the past 30[thinsp]years. Nature 2005;436(7051);686-88.

[35] Muehe D. Aspectos gerais da erosão costeira no Brasil. Mercator - Revista de Geografia da UFC 2005;4(7);97-110.

[36] Nicolodi JL, Pettermann RM. Vulnerability of the Brazilian Coastal Zone in its Environmental, Social, and Technological Aspects. Journal of Coastal Research 2011;SI(64);1372-79.

[37] Ribeiro GP, Rocha CHO, Figueiredo Jr AG, Silva CG, Silva SHF, Moreira PSC, et al. Análise espaço-temporal no suporte à avaliação do processo de erosão costeira em Atafona, São João da Barra (RJ). Revista Brasileira de Cartografia 2004;56(2);129-38.

[38] Van Dolah FM. Marine algal toxins: origins, health effects and their increased occurrence. Environmental Health Perspectives 2000;108(1);133-41.

[39] WHO. Algae and cyanobacteria in coastal and estuarine waters. Guidelines for Safe Recreational Water Environments: Coastal and fresh waters. Geneva: World Health Organization; 2003. p

[40] Shuval HIS, Economic and Social Aspects, Environment otIoPitM, Quantitative oHHAP, (GDB). EotGDB, GESAMP/ World Health Organization W. Scientific, Economic and Social Aspects of the Impact of Pollution in the Marine Environment on Human Health. A Preliminary Quantitative Estimate of the Global Disease Burden (GDB): GESAMP/ World Health Organization - WHO; 1999.

[41] Hudnell HK. Cyanobacterial Harmful Algal Blooms: State of the Science and Research Needs: Springer; 2008.

[42] Allen JI. Marine environment and human health: an overview. In: Hester RE, Harrison RM, eds. Marine Pollution and Human Health - Issues in Environmental Science and Technology. Cambridge: Royal Society of Chemistry; 2011. p

[43] Oliveira JS, Freitas JC. Produtos naturais marinhos: características dos envenenamentos alimentares e substâncias de interesse farmacológico. Higiene alimentar 2001;15(20);22-33.

[44] Manerio E, Rodas VL, Costas E, Hernandez JM. Shellfish consumption: a major risk factor for colorectal cancer. Medical hypotheses 2008;70(2);409-12.

[45] Wallentinus I, Nyberg CD. Introduced marine organisms as habitat modifiers. Marine pollution bulletin 2007;55(7–9);323-32.

[46] Epstein PR, Ford TE, Colwell RR. Marine ecosystems. In: Epstein PR, Sharp D, eds. Health and Climate Change. London: The Lancet; 1994. p14-17.

[47] Teixeira MG, C. CM, L. CV, Pereira MS, E. H. Gastroenteritis epidemic in the area of the Itaparica Dam, Bahia, Brazil. Bulletin of the Pan American Health Organization 1993;27(3);244-53.

[48] Hunter-Cevera J, Karl D, Buckley M. Marine microbial diversity: the key to Earth's habitability. Washington: American Academy of Microbiology; 2005.

[49] US Environmental Protection Agency U. Ambient water quality criteria for bacteria. Cincinnati: USEPA; 1986.

[50] Thompson JR, Marcelino LA, Polz MF. Diversity, Sources, and Detection of Human Bacterial Pathogens in the Marine Environment. In: Belkin S, Colwell RR, eds.

Oceans and Health: Pathogens in the Marine Environment. New York: Springer; 2005. p29-68.

[51] Pereira CS, Amorim SD, Santos AFdM, Siciliano S, Moreno IB, Ott PH, et al. Plesiomonas shigelloides and Aeromonadaceae family pathogens isolated from marine mammals of southern and southeastern Brazilian coast. Brazilian Journal of Microbiology 2008;39749-55.

[52] Shuval H. Estimating the global burden of thalassogenic diseases: human infectious diseases caused by wastewater pollution of the marine environment. Journal of Water and Health 2003;1(2);53-64.

[53] Dorfman M, Rosselot KS. Testing the Waters - A Guide to Water Quality at Vacation Beaches. New York: National Research Defense Council, NRDC; 2009.

[54] Reis EC, Silveira VV-B, Siciliano S. Records of stranded sea turtles on the coast of Rio de Janeiro State, Brazil. Marine Biodiversity Records 2009;2.

[55] Reis EC, Pereira CS, Rodrigues DdP, Secco HKC, Lima LM, Rennó B, et al. Condição de saúde das tartarugas marinhas do litoral centro-norte do estado do Rio de Janeiro, Brasil: avaliação sobre a presença de agentes bacterianos, fibropapilomatose e interação com resíduos antropogênicos. Oecologia Australis 2010;14(3);756- 65.

[56] Moura JF, Sholl TGC, da Silva Rodrigues É, Hacon S, Siciliano S. Marine tucuxi dolphin (Sotalia guianensis) and its interaction with passive gill-net fisheries along the northern coast of the Rio de Janeiro State, Brazil. Marine Biodiversity Records 2009;2doi:10.1017/S1755267209000864.

[57] Moura JF, Siciliano S, Sarcinelli PN, Hacon S. Organochlorine pesticides in marine tucuxi dolphin milk incidentally captured with its calf in Barra de São João, east coast of Rio de Janeiro State, Brazil. Marine Biodiversity Records 2009;2null-null.

[58] Wang JY, Yang SC, Hung SK, Jefferson TA. Distribution, abundance and conservation status of the eastern Taiwan Strait population of Indo-Pacific humpback dolphins, Sousa chinensis. Mammalia 2007157–65.

[59] Ahmed FE. Issues in fishery products safety in the united states. Environmental Toxicology and Water Quality 1993;8(2);141-52.

[60] Van Bressem MF, Raga JA, Di Guardo G, Jepson PD, Duignan PJ, Siebert U, et al. Emerging infectious diseases in cetaceans worldwide and the possible role of environmental stressors. Diseases of Aquatic Organisms 2009;86(2);143-57.

[61] Van Bressem MF, Santos MCdO, Oshima JEdF. Skin diseases in Guiana dolphins (Sotalia guianensis) from the Paranaguá estuary, Brazil: A possible indicator of a compromised marine environment. Marine Environmental Research 2009;67(2);63-68.

[62] Iwamoto M, Ayers T, Mahon BE, Swerdlow DL. Epidemiology of Seafood-Associated Infections in the United States. Clinical Microbiology Reviews 2010;23(2);399–411.

[63] Medeiros DS, Nahuz MAR, da Adr, meio idemep, ponta dádlntpd, 1(2):21p. UEI. Avaliação de risco da introdução de espécies marinhas exóticas por meio

de água de lastro no Terminal Portuário de Ponta Ubu (ES). InterfacEHS 2006;1(2); 21.

[64] Occhipinti-Ambrogi A. Global change and marine communities: Alien species and climate change. Marine pollution bulletin 2007;55(7–9);342-52.

[65] Hewitt CL, Campbell ML. Mechanisms for the prevention of marine bioinvasions for better biosecurity. Marine pollution bulletin 2007;55(7–9);395-401.

[66] Hacon S, Barrocas P, Siciliano S, ambiente Adrpashucpagidse. Avaliação de risco para a saúde humana: uma contribuição para a gestão integrada de saúde e ambiente. Cadenos Saúde Coletiva 2005;13(4);811-35.

[67] Porto MF. Saúde, ambiente e desenvolvimento: reflexões sobre a experiência da COPASAD – Conferência Pan-Americana de Saúde e Ambiente no Contexto do Desenvolvimento Sustentável. Ciência e Saúde Coletiva 1998;3(2);33-46.

[68] Sandifer PA, Holland AF, Rowles TK, Scott GI. The ocean and human health. Environmental Health Perspectives 2004;112(8);A454-A55.

[69] Flores AV, Ribeiro JN, Neves AA, Queiroz ELR. Organoclorados: um problema de saúde pública. Ambiente & Sociedade;7(2);111-25.

[70] Grisolia CK. Agrotóxicos: mutações, reprodução & câncer ; riscos ao homem e ao meio ambiente, pela avaliação de genotoxicidade, carcinogenicidade e efeitos sobre a reprodução. Brasília: Universidade de Brasília; 2005.

[71] Netto ADP, Moreira JC, Dias AEXO, Arbilla G, Ferreira LFV, Oliveira AS, et al. Avaliação da contaminação humana por hidrocarbonetos policíclicos aromáticos (HPAs) e seus derivados nitrados (NHPAs): uma revisão metodológica. Química Nova 2000;23(6);765-73.

[72] Morais S, Costa FG, Pereira ML. Heavy Metals and Human Health. In: Oosthuizen J, ed. Environmental Health - Emerging Issues and Practice. Rijeka: InTech; 2012. p227-46.

[73] Harada M. Minamata Disease: Methylmercury Poisoning in Japan Caused by Environmental Pollution. Critical Reviews in Toxicology 1995;25(1);1-24.

[74] Dewailly E, Ayotte P, Bruneau S, Lebel G, Levallois P, Weber JP. Exposure of the Inuit Population of Nunavik (Arctic Québec) to Lead and Mercury. Archives of Environmental Health 2001;56(4);350-57.

[75] Snedeker SM. Pesticides and breast cancer risk: a review of DDT, DDE, and dieldrin. Environmental Health Perspectives 2001;109(1);35-47.

[76] IARC. Overall Evaluations of Carcinogenicity to Humans: List of all agents, mixtures and exposures evaluated to date. International Agency for Research on Câncer; 2004. p. 32.

[77] Costanza R, d'Arge R, de Groot R, Farber S, Grasso M, Hannon B, et al. The value of the world's ecosystem services and natural capital. Nature 1997;387(15);253-60.

[78] Jack D. Combing the oceans for new therapeutic agents. Lancet 1998;352(9130);794-95.

[79] Bergman W, Feeney RJ. Contributions to the study of marine products. XXXII. The nucleosides of sponges. Journal of Organic Chemistry 1951;16(6);981-87.

[80] McConnell OJ, Longley RE, Koehn FE. The discovery of marine natural products with therapeutic potential. Boston: Butterworth-Heinemann; 1994.

[81] Schwartsmann G. A natureza como fonte de novas drogas anticâncer: a contribuição dos oceanos. Anais da Academia Nacional de Medicina 2000;160(2);95-103.

Permissions

The contributors of this book come from diverse backgrounds, making this book a truly international effort. This book will bring forth new frontiers with its revolutionizing research information and detailed analysis of the nascent developments around the world.

We would like to thank Gbolagade Akeem Lameed, for lending his expertise to make the book truly unique. He has played a crucial role in the development of this book. Without his invaluable contribution this book wouldn't have been possible. He has made vital efforts to compile up to date information on the varied aspects of this subject to make this book a valuable addition to the collection of many professionals and students.

This book was conceptualized with the vision of imparting up-to-date information and advanced data in this field. To ensure the same, a matchless editorial board was set up. Every individual on the board went through rigorous rounds of assessment to prove their worth. After which they invested a large part of their time researching and compiling the most relevant data for our readers. Conferences and sessions were held from time to time between the editorial board and the contributing authors to present the data in the most comprehensible form. The editorial team has worked tirelessly to provide valuable and valid information to help people across the globe.

Every chapter published in this book has been scrutinized by our experts. Their significance has been extensively debated. The topics covered herein carry significant findings which will fuel the growth of the discipline. They may even be implemented as practical applications or may be referred to as a beginning point for another development. Chapters in this book were first published by InTech; hereby published with permission under the Creative Commons Attribution License or equivalent.

The editorial board has been involved in producing this book since its inception. They have spent rigorous hours researching and exploring the diverse topics which have resulted in the successful publishing of this book. They have passed on their knowledge of decades through this book. To expedite this challenging task, the publisher supported the team at every step. A small team of assistant editors was also appointed to further simplify the editing procedure and attain best results for the readers.

Our editorial team has been hand-picked from every corner of the world. Their multi-ethnicity adds dynamic inputs to the discussions which result in innovative outcomes. These outcomes are then further discussed with the researchers and contributors who give their valuable feedback and opinion regarding the same. The feedback is then collaborated with the researches and they are edited in a comprehensive manner to aid the understanding of the subject.

Apart from the editorial board, the designing team has also invested a significant amount of their time in understanding the subject and creating the most relevant covers. They scrutinized every image to scout for the most suitable representation of the subject and create an appropriate cover for the book.

The publishing team has been involved in this book since its early stages. They were actively engaged in every process, be it collecting the data, connecting with the contributors or procuring relevant information. The team has been an ardent support to the editorial, designing and production team. Their endless efforts to recruit the best for this project, has resulted in the accomplishment of this book. They are a veteran in the field of academics and their pool of knowledge is as vast as their experience in printing. Their expertise and guidance has proved useful at every step. Their uncompromising quality standards have made this book an exceptional effort. Their encouragement from time to time has been an inspiration for everyone.

The publisher and the editorial board hope that this book will prove to be a valuable piece of knowledge for researchers, students, practitioners and scholars across the globe.

List of Contributors

F.F. Goulart, B.Q.C. Zimbres and R.B. Machado
Laboratório de Planejamento para a Conservação da Biodiversidade, Departamento de Zoologia, Universidade de Brasília, Campus Darcy Ribeiro, Brasília, DF, Brazil

T.K.B. Jacobson
Campus Planaltina, Universidade de Brasília/Centro de Estudos UnB Cerrado, Universidade de Brasília, Brasília, DF, Brazil

L.M.S. Aguiar
Laboratório de Biologia e Conservação de Mamíferos, Departamento de Zoologia, Universidade de Brasília, Campus Darcy Ribeiro, Brasília, DF, Brazil

G.W. Fernandes
Laboratório de Ecologia Evolutiva e Biodiversidade/DBG, ICB/Universidade Federal de Minas Gerais, Belo Horizonte, MG, Brazil

Dariusz Jaskulski and Iwona Jaskulska
Department of Plant Production and Experimenting, University of Technology and Life Sciences, Bydgoszcz, Poland

Cristina Menta
Department of Evolutionary and Functional Biology, University of Parma, Parma, Italy

Jianming Deng and Qiang Zhang
Key Laboratory of Grass and Agriculture Ecosystem, School of Life Science, Lan Zhou University, Lanzhou, China

Hassan A. I. Ramadan and Nabih A. Baeshen
Department of Biological Sciences, Faculty of Science, King Abdulaziz University, Jeddah, Saudi Arabia

Hassan A. I. Ramadan
Department of Cell Biology, National Research Centre, Dokki, Cairo, Egypt

Ticay-Rivas Jaime R., del Pozo-BañosMarcos, Gutiérrez-RamosMiguel A., Travieso CarlosM. and Jesús B. Alonso
Signals and Communications Department, Institute for Technological Development and Innovation in Communications, University of Las Palmas de Gran Canaria, Campus University of Tafira, 35017, Las
Palmas de Gran Canaria, Las Palmas, Spain

Eberhard William G.
Smithsonian Tropical Research Institute and Escuela de Biologia Universidad de Costa Rica, Ciudad Universitaria, Costa Rica

Florin Vartolomei
Faculty of Geography, SPIRU HARET University, Bucharest, Romania

V. Gergócs
Department of Plant Taxonomy, Ecology and Theoretical Biology, Eötvös Loránd University, Faculty of Science, Budapest, Hungary

R. Homoródi
Department of Mathematics and Informatics, Corvinus University of Budapest, Faculty of Horticulture, Budapest, Hungary

L. Hufnagel
Department of Mathematics and Informatics, Corvinus University of Budapest, Faculty of Horticulture, Budapest, Hungary
Adaptation to Climate Change Research Group of the Hungarian Academy of Sciences, Budapest, Hungary

Stuart A. Harris
Department of Geography, University of Calgary, Canada

Jailson Fulgencio de Moura and Salvatore Siciliano
National School of Public Health – ENSP, Oswaldo Cruz Foundation – FIOCRUZ, Brazil

Emily Moraes Roges, Roberta Laine de Souza and Dalia dos Prazeres Rodrigues
National Reference Laboratory for Bacterial Enteroinfections – LRNEB, Oswaldo Cruz Institute – IOC; Oswaldo Cruz Foundation – FIOCRUZ, Brazil